ZigBee®
Network Protocols and Applications

OTHER COMMUNICATIONS BOOKS FROM AUERBACH

Advances in Biometrics for Secure Human Authentication and Recognition
Edited by Dakshina Ranjan Kisku, Phalguni Gupta, and Jamuna Kanta Sing
ISBN 978-1-4665-8242-2

Anonymous Communication Networks: Protecting Privacy on the Web
Kun Peng
ISBN 978-1-4398-8157-6

Case Studies in Enterprise Systems, Complex Systems, and System of Systems Engineering
Edited by Alex Gorod, Brian E. White, Vernon Ireland, S. Jimmy Gandhi,and Brian Sauser
ISBN 978-1-4665-0239-0

Cyber-Physical Systems: Integrated Computing and Engineering Design
Fei Hu
ISBN 978-1-4665-7700-8

Evolutionary Dynamics of Complex Communications Networks
Vasileios Karyotis, Eleni Stai, and Symeon Papavassiliou
ISBN 978-1-4665-1840-7

Extending the Internet of Things: Towards a Global and Seemless Integration
Antonio J. Jara-Valera
ISBN 9781466518483

Fading and Interference Mitigation in Wireless Communications
Stefan Panic
ISBN 978-1-4665-0841-5

The Future of Wireless Networks: Architectures, Protocols, and Services
Edited by Mohesen Guizani, Hsiao-Hwa Chen, and Chonggang Wang
ISBN 978-1-4822-2094-0

Green Networking and Communications: ICT for Sustainability
Edited by Shafiullah Khan and Jaime Lloret Mauri
ISBN 978-1-4665-6874-7

Image Encryption: A Communication Perspective
Fathi E. Abd El-Samie, Hossam Eldin H. Ahmed, Ibrahim F. Elashry, Mai H. Shahieen, Osama S. Faragallah, El-Sayed M. El-Rabaie, and Saleh A. Alshebeili
ISBN 978-1-4665-7698-8

Intrusion Detection in Wireless Ad-Hoc Networks
Nabendu Chaki and Rituparna Chaki
ISBN 978-1-4665-1565-9

Intrusion Detection Networks: A Key to Collaborative Security
Carol Fung and Raouf Boutaba
ISBN 978-1-4665-6412-1

MIMO Processing for 4G and Beyond: Fundamentals and Evolution
Edited by Mário Marques da Silva and Francisco A.T.B.N. Monteiro
ISBN 978-1-4665-9807-2

Modeling and Analysis of P2P Content Distribution Systems
Yipeng Zhou and Dah Ming Chiu
ISBN 978-1-4822-1920-3

Network Innovation through OpenFlow and SDN: Principles and Design
Edited by Fei Hu
ISBN 978-1-4665-7209-6

Pervasive Computing: Concepts, Technologies and Applications
Guo Minyi, Zhou Jingyu, Tang Feilong, and Shen Yao
ISBN 978-1-4665-9627-6

Physical Layer Security in Wireless Communications
Edited by Xiangyun Zhou. Lingyang Song, and Yan Zhang
ISBN 978-1-4665-6700-9

Security for Multihop Wireless Networks
Shafiullah Khan and Jaime Lloret Mauri
ISBN 978-1-4665-7803-6

Self-Healing Systems and Wireless Networks Management
Junaid Ahsenali Chaudhry
ISBN 978-1-4665-5648-5

Unit and Ubiquitous Internet of Things
Huansheng Ning
ISBN 978-1-4665-6166-3

ZigBee®
Network Protocols and Applications

Edited by
Chonggang Wang • Tao Jiang • Qian Zhang

CRC Press
Taylor & Francis Group
Boca Raton London New York

CRC Press is an imprint of the
Taylor & Francis Group, an **informa** business

ZigBee Alliance, Inc. 2400 Camino Ramon, Suite 375 San Ramon, CA 94583, USA

CRC Press
Taylor & Francis Group
6000 Broken Sound Parkway NW, Suite 300
Boca Raton, FL 33487-2742

First issued in paperback 2019

ISBN-13: 978-1-4398-1601-1 (hbk)
ISBN-13: 978-0-367-37878-3 (pbk)

Visit the Taylor & Francis Web site at
http://www.taylorandfrancis.com

and the CRC Press Web site at
http://www.crcpress.com

Contents

List of Figures

List of Tables

Contributors

Guang Yang
University of Bergen
Bergen, Norway

Yu-Doo Kim
University of Science and Technology
Seoul, South Korea

Il-Young Moon
University of Science and Technology
Seoul, South Korea

M.R. Palattella
University of Luxembourg
Luxembourg

A. Faridi
University of Luxembourg
Luxembourg

G. Boggia
University of Luxembourg
Luxembourg

P. Camarda
University of Luxembourg
Luxembourg

L.A. Grieco
University of Luxembourg
Luxembourg

M. Dohler
University of Luxembourg
Luxembourg

A. Lozano
University of Luxembourg
Luxembourg

Baozhi Chen
Rutgers University
New Brunswick, NJ, USA

Dario Pompili
Rutgers University
New Brunswick, NJ, USA

Al-Khateeb Anwar
Politecnico di Torino
Turin, Italy

Luciano Lavagno
Politecnico di Torino
Turin, Italy

Sudip Misra
Indian Institute of Technology
Kharagpur, India

Sumit Goswami
Indian Institute of Technology
Kharagpur, India

Dan Keun Sung
KAIST, Yuseong-gu
Daejeon, South Korea

Jo Woon Chong
KAIST, Yuseong-gu
Daejeon, South Korea

Su Min Kim
KAIST, Yuseong-gu
Daejeon, South Korea

Pan Zhou
Huazhong University of Science and Technology
Wuhan, China

Tao Jiang
Huazhong University of Science and Technology
Wuhan, China

Chonggang Wang
Huazhong University of Science and Technology
Wuhan, China

Qian Zhang
Huazhong University of Science and Technology
Wuhan, China

Huasong Cao
The University of British Columbia
Vancouver, Canada

Sergio Gonzalez-Valenzuela
The University of British Columbia
Vancouver, Canada

Victor C. M. Leung
The University of British Columbia
Vancouver, Canada

Vincent Tam
University of Hongkong
Hong Kong, China

Johnny Yeung
University of Hongkong
Hong Kong, China

Kumar Padmanabh
Infosys Technologies Ltd.
Bangalore, India

Sougata Sen
Infosys Technologies Ltd.
Bangalore, India

Sanjoy Paul
Infosys Technologies Ltd.
Bangalore, India

Chapter 1

Background Introduction

Guang Yang

CONTENTS

Since wireless technology provides a very convenient way for information transmission, it has become the best contender in most cases. There was a vigorous development in wireless communication technology in these years, such as WiFi for wireless local area network (WLAN), Bluetooth, ZigBee for wireless personal area network (WPAN), WiMAX for wireless metropolitan area network (WirelessMAN) according to IEEE, and near field communication (NFC) technology, they effect every aspect of life. In this chapter, we briefly introduce the WiFi, ZigBee, Bluetooth and NFC from view of their physical layer and MAC layer architecture and function, which are specified by corresponding standards. Relevant security issues are not included in this chapter but by no means are they inessential.

1.1 NFC

Near field communication (NFC) is an open platform technology for a short range (about 10cm) of wireless communication. NFC was approved as an ISO/IEC standard on December 8, 2003 and later was standardized in ECMA-340 and ISO/IEC 18092. It is believed that NFC is an extension of the ISO/IEC 14443 which is a proximity card standard that defies transmission protocol for identification cards in magnetic field. In these standards, modulation schemes, coding, transfer speeds and

Table 1.1: Wireless Communication Technology

Standard	Frequency band	Range	Main feature
WiFi/IEEE 802.11	2.4GHz, 5GHz	20 - 140 m	ease and low cost, low power radio signal
Bluetooth/IEEE 802.15.1	2.4GHz	1m,10m, 100m	designed for low power consumption with a short range based on low - cost transceiver microchips in each device
ZigBee/IEEE 802.15.4	868MHz, 915MHz, 2.4GHz	Many meters	low cost, low power, low data rate and short range
WiMax/IEEE 802.16	2.3GHz, 2.5GHz, 3.5GHz, 3.7GHz, 5.8GHz	Many kilometers	providing wireless data over long distances in a variety of ways
NFC	13.56MHz	Many centimeters	very short range, secure and compatible with RFID

frame format of the Radio frequency (RF) interface of NFC device are specified, also including initialization schemes and conditions required for data collision control during initialization. Furthermore, the standard defines a transport protocol including protocol activation and data exchange methods. Standard ECMA-340 specifies the interface and protocol for simple wireless communication between close coupled devices, it allows but does not specify application in network products and consumer equipment.

1.1.1 *Specifications*

The standard defines NFC working at the center frequency of 13.56 MHz with data rates of 106, 212, and 424 kbps.

1.1.2 *Modulation and Coding*

Amplitude Shift Keying (ASK), in which the amplitude of the carrier frequency is modulated according to the logic of the data to be transmitted. NFC employs two different codings to transfer data. If an active device transfers data at 106 Kbit/s, a modified Miller coding is used. In all other cases Manchester coding is used.

1.1.3 Two Communication Modes

Standard defines both Active and Passive communication modes as follows:

Active mode: In the Active communication mode, both initiator and target use their own RF field to enable communication. The initiator starts the NFCIP-1 communication. The Target responds to an Initiator command in the Active communication mode using self-generated modulation of the self-generated RF field.

Passive mode: In the Passive communication mode, initiator generates the RF field and starts the communication. The Target responds to an Initiator command in the Passive communication mode using a load modulation scheme.

1.1.4 Security

NFC standard does not provide protection against eavesdropping and is also vulnerable to data modifications. Applications have to use higher-layer cryptographic protocols to establish a secure channel.

1.1.5 Applications

As a combination of contactless identification and networking technologies, NFC is a simple, less-expensive, low-power consumer wireless solution and is applied in contactless infrastructure quite often. People use a contactless card when they take public transport, wave a smart card at an entrance when they are going to access it, or have their purchase scanned at a cashier before payment.

1.2 IEEE 802.11 / WiFi

1.2.1 Overview

In most situations, IEEE 802.11 WLAN would firstly come to our mind when we talk about wireless. IEEE802.11 is a set of standards for wireless local area network (WLAN), which was developed by a working group of Institute of Electrical and Electronics Engineers (IEEE).

1.2.1.1 Evolution of IEEE 802.11

The original version of IEEE 802.11 standard was released in 1997. Table below lists the evolution of IEEE 802.11 standard.

IEEE 802.11 The original version of the IEEE 802.11 standard adopts direct sequence spread spectrum (DSSS) and frequency hopping spread spectrum (FHSS) modulation schemes, and specifies two raw data rates that 1 and 2

Table 1.2: Evolution of IEEE 802.11 Standard [2]

Standard	Release Date	Operation Frequency	Typical Throughput	Max.rate	Modulation Technique	Range(in/outdoor)
Legacy	1999	2.4GHz	0.9 Mbps	2 Mbps	DSSS or FHSS	20 / 100 Meters
802.11a	1999	5GHz	23 Mbps	54 Mbps	OFDM	35 / 120 Meters
802.11b	1999	2.4GHz	4.3 Mbps	11 Mbps	DSSS	38 / 140 Meters
802.11g	2003	2.4GHz	19 Mbps	54 Mbps	OFDM	38 / 140 Meters
802.11n	2009	2.4GHz 5GHz	74 Mbps	248 Mbps		70 / 250 Meters
802.11y	2008	3.7GHz	23 Mbps	54 Mbps		50 / 5000 Meters

megabits per second (Mbit/s) to be transmitted in the ISM frequency band at 2.4 GHz. But IEEE 802.11 was rapidly supplemented and popularized by IEEE 802.11b.

IEEE 802.11b IEEE 802.11b adopted the same modulation scheme DSSS as the original version, but the throughput increased to 4.3Mbps and the maximum data rate is 11Mbps. Since it operates in the unlicensed 2.4GHz ISM frequency band, 802.11b devices will coexist with other products which also utilize the same 2.4GHz ISM band, such as Bluetooth devices, ZigBee devices and so on. And the interference issue is what this thesis mainly concerns.

IEEE 802.11g After four years, in 2003, IEEE 802.11g was ratified. It provides a high data rate 54Mbps and uses the same modulation method OFDM as IEEE 802.11a. IEEE 802.11g also faces the interference problem like 802.11b because its operation frequency band is still 2.4GHz.

1.2.1.2 Architecture of IEEE 802.11

1.2.2 IEEE 802.11b

1.2.2.1 Channel Selection

IEEE 802.11b standard defines 14 channels in the 2.4GHz ISM band, and there is 5MHz apart from two adjacent channels. Since the bandwidth of WLAN radio signal is 22MHz, not all channels can be used simultaneously. In fact, only three non-overlapping WLAN channels can be used at the same time, there are Channel 1, 6, 11 for North America and Channel 1, 7, 13 for Europe.

1.2.2.2 Modulation Technique

802.11b extends the DSSS modulation technique which was defined in the original version of 802.11 standards. This extension of the DSSS system builds on the data

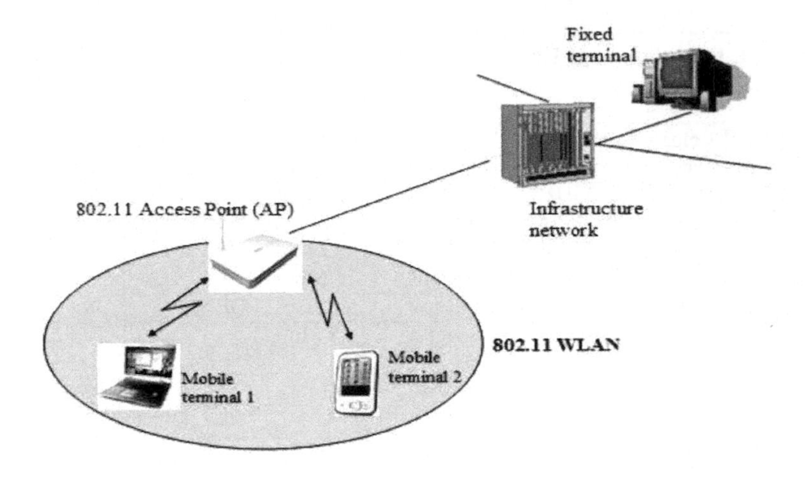

Figure 1.1: 802.11 architecture of infrastructure network.

rate capabilities, to provide 5.5 Mbit/s and 11 Mbit/s payload data rates in addition to the 1 Mbps which is encoded with differential binary phase shift keying (DBPSK) and 2 Mbps rates which is provided using differential quadrature phase shift keying (DQPSK) at the same chip rate. In order to provide the higher rates, quadrature shift keying (QPSK) combined with 8-chip complementary code keying (CCK) is employed as the modulation scheme. The chipping rate is 11 MHz, which is the same as the DSSS system described in IEEE 802.11 standard, 1999 Edition, thus the same occupied channel bandwidth is provided. [3]

- Direct Sequence Spread Spectrum (DSSS) is a modulation technique. It works by modulating a data stream of zeros and ones with a pattern called chipping sequence. In 802.11, Barker code, which is an 11 bits sequence, is used as this chipping sequence. Baker code has certain mathematical properties to make it ideal for modulating radio waves. The basic data stream is XOR with the Barker code to generate a series of data objects called chips. Each bit is encoded by the 11bits Barker code, and each group of 11 chips encodes one bit of data. [4] Rather than using the Barker code, IEEE 802.11b uses 64 complementary code keying (CCK) chipping sequences to achieve 11 Mbps data rate.[4] Different from one bit represented by one Barker symbol used in Barker code, up to 6 bits can be represented by any one particular code word, because there are 64 unique code words that can be used to encode the signal.

- Binary Phase Shift Keying (BPSK) is the simplest form of PSK. It uses two phases which are separated by 180 and so can also be termed 2-PSK. It does not particularly matter exactly where the constellation points are

IEEE 802.11b North American channel selection

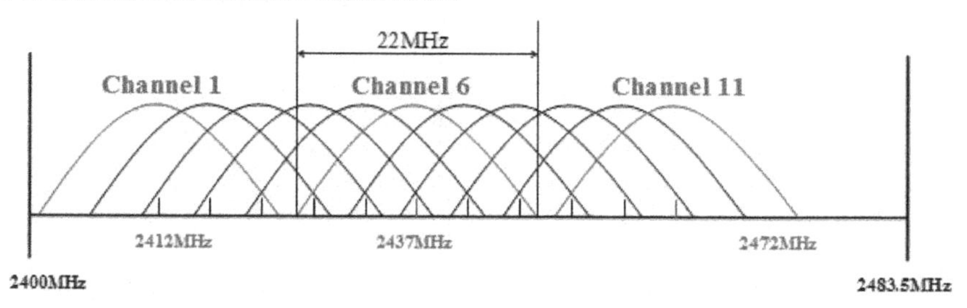

IEEE 802.11b European channel selection

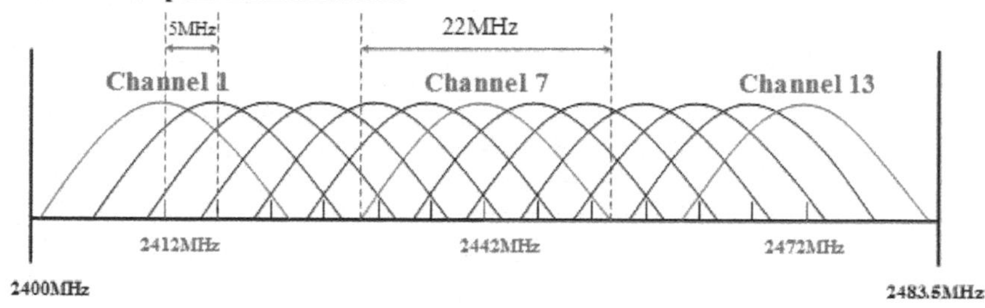

Figure 1.2: WLAN channel selection.

positioned, and in this figure they are shown on the real axis, at 0 and 180. This modulation is the most robust of all the PSKs since it takes serious distortion to make the demodulator reach an incorrect decision. It is, however, only able to modulate at 1 bit per symbol and so is unsuitable for high data-rate applications when bandwidth is limited.

- Quadrature Phase Shift Keying (QPSK) uses four points on the constellation diagram. With four phases, QPSK can encode two bits per symbol, shown in the figure below with 2 bits gray code to minimize the BER, twice the rate of BPSK. That means each adjacent symbol only differs by one bit.

  Analysis shows that QPSK may be used either to maintain the data-rate of BPSK but halve the bandwidth needed or double the data rate compared to a BPSK system while maintaining the bandwidth of the signal. Although QPSK can be viewed as a quaternary modulation, it is easier to see it as two independently modulated quadrature carriers. With this interpretation, the even (or odd) bits are used to modulate the in-phase component of the carrier, while the odd (or even) bits are used to modulate the quadrature-phase component of the carrier.

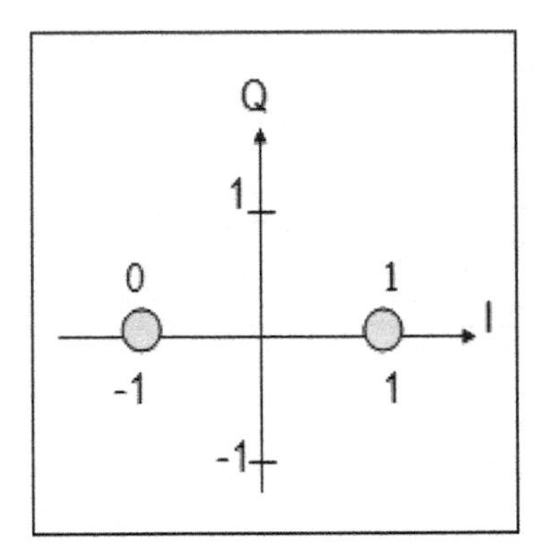

Figure 1.3: Constellation diagram for BPSK.

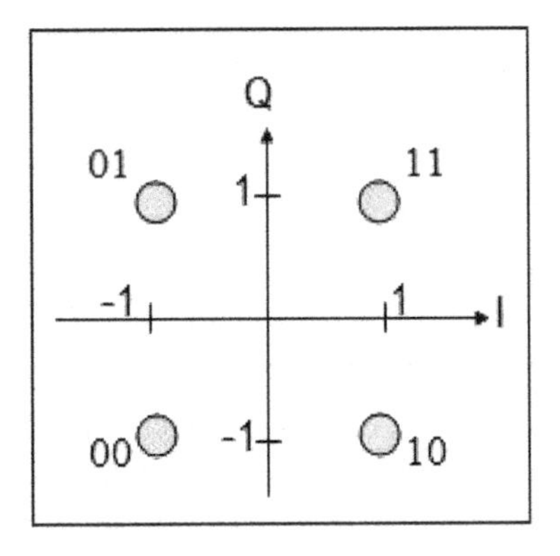

Figure 1.4: Constellation diagram for QPSK.

- Complementary Code Keying (CCK) is an m-ary orthogonal keying modulation where one of M unique (nearly orthogonal) signal code words is chosen for transmission. It allows for multi-channel operation in the 2.4 GHz band using the existing 802.11 DSSS channel structure scheme. The same chipping rate and spectrum shape as the Barker code word used in 802.11 are employed in the spreading. The spreading function allows three non-overlapping channels in the 2.4 to 2.483 GHz band. CCK uses one vector from a set of 64 complex (QPSK) vectors for the symbol and thereby modulates 6 bits (one of 64) on each 8 chips spreading code symbol. Two more bits are sent by QPSK modulating the whole code symbol. This results in modulating 8 bits onto each symbol.[4]

1.2.2.3 Physical Layer Frame Format

Two types of PPDU (PHY protocol data unit) format with different preambles and headers are defined. The format with long preamble is mandatory. Figure 1.5 shows

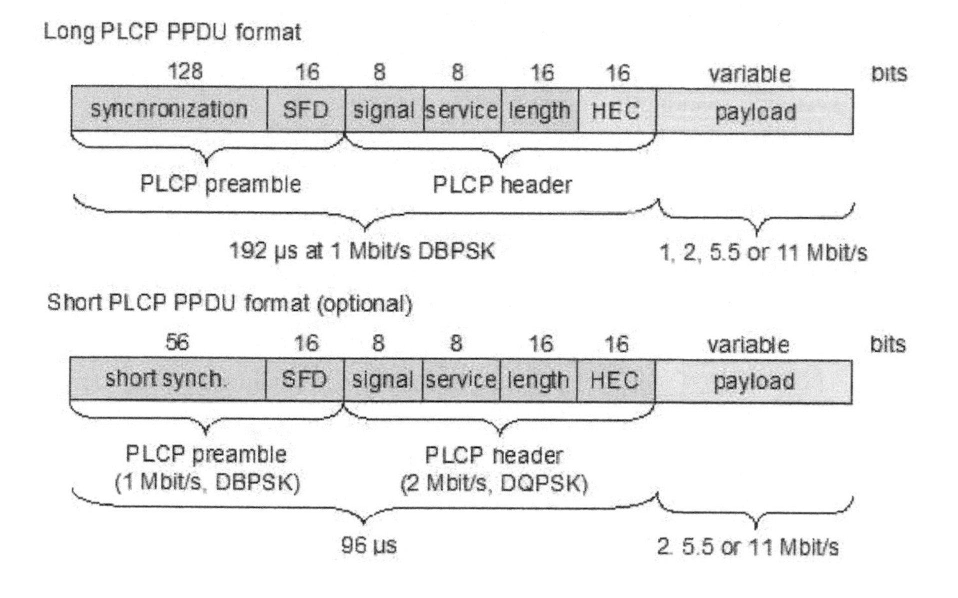

Figure 1.5: 802.11b PLCP PPDU format [3].

that the long PPDU consists of a long PLCP (PHY convergence procedure) preamble, PLCP header and PSDU (PHY service data unit). The preamble includes a long synchronization field with 16 bytes for receiver performing necessary synchronization operations and 2 bytes SFD (start of frame delimiter) provided to indicate the start of PHY-dependent parameters within the PLCP preamble. The modulation which is used for PSDU transmission and reception is indicated by the signal field. In the one byte service field, 3 bits are defined for high rate extension. 2 bytes length field indicates the number of microseconds required to transmit the PSDU. HEC (header error check) is used for signal, service and length protection. Both preamble and header use 1Mbps Barker code spreading with DBPSK, and PSDU is transmitted at 1, 2, 5.5 or 11Mbps.[3] The format with short preamble is defined as optional. Figure shows that the short PPDU includes a short preamble with 9 bytes, 6 bytes header and variable size of PSDU. The preamble uses 1Mbps Barker code spreading with DBPSK modulation while header uses 2Mbps Barker code spreading with DQPSK, and PSDU is transmitted at 2, 5.5 or 11Mbps.[3]

1.2.2.4 Access Method in Media Access Control (MAC) Layer

Carrier sense multiple access with collision avoidance (CSMA/CA) is used in IEEE 802.11 standard.

- Before sending data, station starts sensing the medium, carrier sense based on clear channel assessment (CCA).
- If the media is free for the duration of an Inter-Frame Space (IFS), the station can start sending (IFS depends on service type).
- If the medium is busy, the station has to wait for free IFS, and then the station.
- If another station occupies the medium during the back-off time of the station, the back-off time stops (fairness).
- SIFS (Short Inter Frame Spacing): highest priority, for ACK, CTS, polling response
- PIFS (PCF, Point Coordinator Function IFS): medium priority, for time-bounded service using PCF
- DIFS (DCF, Distributed Coordination Function IFS): lowest priority, for asynchronous data service

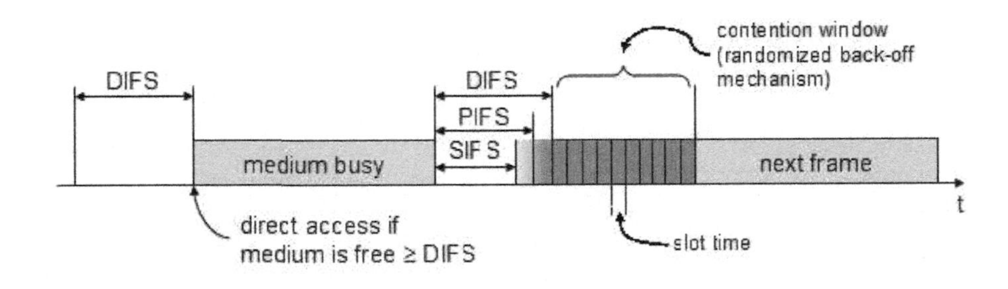

Figure 1.6: CSMA/CA mechanism [3].

1.2.3 IEEE 802.11g

IEEE 802.11g is an amendment of the IEEE 802.11 specification that extended throughput to up to 54 Mbps using the same 2.4 GHz band as 802.11b. Extended Rate PHY (ERP) is defined in it. The channel selection and MAC layer access method used in 802.11g are the same as 802.11b.

1.2.3.1 Modulation Technique

The modulation scheme used in 802.11g is orthogonal frequency-division multiplexing (OFDM) copied from 802.11a with data rates of 6, 9, 12, 18, 24, 36, 48, and 54 Mbit/s, and reverts to CCK (like the 802.11b standard) for 5.5 and 11 Mbit/s and DBPSK/DQPSK+DSSS for 1 and 2 Mbit/s. Even though 802.11g operates in the same frequency band as 802.11b, it can achieve higher data rates because of its heritage to 802.11a. In order to provide 6, 9, 12, 18, 24, 36, 48, and 54 Mbit/s payload data rates while reusing the long and short preambles described in 802.11b, the modulation technique called DSSS-OFDM was used.

1.2.3.2 Physical Layer Frame Format

Besides long preamble PPDU formats (based on 802.11b), short preamble PPDU format (which is optional in 802.11b) and ERP-OFDM preamble PPDU format (based on 802.11a), ERP defined in 802.11g provides two optional PPDU formats to support the optional DSSS-OFDM modulation rates. The major change is the format of the

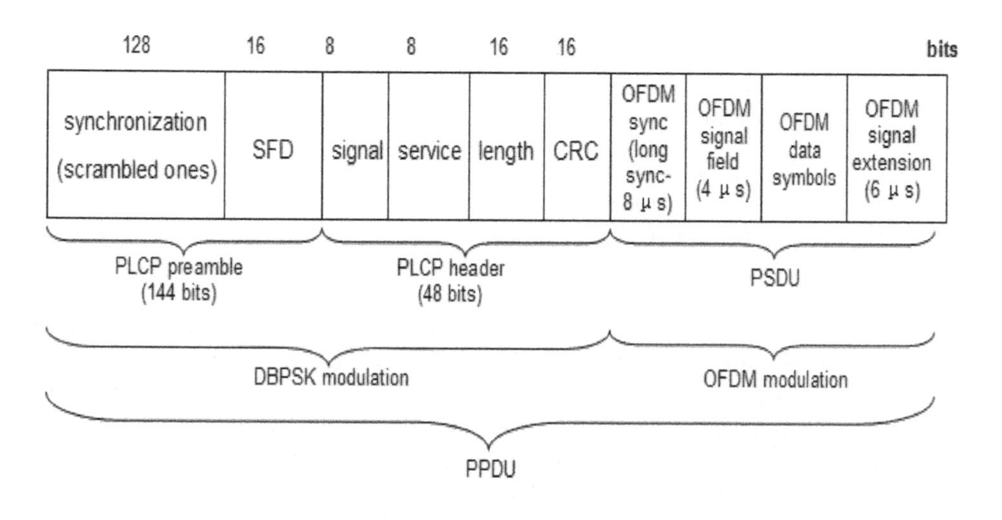

Figure 1.7: Long preamble PPDU format for DSSS-OFDM [5].

PSDU. The single carrier PSDU defined in 802.11b is replaced by a PSDU which consists of OFDM synchronization, OFDM signal, OFDM data symbol and OFDM signal extension.[5]

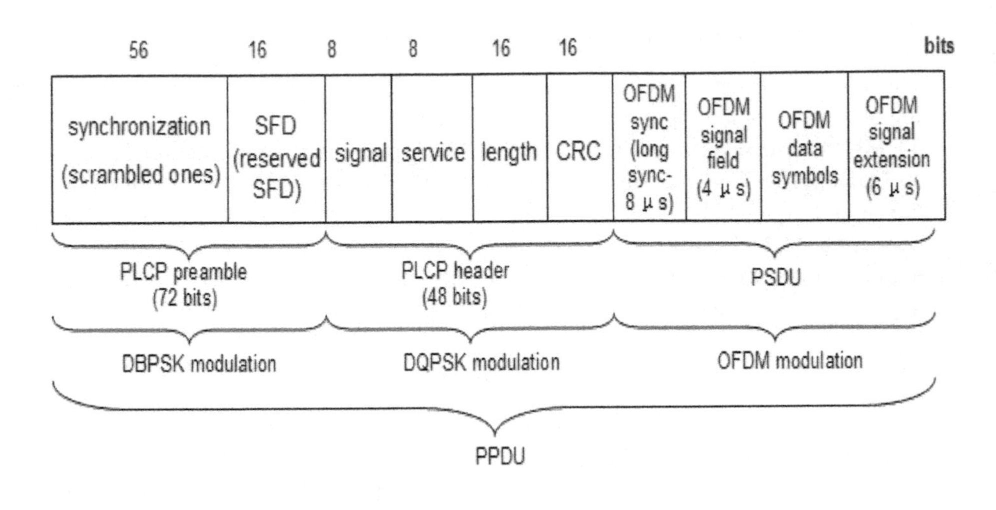

Figure 1.8: Short preamble PPDU format for DSSS-OFDM [5].

1.3 802.15.1/Bluetooth

1.3.1 Overview

Bluetooth wireless technology is a short-range communications technology intended to replace the cables connecting portable and/or fixed devices while maintaining high levels of security. The key features of Bluetooth technology are robustness, low power, and low cost. The Bluetooth specification defines a uniform structure for a wide range of devices to connect and communicate with each other. [6]

Bluetooth technology has achieved global acceptance such that any Bluetooth enabled device, almost everywhere in the world, can connect to other Bluetooth enabled devices in proximity. Bluetooth enabled electronic devices connect and communicate wirelessly through short-range, ad hoc networks known as piconets. Each device can simultaneously communicate with up to seven other devices within a single piconet. Each device can also belong to several piconets simultaneously. Piconets are established dynamically and automatically as Bluetooth enabled devices enter and leave radio proximity.[6]

A fundamental Bluetooth wireless technology strength is the ability to simultaneously handle both data and voice transmissions. This enables users to enjoy a variety of innovative solutions such as a hands-free headset for voice calls, printing and fax capabilities, and synchronizing PDA, laptop, and mobile phone applications to name a few.[6]

Bluetooth wireless technology is geared towards voice and data applications and able to penetrate solid objects. It is omni-directional and does not require line-of-sight positioning of connected devices.[6]

1.3.2 PHY Layer Specifications and Features

1.3.2.1 Frequency Band and Channel

Bluetooth operates in the 2.4GHz frequency band, it defines 79 channels (Channel 0: 2402MHz, Channel 78: 2480MHz) and each channel has 1MHz bandwidth.

1.3.2.2 Modulation Technique : GFSK

Gaussian frequency shift keying (GFSK) is a type of frequency shift keying (FSK) that utilizes a Gaussian filter to smooth positive/negative frequency deviations, which represent a binary 1 or 0. For Bluetooth the minimum deviation is 115kHz.

1.3.2.3 Power Class

It is primarily designed for low power consumption with a short range. The transmission distance is 1 meter, 10 meter, 100 meter depending on the Bluetooth device class. Each device is classified into three power classes.

Class 1: Designed for the range of approximate 100 meter, Maximum permitted power is 100 mW (20 dBm)

Class 2: Designed for the range of approximate 10 meter, Maximum permitted power is 2.5 mW (4 dBm)

Class 3: Designed for the range of approximate 1 meter, Maximum permitted power is 1 mW (0 dBm)

1.3.2.4 Using FHSS and TDD

As we know a lot of technologies utilize the unlicensed 2.4GHz ISM frequency band, Bluetooth adopts another spread spectrum technique which is different from DSSS used by 802.11b, called frequency hopping spread spectrum (FHSS) to resist the interference from other devices using the same frequency band.

1.3.2.5 Physical Links

1. Data Link : Asynchronous Connectionless (ACL)
    - Primarily used for data transmission
    - Data rate : asymmetric data rate 723.2 kbps (one way), 57.6 kbps (in the rcturn) symmetric data rate 433.9 kbps
    - Asynchronous
    - Fast acknowledge
    - Point - to - multipoint
    - Packet switched
2. Voice Link : Synchronous Connection Oriented (SCO)

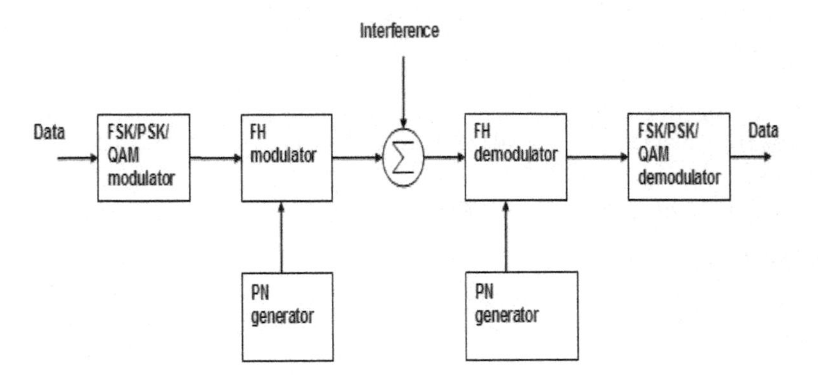

Figure 1.9: Frequency hopping spread spectrum.

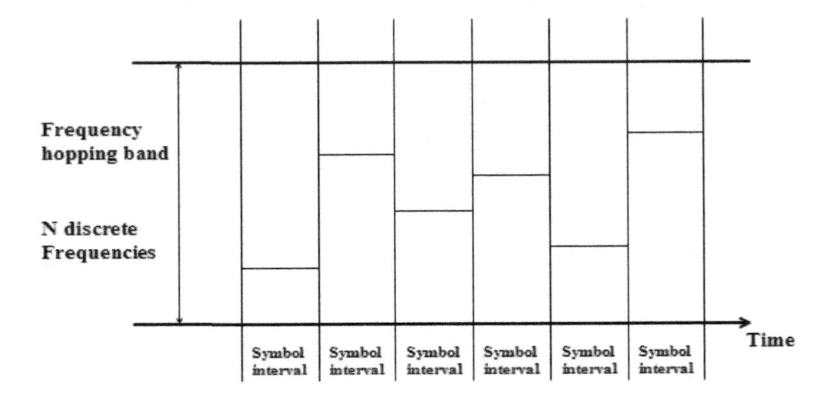

Figure 1.10: Frequency hopping.

- Used for voice transmission
- Voice channels operated with 64 kbps duplex
- Forward error correction (FEC)
- No retransmission
- Point - to -Point
- Circuit switched

1.3.3 Network Topology

The following figure shows the network topology of Bluetooth. Piconet is the connection of devices connected in an ad hoc fashion. One unit acts as master and the

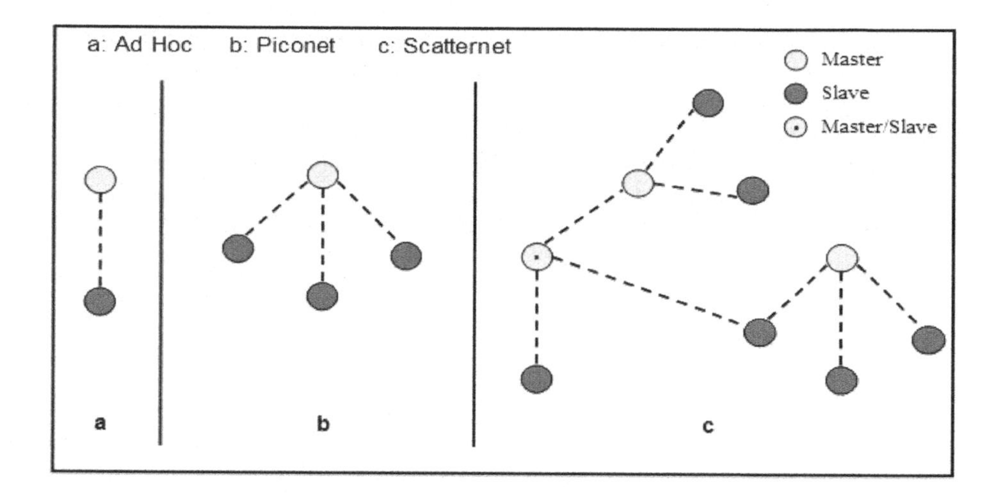

Figure 1.11: Bluetooth network topology [7].

others as slaves for the lifetime of the piconet. Master determines the hopping pattern. And each piconet has a unique hopping pattern which is determined by device ID the master gives. Slaves have to synchronize by the system clock of the master. When a new unit wants to participate in a piconet, synchronization to hopping sequence should be done first. Each piconet has one master and up to 7 active (simultaneous) slaves or over 200 passive slaves. Figure a and b illustrate the point to point connection (ad hoc) and point to multipoint connection piconet respectively. Scatternet consists of multiple co-located piconets through the sharing of common master or slave devices. Devices can be slave in one piconet and master of another. They also can jump back and forth between the piconets. Scatternet is high capacity system, which has minimal impact with up to 10 piconets within range. Figure c shows an example of scatternet.

1.4 802.15.4 Low-Rate WPAN

A wireless personal area network (WPAN), which concentrates on personal environment network solution, can be represented as a network for interconnecting personal devices by using wireless connections around an individual workspace.[8]

LR-WPANs address wireless networking and mobile computing devices have been making an impact in various fields. More or less, they have already changed our study and business modality, and will continue to. PCs, PDAs, peripherals, cell phones, Bluetooth earphone, wireless mouse, exist in our daily life, and for certain there will be more WPAN devices present tomorrow.

WPAN is a really popular technology nowadays. It emphasizes low cost and low power consumption in transmission. While, those attractive advantages usually make a sacrifice of transmission range and rate. That is why it is also called as short distance wireless networks. This is just like going an opposite way of the WLAN technologies which emphasize higher rate and longer range at the expense of cost and power consumption.

Some IEEE standards, especially IEEE 802.15 serial standards are referenced in our study through entire process, including theoretical analysis part, simulation part and test part. IEEE 802.15 serial standards are established by IEEE 802.15 Working Group for Personal Area Network or short distance wireless networks. Here, we introduce them respectively.[8]

- IEEE 802.15.1-2002 Standard primarily defines the lower layer transport layer (L2CAP, LMP, Baseband, and radio) of the Bluetooth wireless technology. It also has reviewed and provided a standard adaptation of the Bluetooth Specification v1.1 Foundation MAC (L2CA, PLMP, and Baseband) and PHY (Radio), and specifies other related aspects. It mainly established Bluetooth device implementations.

- IEEE 802.15.2-2003 is established for coexistence analysis of Wireless Personal Area Network and Wireless Local Area Network (802.11). The IEEE 802.15.2 working group developed a coexistence model to quantify the mutual interference of a WLAN and a WPAN. The working group also developed a set of Coexistence Mechanisms to facilitate coexistence of WLAN and WPAN devices. We use the BER analysis model provided by IEEE 802.15.2, also the algorithms of BER under different transmission types.

- IEEE 802.15.4 specifies wireless medium access control (MAC) sub layer and physical layer (PHY) specifications for low-rate wireless personal area networks. It also explores coexistence of WLAN and WPAN in its Annex. ZigBee is built on the IEEE 802.15.4. The two lower layers: the physical (PHY) layer and the medium access control (MAC) sub-layer of ZigBee stack architecture is specified in IEEE 802.15.4. We will go into more details in following chapters.

1.4.1 IEEE 802.15.4 Features

- Data rates of 250 kbps, 40 kbps, and 20 kbps. Symbol rate is 62.5 ksymbol/sppm.

- Two addressing modes; 16-bits short (short address) and 64-bit IEEE addressing (long address).

- Optional use Star-topology or Peer to Peer topology, and also supposes Cluster Tree nowadays.

- CSMA-CA channel access.

- Automatic network establishment by the coordinator.
- Full handshake protocol for transfer reliability.
- Power management to ensure low power consumption.
- 16 channels in the 2.4GHz ISM band, 10 channels in the 915MHz and one channel in the 868MHz band.
- Optional to use Acknowledgement packet.
- Transmit Power : About 1mW transmit power.
- RSSI (Received signal strength indication) measurement

1.4.2 Type of Device

IEEE 802.15.4 specifies components based on its transmission mechanism. The most basic components are devices. The LR-WPAN networks consist of devices. There are two types of device: FFD and RFD. FFD is full-function device while RFD is reduced-function device. As their names indicate, a FFD has more functionalities than a RFD. Generally, FFD can operate in three modes that are serving as a personal area network (PAN) coordinator, a coordinator, or a device. FFDs take charge of main data source transmission in a network, are able to talk to any other devices. RFDs act as end device that only can associate with one FFD at one time. IEEE 802.15.4 network need at least one FFD in the network to act as a coordinator. All devices should have 64 bits extended address or use 16 bits short address allocated by the coordinator instead. Commonly, RFDs use battery power while FFDs use line power.

1.4.3 Topology

The LR-WPAN may operate in either of two topologies: star topology or peer-to-peer topology. As Figure 1.12 shows, star topology contains devices and a PAN coordinator which acts as a central controller. The PAN coordinator initiates and terminates the network communication, it also routes communication packets. Generally, the coordinator is a FFD, which could establish a network and identify its network. Other available device, no matter FFD or RFD could join the network. Every star topology network is independent. In star topology, end device cannot communicate with each other without the coordinator. These types of networks are suitable for simple WPAN requirements. This topology, on one hand, can reduce the possibility that any end device causes the connection to fail. But, on the other hand, if it depends on the central coordinator too much it may cause the network incurable when the central coordinator collapses down. In star topology, addressing mode uses network and device identifier.

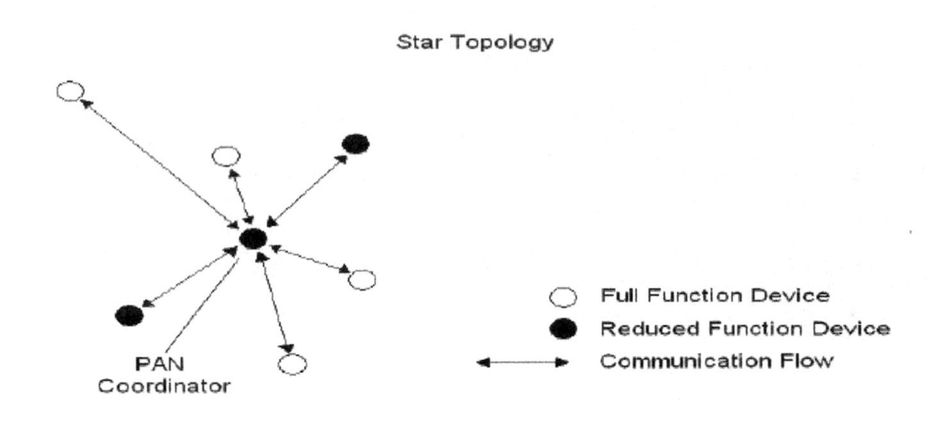

Figure 1.12: Star topology [9].

The peer-to-peer topology is more complex, it allows end devices to communicate with each other within its radio sphere of influence. It takes use of a PAN coordinator with more network functions comparing with the coordinator in star topology. This topology makes it possible to achieve more complex mission and extendable. In peer-to-peer topology, addressing mode uses source/destination identifier.

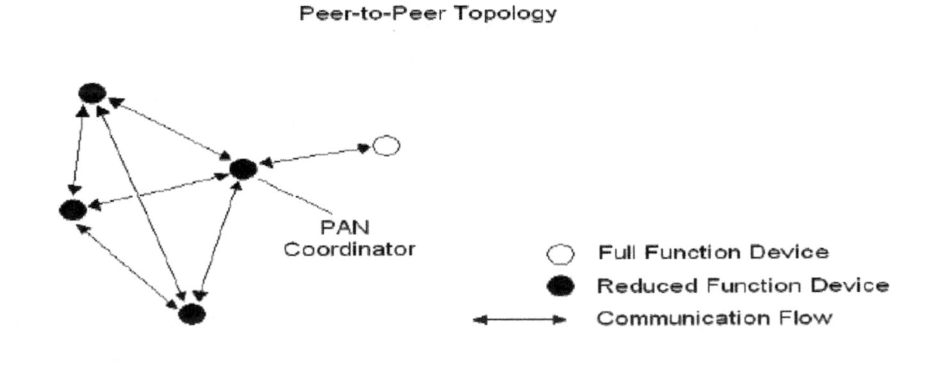

Figure 1.13: Peer - to - peer topology [9].

1.4.4 PHY Layer Specification

In IEEE 802.15.4, the PHY layer provides PHY data service and PHY management service interfacing to the physical layer management entity (PLME). The PHY data services enable the transmission and reception of PHY protocol data units (PPDUs) across the physical radio channel. IEEE 802.15.4 PHY layer is responsible for:

- Activation and deactivation of the radio transceiver
- Energy Detection within the current channel
- Link quality indication for received packets
- CCA for CSMA/CA
- Channel frequency selection
- Data transmission and reception

The standard specifies three license-free bands, which are: 868-868.9MHz, 902-928MHz and 2400-2483.5MHz. Different frequency bands have their specified transmission rates and modulation modes. 27 channels are available across the frequency bands, from number 0 to 26. We focus on analysis of transmission quality over the 2450MHz frequency band since it is an unlicensed ISM band, which is also utilized by WLAN transmission. 16 Channels distribute over the 2450 MHz frequency band from 2405MHz to 2480MHz, which can be obtained as:

$$Fc = 2405 + 5(k - 11) inmegahertz, fork = 11,1226[9] \tag{1.1}$$

where k is the channel number

1.4.4.1 Spread and Modulation

In 2450MHz, PHY use Direct Sequence Spread Spectrum (DSSS), and modulation type is OQPSK with 32 PN-code lengths. An RF bandwidth occupies 2MHz

DSSS used in IEEE 802.15.4 Direct-sequence spread-spectrum is one kind of spread spectrum technique. In short, spread spectrum technologies make signals that take up wider frequency bandwidth by using various pseudorandom sequences. Usually speaking, of spread spectrum are the enhance the signal's resistance to noise or interference and prevent malicious detection. At the receiver, the same pseudorandom sequence shall be used for de-spread. In this scenario, the DSSS processes employ PN sequences to represent each symbol. All bytes contained in the PPDU were split into 4 LSBs and 4 MSBs, and each of them shall map into one data symbol. There are 42 symbols as shown in the following table: 0000,1000,0100,1100,0010,1010, 0110,1110,0001,1001,0101,1101,0011,1011,0111,1111. Each data symbol shall be mapped into a 32-chip PN sequence (C0, C1 C30, C31) as specified. In de-spread process, the 32-chip PN shall map back to 4 LSBs or MSBs.

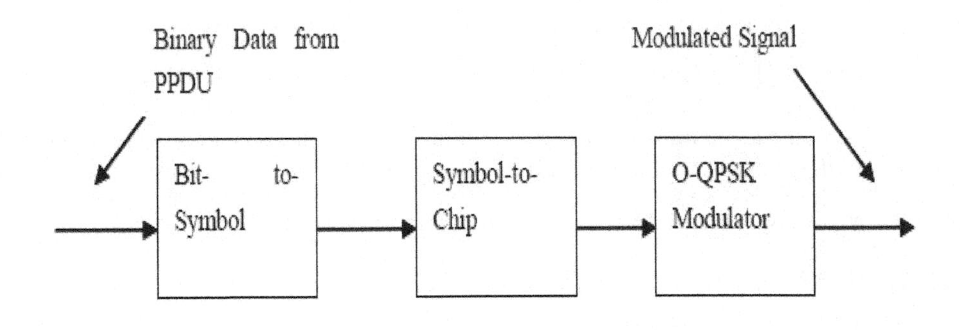

Figure 1.14: Spread and modulation functions[9].

Logically, it is possible that those spreaded signals could not find exactly the same chip used to map back since they were transmitted under noise and interference. There are some prescribe of receiver sensitivity, but this paper did not come into that part.

OQPSK is offset quadrature phase-shift keying modulation scheme. It divides signal into two portions which are in-phase (I) and quadrature-phase (Q), and transmitted shifted by half symbol duration. There is no phase shifts by 180 compared with QPSK. [10] The following figure illustrates that the chip sequences representing each data symbol which we describe in DSSS scenario are modulated onto the carrier using OQPSK with half sine pulse shaping. Even-indexed chips are modulated onto the in-phase (I) carrier and odd-indexed chips are modulated onto the quadrature-phase (Q).

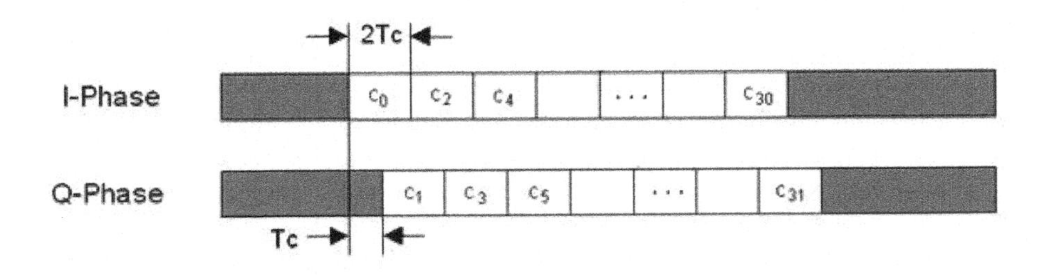

Figure 1.15: OQPSK chip offset [9].

Table 1.3: Symbol-to-Chip Mapping in DSSS [9]

Data Symbol (decimal)	Data Symbol (binary)$b_0 b_1 b_2 b_3$	Chip values ($c_0 c_1 \ldots c_{30} c_{31}$)
0	0000	11011001110000110101001000101110
1	1000	11101101100111000011010100100010
2	0100	00101110110110011100001101010010
3	1100	00100010111011011001110000110101
4	0010	01010010001011101101100111000011
5	1010	00110101001000101110110110011100
6	0110	11000011010100100010111011011001
7	1110	10011100001101010010001011101101
8	0001	10001100100101100000011101111011
9	1001	10111000110010010110000001110111
10	0101	01111011100011001001011000000111
11	1101	01110111101110001100100101100000
12	0011	00000111011110111000110010010110
13	1011	01100000011101111011100011001001
14	0111	10010110000001110111101110001100
15	1111	11001001011000000111011110111000

1.4.4.2 PPDU Frame Format

Usually, PPDU format is used to specify transmission data packets over PHY layer. The PPDU structure can be illustrated as following figure: Where, the synchronization header (SHR) includes Preamble (32 bits) and Start of Packet Delimiter (8 bits). The PHY header (PHR) has 8bits that first 7bits indicate length of PSDU and reserved 1 bit indicates whether the packet is received. Payload is data field from 0 to 127 bytes.

1.4.4.3 Clear Channel Assessment (CCA)

[9] CCA is an algorithm used to judge whether a channel is busy or idle by detection of the channel energy. IEEE 802.15.4 PHY layer offer three modes of CCA algorithm: Mode1: define an energy threshold, if energy of the channel is above the threshold, the channel is judged as busy, otherwise it is idle. Mode2: use carrier sense. The channel is judged as busy when a signal with modulation and spreading characteristics of IEEE 802.15.4 are detected. Mode3: use the carrier sense when detected energy above threshold. Report the channel is busy when a signal with the modulation and spread characteristics of IEEE 802.15.4 detected while with the channel energy above the ED threshold.

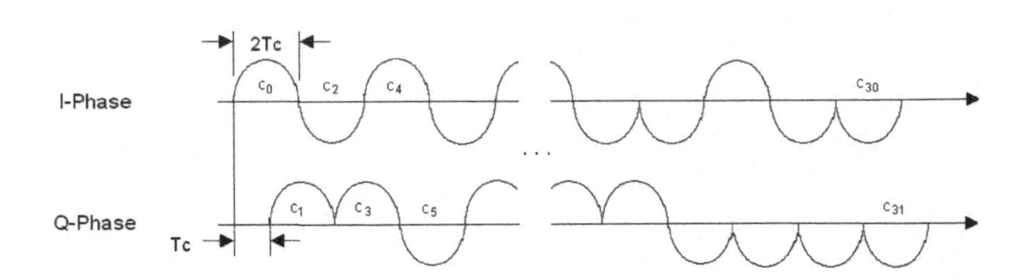

Figure 1.16: Sample baseband chip sequences with shaping[9].

Octets:4	1	1		variable
Preamble	SFD	Frame length (7 bites)	Reserved (1 bit)	PSDU
SHR		PHR		PHY payload

Figure 1.17: 802.15.4 PPDU format[9].

1.4.5 MAC Specification

MAC sublayer handles all access to the physical radio channel and is responsible for:

- Provide a reliable link between two peer MAC entities
- Coordinator generates of network beacons
- Support PAN association and disassociation
- Employ CSMA/CA mechanism for channel access
- Handling and maintaining guaranteed time slot mechanism

1.4.5.1 Superframe Structure

Superframe structure is used for channel bonding, but not limited to. (More information can be found in IEEE 802.22[11].) It is defined and sent by coordinator in WPAN; it can optionally bond its channel. If a WPAN does not wish to use the superframe structure, coordinator of this network shall not transmit beacons; all transmission shall use an un-slotted CSMA-CA mechanism to access the channel.[12] A

WPAN, which is beacons-enabled, wishes to use the superframe structure, and the superframe is sent bounded by network beacon frame. It is divided into 16 equally sized slots as the following Figure 2-16 shows: In IEEE 802.15.4 takes more specification

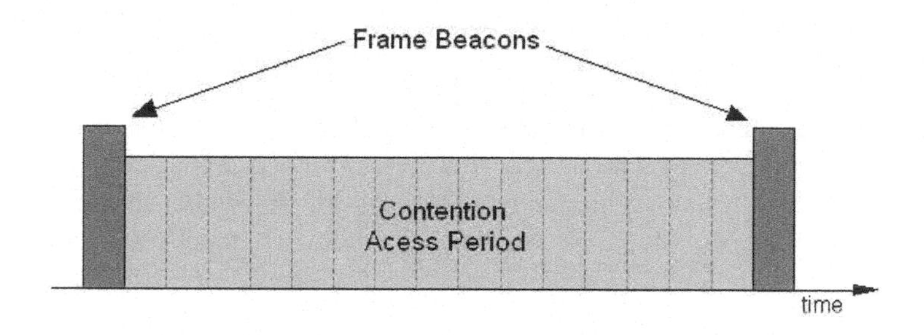

Figure 1.18: Superframe is sent bounded by network beacon frame[9].

on this point that "For low-latency applications or applications requiring specific data bandwidth, the PAN coordinator may dedicate portions of the active superframe to that application. These portions are called guaranteed time slots (GTSs)." [9] In short, GTSs provide a specific duration of time for the superframe without contention or latency. GTSs are set in Contention Free Period (CFP), continually following the end

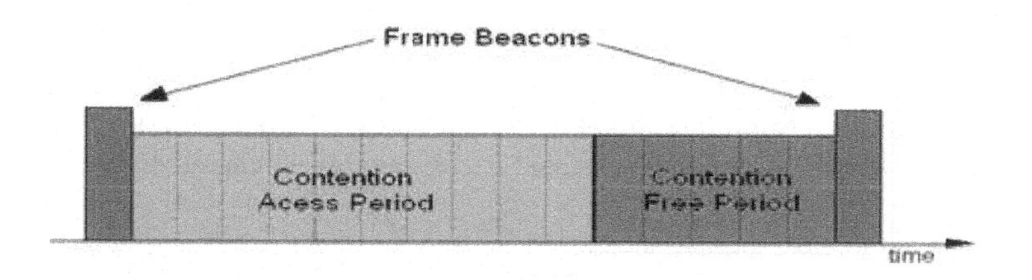

Figure 1.19: Superframe with GTSs[9].

slot of Contention Access Period (CAP). A device transmitting in GTSs shall ensure that the transmissions are completed before the next GTS or the end of this CFP. A WPAN coordinator may allocate up to 7 GTSs which means 7 devices at maximum could use GTS in one network. As coordinator, it must keep record of: starting slot, length of superframe slot, direction and associated device address. As the device associated with GTS must keep record of: starting slot, length of slot, direction. The data frames which use GTS should use short address (16 bits) for transmission.

1.4.5.2 Transmission Mechanisms

IEEE 802.15.4 specifies three types of data transactions: a device transfer data to a coordinator, a coordinator transfer data to a device, and data transfer between two peer devices. All three transmission mechanisms can be used in peer-to-peer topology. Start topology is limited to provide the third one since it does not support communication between two peer devices.

Data transfer to a coordinator from a device In non beacon-enabled network, if a device wishes to send data, it simply sends a data frame using an un-slotted CSMA-CA to the coordinator of the network. As shown in Figure 1.20, the coordinator responds with an optional acknowledgment frame. In a beacon-enabled network, when a device wishes to send data to the coordinator, it synchronizes to the superframe structure after it finds the network beacon. Then it sends its data using slotted CSMA-CA to the coordinator. The coordinator responds a with an optional acknowledgment frame.

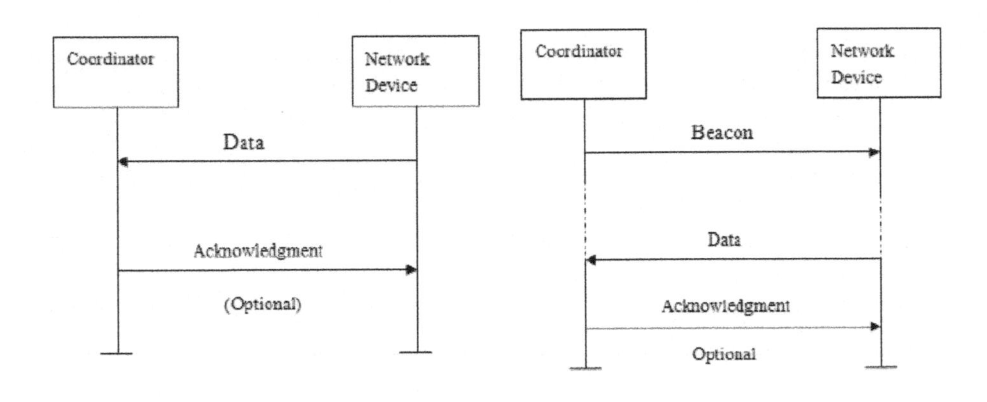

Figure 1.20: Communication to a coordinator in a nonbeacon-enabled network[9].

Figure 1.21: Communication to a coordinator in a beacon-enabled network[9].

Data transfer from a coordinator When a coordinator wishes to send data to a device in a non-enabled network, firstly the device requests the data that is stored in the coordinator. Since the coordinator stores the data for different device, a device may use a MAC command requesting the data by using un-slotted CSMA-CA to the coordinator at an application-defined rate. If the data are pending, the coordinator using un-slotted CSMA-CA to send the data frames. Otherwise, if the data are not pending, the coordinator sends the data with a zero-length payload to denote that. Then the coordinator responds with an acknowledgment frame and sends data frame sequentially. The coordinator sends out an acknowledgment frame when it finishes the data frame receive. In

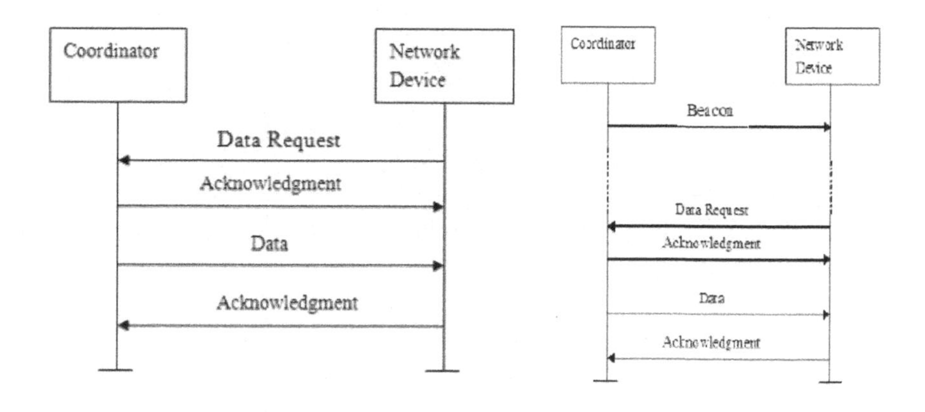

Figure 1.22: Communication from a coordinator in a nonbeacon-enabled network[9].

Figure 1.23: Communication from a coordinator in a beacon-enabled network[9].

a beacon-enabled network, when a coordinator wishes to send date to a device, it first composite network beacon with the data message is pending. The device may listen to this beacon and send a MAC command requesting the data using CSMA-CA. Then, the coordinator receives this request, it sends out an acknowledgment and sent the data using CSMA-CA sequentially. When the data is transmitted, the device sends out an acknowledgment frame. At this time, the message indicates that the pending data is removed from the pending message list.

1.4.5.3 MAC Layer Frame Format[9]

Four MAC frame formats are defined in IEEE 802.15.4. There are beacon, MAC command, data and acknowledgement frame. Beacon frame has $7+(4or10)+k+m+n$ as MAC sublayer frame, and totally has $13+(4or10)+k+m+n$bytes as PPDU in PHY layer. Where, k is GTS Fields value, m is pending address fields and n is Beacon payload. MAC command frame has $6+(4or20)+n$ as MAC sublayer frame, and totally has $12+(4to20)+n$bytes as PPDU. n is command payload. Data frame has $5+(4to20)+n$as MAC sublayer frame and has $11+(4to20)+n$ as PPDU in PHY layer. 4 to 20 are address information. n is data payload Acknowledgement frame has 5 bytes as MAC sublayer frame and has 11 bytes as PPDU in PHY layer.

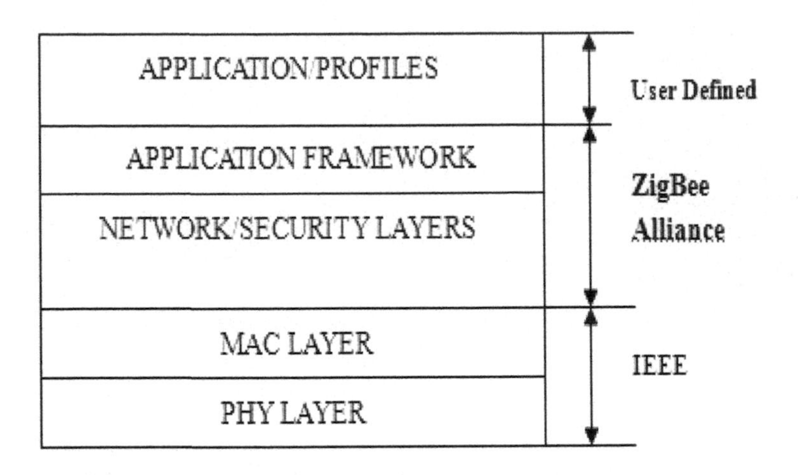

Figure 1.24: ZigBee layers [13].

1.4.6 ZigBee

1.4.6.1 Overview

When ZigBee networks began to gain consideration in 1998, people realized that besides those high data rate required network, there are many wireless networks that require low latency and low energy consumption but not high data rate, such as control or sensor network. The ZigBee 1.0 specification was released on December 14, 2004, to appease those requirements. The ZigBee network is actually a standard specified by two organizations that are IEEE 802.15 WPAN task group 4 and ZigBee Alliance. IEEE 802.15.4, which was established in May 2003, scopes on definition of physical layer (PHY) and media access control (MAC). ZigBee Alliance defined application support sub-layer (APS), ZigBee device object (ZDO), ZigBee device profile (ZDP), application framework, network layer (NWK) and ZigBee security services. ZigBee Alliance also publishes application profiles. The IEEE 802.15.4 PHY and MAC along with ZigBee network and application support layer provide low cost, low power consumption, short range operation, easy to implement and have appropriate level of security communication approach. ZigBee is typically is used for industrial control, embedded sensing, medical data collection, smoke and intruder warning, building automation, home automation, etc.

1.4.6.2 Architecture

The ZigBee stack as illustrated as Figure 1.24, the PHY and MAC layer are specified in IEEE 802.15.4 standard as a previous section introduced. The ZigBee Alliance builds on this foundation, providing network (NWK) layer and the framework for

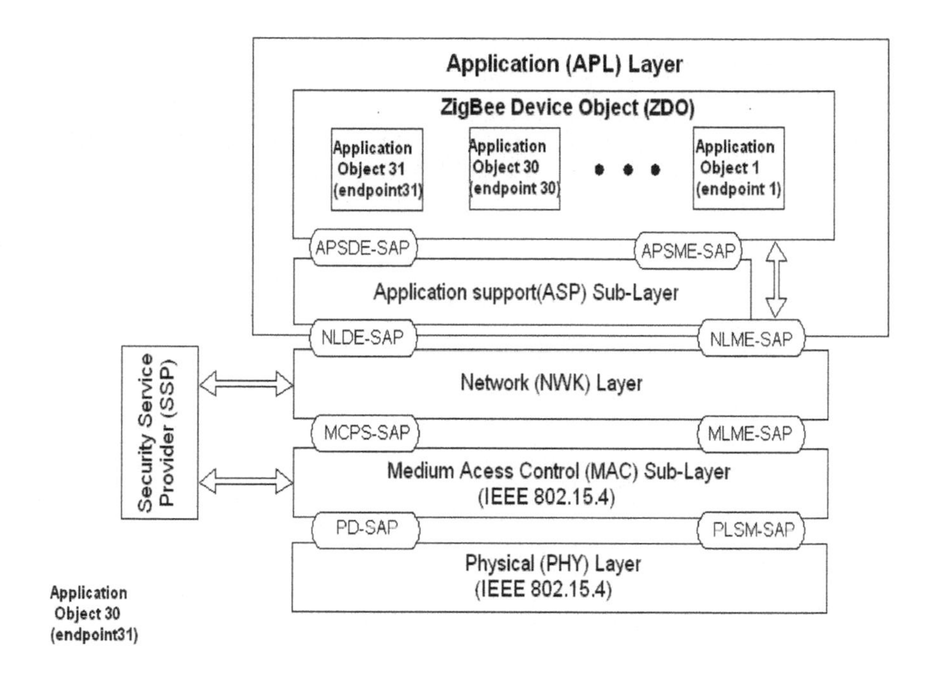

Figure 1.25: ZigBee stack[14].

application layer standards, which includes the application support sub-layer (APS), the ZigBee device objects (ZDO) and the manufacturer-defined application objects.

- Network layer: "The network layer builds upon the IEEE 802.15.4 MAC's features to allow extensibility of coverage. Additional cluster can be added; networks can be consolidated or split up."[14] The ZigBee NWK layer mainly takes charge of establishing a new network Joining and leaving a network, configuring the stack for operation when a new device joins the network, assigning address to device which is joining the network, this operation is carried by coordinator, routing frame to their destinations, enabling a device to synchronization with another device either through tracking beacons or by polling, and applying security operations.

- Application layer: The application layer consists of APS sub-layer and ZDO. It logically includes manufacturer-defined applications, which can be hardware or software. APS sub-layer provides discovery and binding services. Discovery is used for detecting devices which are working in range of a device. Binding is used to match two or more devices together and forwarding messages between bound devices.

1.4.6.3 ZigBee Network

ZigBee networks support three network topologies: star topology, peer-to-peer topology and cluster tree topology. Figure 1.26 shows those three topologies. Star net-

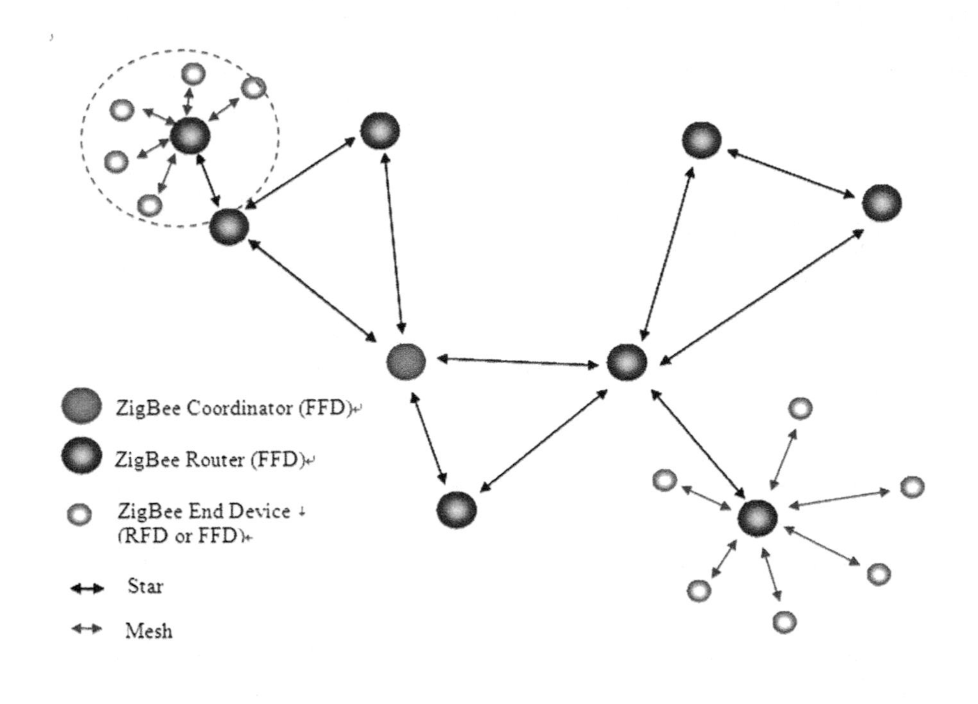

Figure 1.26: ZigBee network model[14].

works are suitable for simple requirement with low power consumption. Peer-to-peer networks have the capability of high level reliability and provides various paths in the network. Cluster tree topology actually just utilizes a hybrid star and peer-to-peer topology, benefits both for high level of reliability and support for battery power nodes.[13]

References

[1] Official IEEE 802.11 working group project timelines (18/11/07), retrieved on 2007-11-18.

[2] IEEE Std. 802.11TM-2007, IEEE Standard for Information technology - Telecommunications and information exchange between systems - Local and metropolitan area networks - Specific requirements Part 11: Wireless LAN Medium Access Control (MAC) and Physical Layer (PHY) specifications

[3] IEEE Std.802.11b-1999, IEEE standard for Wireless LAN Medium Access Control (MAC) and Physical Layer (PHY) specifications: Higher-Speed Physical Layer Extension in the 2.4 GHz Band

[4] IEEE 802.11b White Paper, VOCAL Technologies, Ltd. 2003-10.

[5] IEEE Std. 802.11gTM-2003, IEEE Standard for Information technology - Telecommunications and information exchange between systems - Local and metropolitan area networks - Specific requirements Part 11: Wireless LAN Medium Access Control (MAC) and Physical Layer (PHY) specifications, Amendment 4: Further Higher Data Rate Extension in the 2.4 GHz Band

[6] Bluetooth homepage: Available:*http : //www.bluetooth.com*

[7] IEEE Std. 802.15.1-2002, IEEE Standard for Information technology - Telecommunications and information exchange between systems - Local and metropolitan area networks - Specific requirements Part 15.1: Wireless Medium Access Control (MAC) and Physical Layer (PHY) Specifications for Wireless Personal Area Networks (WPANs)"

[8] IEEE 802.15 Working Group for WPAN homepage.
Available:*http : //www.ieee*802.*org*/15/

[9] IEEE Std.802.15.4: IEEE standard for Wireless Medium Access Control (MAC) and Physical Layer (PHY) Specifications for Low-Rate Wireless Personal Area Networks (LR-WPANs), 2003

[10] OQPSK introduction, waveform description
Available:*http : //www.hoka.com/th/*TH_w*aveformdescription.pdf*

[11] IEEE Std.P802.22/D0.1: Draft Standard for Wireless Regional Area Networks Part 22: Cognitive Wireless RAN Medium Access Control (MAC) and Physical Layer (PHY) specifications: Policies and procedures for operation in the TV Bands.

[12] MAC functional-Superframe Structure
Available:*http : //alkautsarpens.wordpress.com*/2008/01/28/*superframe − structure* − 2/

[13] ZigBee: Wireless Control That Simply Works, William C.Craig, Program Manager Wireless Communication, ZMD America, Inc

[14] ZigBee Technology: Wireless Control that Simply Works, Patrick Kinney, Kinney Consulting LLC, Chair of IEEE 802.15.4 Task Group, Secretary of ZigBee BoD, Chair of ZigBee Building Automation Profile WG

Chapter 2

ZigBee and IEEE 802.15.4 Standards

Yu-Doo Kim, Il-Young Moon

CONTENTS

2.1 Introduction

ZigBee is one of the wireless network technologies which are widely used from the low power environment. The low cost allows the technology to be widely deployed in wireless control and monitoring applications, the low power-usage allows longer life with smaller batteries, and the mesh networking provides high reliability and larger range. This technology defined by the ZigBee specification is intended to be simpler and less expensive than other WPAN, such as Bluetooth. ZigBee is targeted at radio-frequency applications that require a low data rate, long battery life, and secure networking. So it was organized as a very simple structure and offered at a low price. Current wireless network environment is separating that broadband network and low speed and low cost network. Therefore ZigBee is one important technology for low cost wireless networks. ZigBee technology is separated by two parts. First the application layer designate by ZigBee alliance which is association of companies. It is defining the network, security and application software layers. Next, the physical and medium access control layers are making standards by IEEE 802.15.4. MAC layer was access to wireless channel through CSMA/CA mechanism. Application and Network layer had been constituted ZigBee alliance. Application layer is made up APS (Application Support Sublayer), ZDO (ZigBee Device Object) and application framework. Network layer does exchange the information between terminal nodes. In this chapter, we explain the ZigBee Protocols and IEEE 802.15.4 standards through three parts. First, It explain about what is ZigBee. ZigBee chapter is included introduction of ZigBee architecture , application layer, network layer and security. In

Next chapter, it shows IEEE 802.15.4 standards. IEEE 802.15.4 group decide network topology, architecture, frame structure and channel access mechanism. In the final chapter, we will examine current research of ZigBee. Especially, it is explained related MAC standards such as RFID, smart utility.

2.2 ZigBee

The ZigBee architecture is separated by layers. Each layer performs a specific service set for the layer above. A data entity provides a data transmission service and a management entity provides all other services. Each service entity exposes an interface to the upper layer through a SAP (service access point), and each SAP supports a number of service primitives to achieve the required functionality. The IEEE 802.15.4 defines the two lower layers. First, the PHY(Physical) layer and the MAC (medium access control) sub-layer. The ZigBee Alliance builds on this foundation by providing the network layer and the framework for the application layer. The application layer framework consists of the APS (application support sub-layer) and the ZDO (ZigBee device objects). Application objects defined by the manufacturer use the framework and share APS and security with the ZDO. IEEE 802.15.4 has two PHY layers that operate in two separate frequency ranges, 868/915 MHz and 2.4 GHz. The lower frequency PHY layer covers both the 868 MHz European band and the 915 MHz band, used in countries such as the US and Australia. The higher frequency PHY layer is used virtually worldwide. The IEEE 802.15.4 MAC controls access to the radio channel using a CSMA-CA mechanism. Its responsibilities may also include transmitting beacon frames, synchronization, and providing a reliable transmission mechanism.

2.2.1 ZigBee Architecture

The ZigBee network layer supports star, tree, and mesh topologies. In a star topology, the network is controlled by one single device. The ZigBee coordinator is responsible for initiating and maintaining the devices on the network. All other devices, known as end devices, directly communicate with the ZigBee coordinator. In mesh and tree topologies, the ZigBee coordinator is responsible for starting the network and for choosing certain key network parameters, but the network may be extended through the use of ZigBee routers. In tree networks, routers move data and control messages through the network using a hierarchical routing strategy. Tree networks may employ beacon-oriented communication as described in the IEEE 802.15.4 specification. Mesh networks allow full peer-to-peer communication. ZigBee routers in mesh networks do not currently emit regular IEEE 802.15.4 beacons. This specification describes only intra-PAN networks, that is, networks in which communications begin and terminate within the same network.

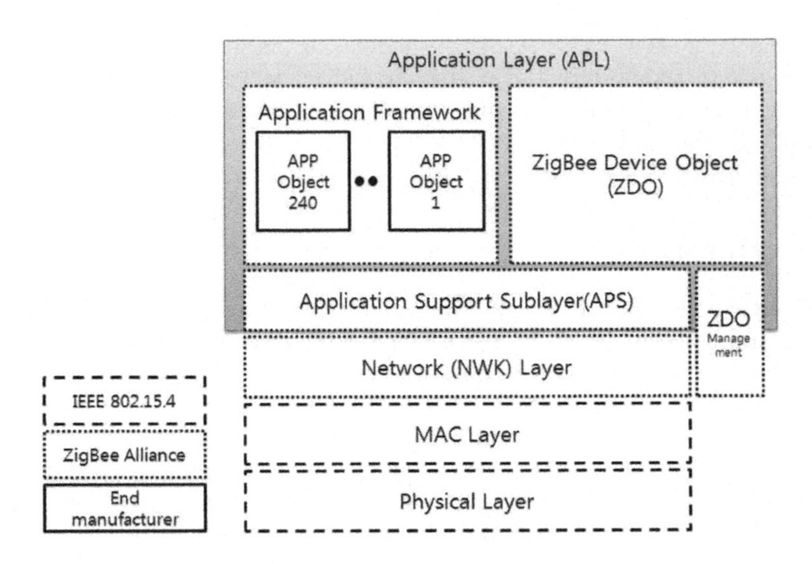

Figure 2.1: Structure of ZigBee and IEEE 802.15.4 architecture.

2.2.2 Application Layer

The ZigBee stack architecture includes a number of layered components including the IEEE 802.15.4 Medium Access Control layer, Physical layer, and the ZigBee Network layer. Each component provides an application with its own set of services and capabilities. Although this chapter may refer to other components within the ZigBee stack architecture, its primary purpose is to describe the component labeled APL (Application) Layer. The ZigBee application layer consists of the APS sublayer, the ZDO and the defined by manufacturer application objects.

2.2.2.1 APS (Application Support Sub-layer)

The application support sub-layer provides the interface between the network layer and the application layer through a general set of services for use by both the ZigBee device object and the defined by manufacturer application objects. These services are offered via two entities. The data service and the management service. The APSDE (APS data entity) provides the data transmission service via its associated SAP, the APSDE-SAP. The APSME (APS management entity) provides the management service via its associated SAP, the APSME-SAP, and maintains a database of managed objects known as the AIB (APS information base).

- APSDE : The APSDE shall provide a data service to the network layer and both ZDO and application objects to enable the transport of application PDUs between two or more devices. The devices themselves must be located on the same network.

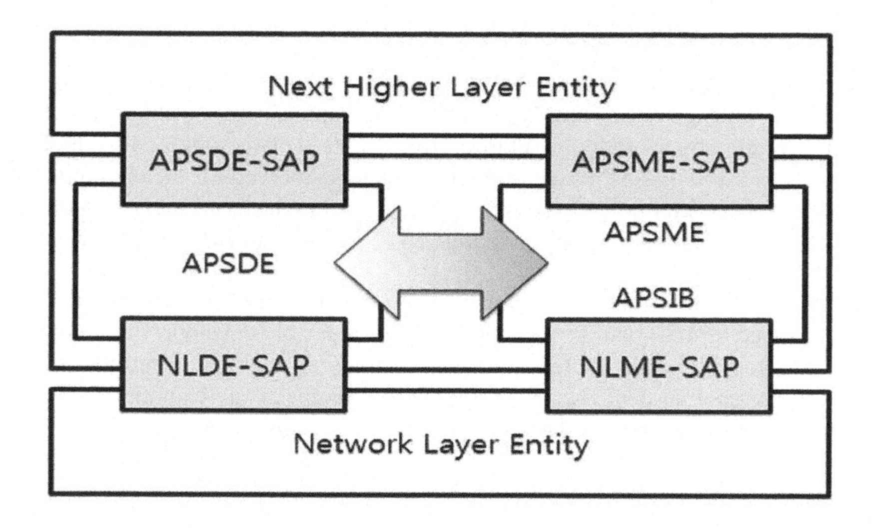

Figure 2.2: APS sub-layer model.

- APSME : The APSME shall provide a management service to allow an application to interact with the stack. The APSME shall provide the ability to match two devices together based on their services and their needs. This service is called the binding service, and the APSME shall be able to construct and maintain a table to store this information.

The APS sub-layer provides an interface between a NHLE (next higher layer entity) and the network layer. The APS sub-layer conceptually includes a management entity called the APSME (APS sub-layer management entity). This entity provides the service interfaces through which sub-layer management functions may be invoked. The APSME is also responsible for maintaining a database of managed objects pertaining to the APS sub-layer. This database is referred to as the APS sub-layer information base. The APS sub-layer provides two services, accessed through two SAP (service access points). These are the APS data service, accessed through the APSDE-SAP (APS sublayer data entity SAP), and the APS management service, accessed though the APSME-SAP (APS sub-layer management entity SAP). These two services provide the interface between the NHLE and the network layer, via the NLDE-SAP and, to a limited extent, NLME-SAP interfaces. The NLME-SAP interface between the network layer and the APS sub-layer supports only the NLME-GET and NLME-SET primitives; all other NLME-SAP primitives are available only via the ZDO. In addition to these external interfaces, there is also an implicit interface between the APSME and the APSDE that allows the APSME to use the APS data service.

2.2.2.2 *Application Framework*

The ZigBee application framework is the environment in which application objects are hosted on ZigBee devices. Up to 240 distinct application objects can be defined, each identified by an endpoint address from 1 to 240. Two additional endpoints are defined for APSDE-SAP usage. The endpoint 0 is reserved for the data interface to the ZDO, and endpoint 255 is reserved for the data interface function to broadcast data to all application objects. Endpoints 241-254 are reserved for future use.

- Application Profiles : Application profiles are agreements for messages, message formats, and processing actions that enable developers to create an interoperable, distributed application employing application entities that reside on separate devices. These application profiles enable applications to send commands, request data, and process commands and requests.
- Clusters : Clusters are identified by a cluster identifier, which is associated with data flowing out of, or into, the device. Cluster identifiers are unique within the scope of a particular application profile.

2.2.2.3 *ZDO (ZigBee Device Objects)*

The ZigBee device objects represent a base class of functionality that provides an interface between the application objects, the device profile, and the APS. The ZDO is located between the application framework and the application support sub-layer. It satisfies common requirements of all applications operating in a ZigBee protocol stack. The ZDO is responsible for the following:

- Initializing the application support sub-layer, the network layer, and the Security Service Provider.
- Assembling configuration information from the end applications to determine and implement discovery, security management, network management, and binding management.

The ZDO presents public interfaces to the application objects in the application framework layer for control of device and network functions by the application objects. The ZDO interfaces with the lower portions of the ZigBee protocol stack, on endpoint 0, through the APSDE-SAP for data, and through the APSME-SAP and NLME-SAP for control messages. The public interface provides address management of the device, discovery, binding, and security functions within the application framework layer of the ZigBee protocol stack.

2.2.3 Network Layer

The network layer is required to provide functionality to ensure correct operation of the IEEE 802.15.4 MAC and to provide a suitable service interface to the application layer. To interface with the application layer, the network layer conceptually includes two service entities that provide the necessary functionality. These service entities are the data service and the management service. The NLDE (network layer data entity) provides the data transmission service via its associated SAP, the NLDE-SAP, and the NLME (network layer management entity) provides the management service via its associated SAP, the NLME-SAP. The NLME utilizes the NLDE to achieve some of its management tasks and it also maintains a database of managed objects known as the NIB.

2.2.3.1 NLDE (Network Layer Data Entity)

The NLDE shall provide a data service to allow an application to transport APDU (application protocol data units) between two or more devices. The devices themselves must be located on the same network. The NLDE will provide the following service,

- NPDU(Generation of the Network level PDU) : The NLDE shall be capable of generating an NPDU from an application support sub-layer PDU through the addition of an appropriate protocol header.
- Topology-specific routing : The NLDE shall be able to transmit an NPDU to an appropriate device that is either the final destination of the communication or the next step toward the final destination in the communication chain.
- Security: The ability to ensure both the authenticity and confidentiality of a transmission.

2.2.3.2 NLME (Network Layer Management Entity)

The NLME shall provide a management service to allow an application to interact with the stack. The NLME shall provide the following services,

- Configuring a new device : this is the ability to sufficiently configure the stack for operation as required. Configuration options include beginning an operation as a ZigBee coordinator or joining an existing network.
- Starting a network : this is the ability to establish a new network.
- Joining, rejoining and leaving a network : this is the ability to join, rejoin or leave a network as well as the ability of a ZigBee coordinator or ZigBee router to request that a device leave the network.
- Addressing : this is the ability of ZigBee coordinators and routers to assign addresses to devices joining the network.
- Neighbor discovery : this is the ability to discover, record, and report information pertaining to the one-hop neighbors of a device.

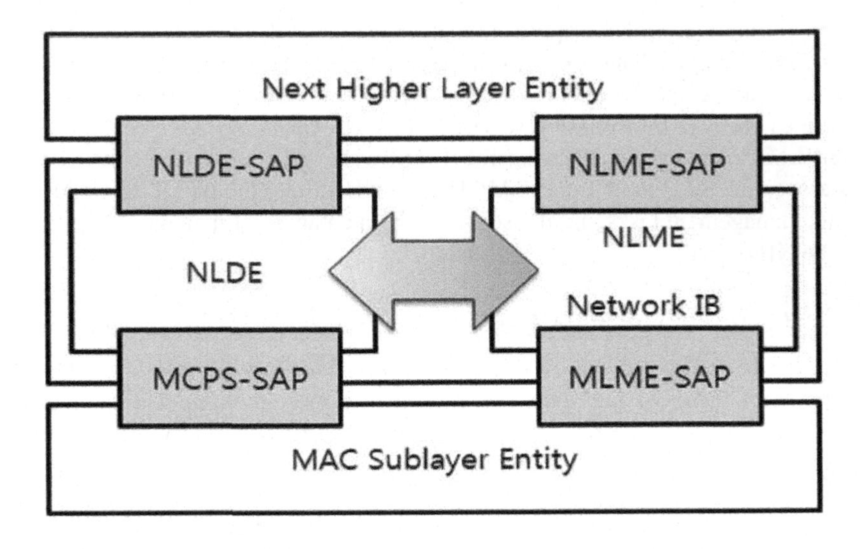

Figure 2.3: Network layer model.

- Route discovery : this is the ability to discover and record paths through the network, whereby messages may be efficiently routed.
- Reception control : this is the ability for a device to control when the receiver is activated and for how long, enabling MAC sub-layer synchronization or direct reception.
- Routing : this is the ability to use different routing mechanisms such as unicast, broadcast, multicast or many to one to efficiently exchange data in the network.

2.2.3.3 Service Specification

The network layer provides two services, accessed through two SAP (service access points). These are the network data service, accessed through the NLDE-SAP (network layer data entity SAP), and the network management service, accessed through the NLME-SAP (network layer management entity SAP). These two services provide the interface between the application and the MAC sub-layer, via the MCPS-SAP and MLME-SAP interfaces. In addition to these external interfaces, there is also an implicit interface between the NLME and the NLDE that allows the NLME to use the NWK data service.

SYNC	PHY HDR	MAC HDR	NWK HDR	Auxiliary HDR	Encrypted NWK Payload	MIC

Figure 2.4: Network layer security framework.

2.2.4 *Security*

The ZigBee security provided includes methods for key establishment, key transport, frame protection, and device management. These services form the building blocks for implementing security policies within a ZigBee device.

2.2.4.1 Architecture

The security architecture includes security mechanisms at two layers of the protocol stack. The network and APS layers are responsible for the secure transport of their respective frames. Furthermore, the APS provides services for the establishment and maintenance of security relationships. The ZDO manages the security policies and the security configuration of a device. The security mechanisms provided by the APS and network layers are described in this version of the specification.

2.2.4.2 Network Layer Security

When a frame originating at the network layer needs to be secured, or when a frame originates at a higher layer and the nwkSecureAllFrames attribute in the NIB is TRUE, ZigBee shall use the frame-protection mechanism, unless the SecurityEnable parameter of the NLDE-DATA.request primitive is FALSE, explicitly prohibiting security. The network layer's frame-protection mechanism shall make use of the AES (Advanced Encryption Standard) and use CCM as specified. The security level applied to a network frame shall be determined by the nwkSecurityLevel attribute in the NIB. Upper layers manage network layer security by setting up active and alternate network keys and by determining which security level to use.

2.2.4.3 APL Layer Security

When a frame originating at the APL layer needs to be secured, the APS sublayer shall handle security. The APS layer allows frame security to be based on link keys or the network key. Figure 2.5 shows an example of the security fields that may be included in an APL frame. The APS layer is also responsible for providing applications and the ZDO with key establishment, key transport, and device management services.

2.2.4.4 Trust Center Role

For security purposes, ZigBee defines the role of Trust Center. The Trust Center is the device trusted by devices within a network to distribute keys for the purpose of

Figure 2.5: APL layer security framework.

network and end-to-end application configuration management. All members of the network shall recognize exactly one Trust Center, and there shall be exactly one Trust Center in each secure network. In high-security, commercial applications a device can be pre-loaded with the Trust Center address and initial master key. Alternatively, if the application can tolerate a moment of vulnerability, the master key can be sent via an in-band unsecured key transport. If not pre-loaded, a device's Trust Center defaults to the ZigBee coordinator or a device designated by the ZigBee coordinator. In low-security, residential applications a device securely communicates with its Trust Center using the current network key, which can be preconfigured or sent via an in-band unsecured key transport. For purposes of trust management, a device accepts an initial master or active network key originating from its Trust Center via unsecured key transport. For purposes of network management, a device accepts an initial active network key and updated network keys only from its Trust Center. For purposes of configuration, a device accepts master keys or link keys intended for establishing end-to-end security between two devices only from its Trust Center. Aside from the initial master key or network key, additional link, master, and network keys are generally only accepted if they originate from a device's Trust Center via secured key transport.

2.3 IEEE 802.15.4 Standards

WPAN (Wireless personal area networks) are used to convey information over relatively short distances. Unlike WLAN (wireless local area networks), connections effected via WPANs involve little or no infrastructure. This feature allows small, power-efficient, inexpensive solutions to be implemented for a wide range of devices. The IEEE 802.15.4 standards' scope is to define the physical layer and medium access control sublayer specifications for low data rate wireless connectivity with fixed, portable, and moving devices with no battery or very limited battery consumption requirements typically operating in the POS (personal operating space) of 10m. It is foreseen that, depending on the application, a longer range at a lower data rate may be an acceptable trade-off. It is the intent of this project to work toward a level of coexistence with other wireless devices in conjunction with Coexistence Task Groups, such as 802.15.2 and 802.11/ETSI-BRAN/MMAC 5GSG.

2.3.1 Overview

A LR-WPAN is a simple, low-cost communication network that allows wireless connectivity in applications with limited power and relaxed throughput requirements. The main objectives of an LR-WPAN are ease of installation, reliable data transfer, short-range operation, extremely low cost, and a reasonable battery life, while maintaining a simple and flexible protocol. Some of the characteristics include:

- Over-the-air data rates of 250 kb/s, 40 kb/s, and 20 kb/s
- Star or peer-to-peer operation
- Allocated 16 bit short or 64 bit extended addresses
- Allocation of GTS (guaranteed time slots)
- CSMA-CA (Carrier sense multiple access with collision avoidance channel access)
- Fully acknowledged protocol for transfer reliability
- Low power consumption
- ED (Energy detection)
- LQI (Link quality indication)
- 16 channels in the 2450 MHz band, 10 channels in the 915 MHz band, and 1 channel in the 868 MHz band

Two different device types can participate in an LR-WPAN network. A FFD (full-function device) and a reduced-function device. The FFD can operate in three modes serving as a PAN (personal area network) coordinator, a coordinator, or a device. An FFD can talk to RFDs or other FFDs, while an RFD can talk only to an FFD. An RFD is intended for applications that are extremely simple, such as a light switch or a passive infrared sensor. They do not need to send large amounts of data and may only associate with a single FFD at a time. Consequently, the RFD can be implemented using minimal resources and memory capacity.

2.3.2 Network Topology

Depending on the application requirements, the LR-WPAN may operate in either of two topologies: the star topology or the peer-to-peer topology. In the star topology the communication is established between devices and a single central controller, called the PAN coordinator. A device typically has some associated application and is either the initiation point or the termination point for network communications. A PAN coordinator may also have a specific application, but it can be used to initiate, terminate, or route communication around the network. The PAN coordinator is the

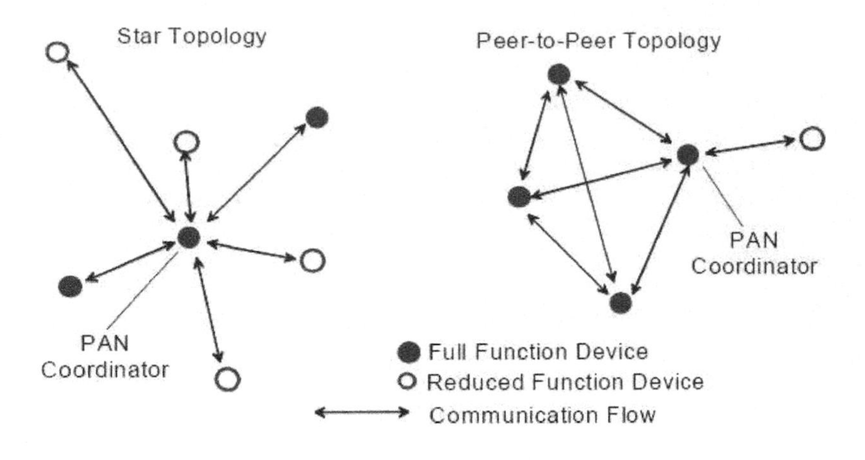

Figure 2.6: Network topology of IEEE 802.15.4.

primary controller of the PAN. All devices operating on a network of either topology shall have unique 64 bit extended addresses. This address can be used for direct communication within the PAN, or it can be exchanged for a short address allocated by the PAN coordinator when the device associates. The PAN coordinator may be main powered, while the devices will most likely be battery powered. Applications that benefit from a star topology include home automation, personal computer (PC) peripherals, toys and games, and personal health care. The peer-to-peer topology also has a PAN coordinator. But, it differs from the star topology in that any device can communicate with any other device as long as they are in range of one another. Peer-to-peer topology allows more complex network formations to be implemented, such as mesh networking topology. Applications such as industrial control and monitoring, wireless sensor networks, asset and inventory tracking, intelligent agriculture, and security would benefit from such a network topology. A peer-to-peer network can be ad hoc, self-organizing and self-healing. It may also allow multiple hops to route messages from any device to any other device on the network. Such functions can be added at the network layer, but are not part of this standard. Each independent PAN will select a unique identifier. This PAN identifier allows communication between devices within a network using short addresses and enables transmissions between devices across independent networks.

2.3.2.1 Star Network

The star network's basic structure can be seen Figure 2.6. After an FFD is activated for the first time, it may establish its own network and become the PAN coordinator. All star networks operate independently from all other star networks currently in operation. This is achieved by choosing a PAN identifier, which is not currently used by any other network within the radio sphere of influence. Once the PAN identifier is chosen, the PAN coordinator can allow other devices to join its network. Both FFDs and RFDs may join the network.

2.3.2.2 Peer-to-Peer Network

In a peer-to-peer topology, each device is capable of communicating with any other device within its radio sphere of influence. One device will be nominated as the PAN coordinator, for instance, by virtue of being the first device to communicate on the channel. Further network structures can be constructed out of the peer-to-peer topology and may impose topological restrictions on the formation of the network. An example of the use of the peer-to-peer communications topology is the cluster-tree. The cluster-tree network is a special case of a peer-to-peer network in which most devices are FFDs. An RFD may connect to a cluster tree network as a leave node at the end of a branch, because it may only associate with one FFD at a time. Any of the FFDs may act as a coordinator and provide synchronization services to other devices or other coordinators. Only one of these coordinators can be the overall PAN coordinator, which may have greater computational resources than any other device in the PAN. The PAN coordinator forms the first cluster by establishing itself as the CLH (cluster head) with a CID (cluster identifier) of zero, choosing an unused PAN identifier, and broadcasting beacon frames to neighboring devices. A candidate device receiving a beacon frame may request to join the network at the CLH. If the PAN coordinator permits the device to join, it will add the new device as a child device in its neighbor list. Then the newly joined device will add the CLH as its parent in its neighbor list and begin transmitting periodic beacons. Other candidate devices may then join the network at that device. If the original candidate device is not able to join the network at the CLH, it will search for another parent device. The simplest form of a cluster tree network is a single cluster network, but larger networks are possible by forming a mesh of multiple neighboring clusters. Once predetermined application or network requirements are met, the PAN coordinator may instruct a device to become the CLH of a new cluster adjacent to the first one. Other devices gradually connect and form a multicluster network structure.

2.3.3 Architecture

The LR-WPAN architecture is defined in terms of a number of blocks in order to simplify the standard. These blocks are called layers. Each layer is responsible for one part of the standard and offers services to the higher layers. The interfaces between the layers serve to define the logical links that are described in this standard. An LR-WPAN device comprises a PHY, which contains the RF (radio frequency) transceiver along with its low-level control mechanism, and a MAC sublayer that provides access to the physical channel for all types of transfer. Figure 2.7 shows these blocks in a graphical representation. The upper layers consist of a network layer, which provides network configuration, manipulation, and message routing, and an application layer, which provides the intended function of the device. The definition of these upper layers is outside the scope of this standard. An IEEE 802.2 Type 1 LLC

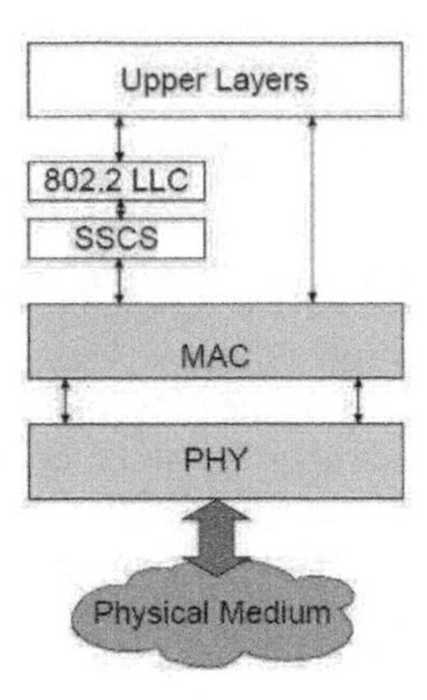

Figure 2.7: WPAN architecture.

(logical link control) can access the MAC sublayer through the SSCS (service specific convergence sublayer). The LR-WPAN architecture can be implemented either as embedded devices or as devices requiring the support of an external device such as a PC.

2.3.3.1 *PHY Layer*

The PHY provides two services: the PHY data service and the PHY management service interfacing to the PLME (physical layer management entity). The PHY data service enables the transmission and reception of PPDU (PHY protocol data units) across the physical radio channel. The features of the PHY are activation and deactivation of the radio transceiver, ED (Receiver Energy Detection), LQI, channel selection, CCA (clear channel assessment), and transmitting as well as receiving packets across the physical medium. The radio shall operate at one of the following license-free bands,

- 868 868.6 MHz (Europe)
- 902 928 MHz (North America)
- 2400 2483.5 MHz (Worldwide)

The receiver ED measurement is intended for use by a network layer as part of channel selection algorithm. It is an estimate of the received signal power within the bandwidth of an IEEE 802.15.4 channel. No attempt is made to identify or decode signals on the channel. The ED time should be equal to 8 symbol periods. Upon reception of a packet, the PHY sends the PSDU length, PSDU itself and link quality in the PD-DATA.indication primitive. The LQI measurement is a characterization of the strength and/or quality of a received packet. The measurement may be implemented using receiver ED, a signal-to-noise estimation or a combination of these methods.

The use of LQI result is up to the network or application layers. CCA is performed according to at least one of the following three methods.

- Energy above threshold. CCA shall report a busy medium upon detecting any energy above the ED threshold.
- Carrier sense only. CCA shall report a busy medium only upon the detection of a signal with the modulation and spreading characteristics of IEEE 802.15.4. This signal may be above or below the ED threshold.
- Carrier sense with energy above threshold. CCA shall report a busy medium only upon the detection of a signal with the modulation and spreading characteristics of IEEE 802.15.4 with energy above the ED threshold.

2.3.3.2 MAC Layer

The MAC sublayer provides two services: the MAC data service and the MAC management service interfacing to the MLME-SAP. The MAC data service enables the transmission and reception of MAC protocol data units across the PHY data service. The features of the MAC sublayer are beacon management, channel access, GTS management, frame validation, acknowledged frame delivery, association, and disassociation. In addition, the MAC sublayer provides hooks for implementing application appropriate security mechanisms. The slotted CSMA-CA shall be used if superframe structure is used in the PAN. If beacons are not being used in the PAN or a beacon cannot be located in a beacon-enabled network, an unslotted CSMA-CA algorithm is used.

- Slotted CSMA-CA : The backoff period boundaries of every device in the PAN are aligned with the superframe slot boundaries of the PAN coordinator. Each time a device wishes to transmit data frames during the CAP, it shall locate the 11 boundary of the next backoff period.
- Unslotted CSMA-CA : The backoff periods of one device do not need to be synchronized to the backoff periods of another device.

2.3.4 Frame Structure

2.3.4.1 Super Frame

The LR-WPAN standard allows the optional use of a superframe structure. The format of the superframe is defined by the coordinator. The superframe is bounded by network beacons, is sent by the coordinator, and is divided into 16 equally sized slots. The beacon frame is transmitted in the first slot of each superframe. If a coordinator does not wish to use a superframe structure, it may turn off the beacon transmissions. The beacons are used to synchronize the attached devices, to identify the PAN, and to describe the structure of the superframes. Any device wishing to communicate

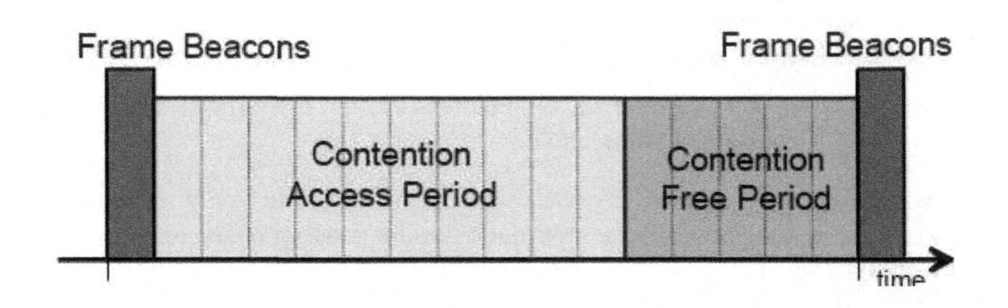

Figure 2.8: Superb frame structure.

during the CAP (contention access period) between two beacons shall compete with other devices using a slotted CSMA-CA mechanism. All transactions shall be completed by the time of the next network beacon. The superframe can have an active and an inactive portion. During the inactive portion, the coordinator shall not interact with its PAN and may enter a low-power mode.

2.3.4.2 Beacon Frame

Figure 2.9 shows the structure of the beacon frame, which originates from the MAC sublayer. A coordinator can transmit network beacons in a beacon-enabled network. The MSDU (MAC service data unit) contains the superframe specification, pending address specification, address list, and beacon payload fields (see Section 7.2.2.1). The MSDU is prefixed with a MHR (MAC header) and appended with a MFR (MAC footer). The MHR contains the MAC frame control fields, BSN (beacon sequence number), and addressing information fields. The MFR contains a 16 bit FCS(frame check sequence). The MHR, MSDU, and MFR together form the MAC beacon frame. The MPDU is then passed to the PHY as the PHY beacon packet payload. The PSDU is prefixed with a SHR (synchronization header), containing the preamble sequence and SFD (start-of frame delimiter) fields, and a PHR containing the length of the PSDU in octets. The preamble sequence enables the receiver to achieve symbol synchronization. The SHR, PHR, and PSDU together form the PHY beacon packet.

2.3.4.3 Data Frame

The data payload is passed to the MAC sublayer and is referred to as the MSDU. The MSDU is prefixed with an MHR and appended with an MFR. The MHR contains the frame control, sequence number, and addressing information fields. The MFR is composed of a 16 bit FCS. The MHR, MSDU, and MFR together form the MAC data frame. The MPDU is passed to the PHY as the PHY data frame payload. The PSDU is prefixed with an SHR, containing the preamble sequence and SFD fields, and a PHR containing the length of the PSDU in octets. The preamble sequence and the data SFD enable the receiver to achieve symbol synchronization. The SHR, PHR, and PSDU together form the PHY data packet.

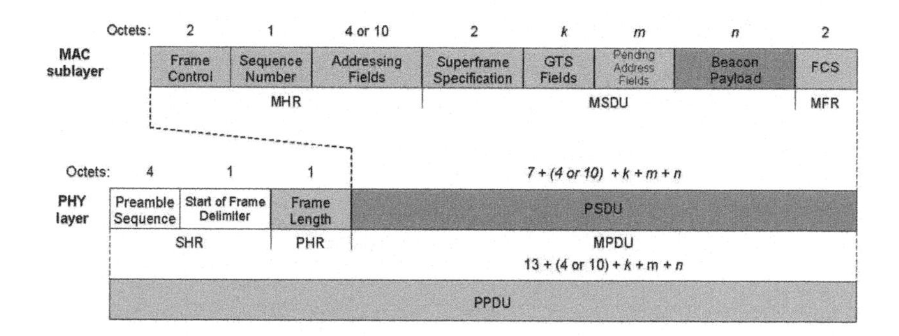

Figure 2.9: Beacon frame structure.

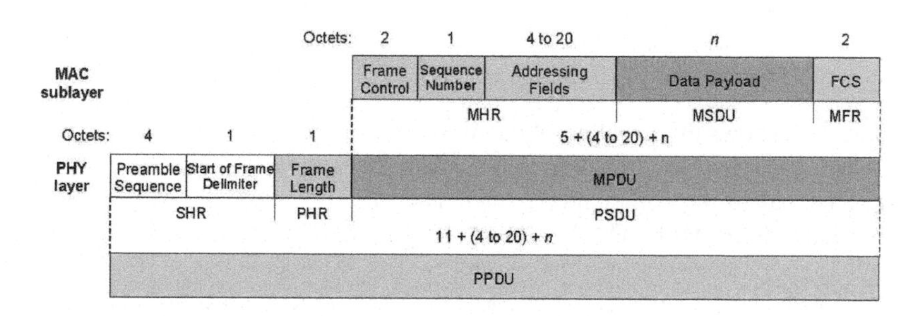

Figure 2.10: Data frame structure.

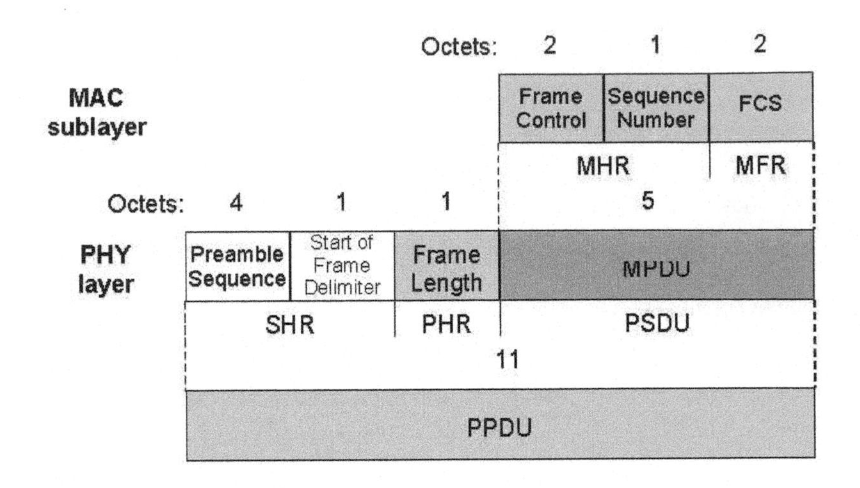

Figure 2.11: Acknowledgement frame structure.

2.3.4.4 Acknowledgement Frame

Figure 2.11 shows the structure of the acknowledgment frame, which originates from the MAC sublayer. The MAC acknowledgment frame is constructed from an MHR and an MFR. The MHR contains the MAC frame control and data sequence number fields. The MFR is composed of a 16 bit FCS. The MHR and MFR together form the MAC acknowledgment frame. The MPDU is passed to the PHY as the PHY acknowledgment frame payload. The PSDU is prefixed with the SHR, containing the preamble sequence and SFD fields, and the PHR containing the length of the PSDU in octets. The SHR, PHR, and PSDU together form the PHY acknowledgment packet.

2.3.4.5 MAC Command Frame

Figure 2.12 shows the structure of the MAC command frame, which originates from the MAC sublayer. The MSDU contains the command type field and command specific data, called the command payload. The MSDU is prefixed with an MHR and appended with an MFR. The MHR contains the MAC frame control, data sequence number, and addressing information fields. The MFR contains a 16 bit FCS. The MHR, MSDU, and MFR together form the MAC command frame. The MPDU is then passed to the PHY as the PHY command frame payload. The PSDU is prefixed with an SHR, containing the preamble sequence and SFD fields, and a PHR containing the length of the PSDU in octets. The preamble sequence enables the receiver to achieve symbol synchronization. The SHR, PHR, and PSDU together form the PHY command packet.

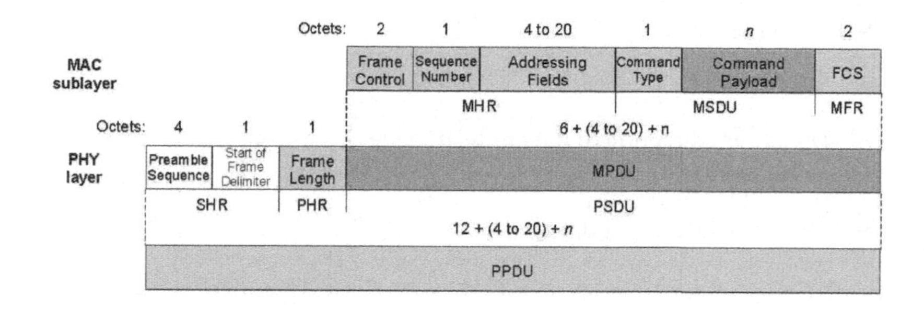

Figure 2.12: MAC command frame structure.

2.3.5 Channel Access Mechanism

The LR-WPAN uses two types of channel access mechanism, depending on the network configuration. Non beacon-enabled networks use an unslotted CSMA-CA channel access mechanism. Each time a device wishes to transmit data frames or MAC commands, it shall wait for a random period. If the channel is found to be idle, following the random backoff, the device shall transmit its data. If the channel is found to be busy, following the random backoff, the device shall wait for another random period before trying to access the channel again. Acknowledgment frames shall be sent without using a CSMA-CA mechanism. Beacon-enabled networks use a slotted CSMA-CA channel access mechanism, where the backoff slots are aligned with the start of the beacon transmission. Each time a device wishes to transmit data frames during the CAP, it shall locate the boundary of the next backoff slot and then wait for a random number of backoff slots. If the channel is busy, following this random backoff, the device shall wait for another random number of backoff slots before trying to access the channel again. If the channel is idle, the device can begin transmitting on the next available backoff slot boundary. Acknowledgment and beacon frames shall be sent without using a CSMA-CA mechanism.

2.3.6 GTS Allocation and Management

The GTS allows a device to operate on the channel within a portion of the superframe. It is dedicated exclusively to that device. If it is currently tracking the beacons, a device shall attempt to allocate and use a GTS only. The GTS shall be allocated only by the PAN coordinator and it shall be used only for communications between the PAN coordinator and a device. Single GTS can extend over one or more superframe slots. The PAN coordinator may allocate up to seven GTSs at the same time, provided there is sufficient capacity in the superframe. The PAN coordinator must be stored this information, these parameters are included in the GTS request command. In radical state, PAN Coordinator can deallocate one or more GTS to keep minimum CAP length.

- Starting slot
- Length
- Direction : direction of data flow, from device to coordinator, from coordinator to device
- Associated device address

Each device request can transmit GTS or receive GTS and the device can allocate and use GTS when tracking beacon. If synchronization with all PAN Coordinators is broken, all GTS allocations are lost, too.

2.4 Related MAC Standards

In this chapter, we see the trends of current ZigBee research. Especially, there are several important standards and working groups. First, IEEE 802.15.5 is researched for mesh network. Next there are many task group. 802.15.4e research WPAN enhancements, 802.15.4f task group made for RFID, 802.15.4g is working for smart utility, etc.

The IEEE 802.15 Task group 5 is chartered to determine the necessary mechanisms that must be present in the PHY and MAC layers of WPANs to enable mesh networking.

2.4.0.1 IEEE 802.15.5

The mesh network is a PAN that employs one of two connection arrangements, full mesh topology or partial mesh topology. In the full mesh topology, each node is connected directly to each of the others. In the partial mesh topology, some nodes are connected to all the others, but some of the nodes are connected only to those other nodes with which they exchange the most data. Mesh networks have the capability to provide:

- Extension of network coverage without increasing transmit power or receive sensitivity
- Enhanced reliability via route redundancy
- Easier network configuration
- Better device battery life due to fewer retransmissions

2.4.0.2 IEEE 802.15.4e WPAN Enhancement

The IEEE 802.15.4e is chartered to define a MAC enhancement to the existing standard 802.15.4-2006. The intent of this amendment is to enhance and add functionality to the 802.15.4-2006 MAC to better support the industrial markets and permit compatibility with modifications being proposed within the Chinese WPAN. This group has defined the application spaces that it will address along with the MAC behavior changes and additions that are required to enable those application spaces. These application spaces are Factory Automation, Process Automation, Asset Tracking, General Sensor Control, Home Medical Health and Monitor, Telecom Application, Neighborhood Area Networks, Audio.

2.4.0.3 IEEE 802.15.4f for RFID

The IEEE 802.15.4f is chartered to define new wireless Physical layer and enhancements to the 802.15.4-2006 standard MAC layer which are required to support new PHY for Active RFID System. An Active RFID tag is a device which is typically attached to an asset or person with a unique identification and the ability to produce its own radio signal not derived from an external radio signal. Tag applications include wireless sensor telemetry, control, and location determination. To generate a radio signal, Active RFID tags must employ some source of power. Traditionally this has been accomplished by integrated batteries, although designs exist for such devices that harvest ambient energy from the surrounding environment.

2.4.0.4 IEEE 802.15.4g for Smart Utility

The role of IEEE 802.15 Smart Utility Networks is to create a PHY amendment to 802.15.4 to provide a global standard that facilitates very large scale process control applications such as the utility smart-grid network capable of supporting large, geographically diverse networks with minimal infrastructure, with potentially millions of fixed endpoints.

There are other task groups. TG6 (Task Group 6) is researching for body area networks, TG7 was made for visible light communication and IGthz has been chartered to explore the feasibility of Terahertz for wireless communications.

2.5 Conclusion

In this chapter, we explain ZigBee and IEEE 802.15.4 standards. The ZigBee is separated into two parts. Application and network layer defined by ZigBee Alliance. Then, physical and medium access control layer defined by IEEE 802.15.4. The ZigBee has many advantages: low cost, low power, and small size. So it is widely used. Especially, it is very useful in sensor networks because it is composed of small device and has a low-power battery. If you use ZigBee in sensor networks, you can have many advantages in your research.

References

[1] IEEE Std 802.15.4-2003 Part 15.4: Wireless Medium Access Control (MAC) and Physical Layer (PHY) Specifications for Low-Rate Wireless Personal Area Networks (LR-WPANs), 2003.

[2] Jose A. Gutierrez, "IEEE 802.15.4 Tutorial", Document of IEEE 802.15-03/036r0, Jan. 2003.

[3] ZigBee Alliance, ZigBee Specification, 2008.

[4] ZigBee Device Object, "ZigBee document 03525r5ZB", ZigBee Alliance, March 2004.

[5] Yu-Doo Kim and Il-Young Moon, "Improved AODV Routing Protocol for Wireless Sensor Network based on ZigBee", The 11th International Conference On Advanced Communication Technology, 2009.

[6] Ran Peng, Sun Mao-heng, Zou You-min,"ZigBee Routing Selection Strategy Based on Data Services and Energy-balanced ZigBee Routing", Proceedings of the 2006 IEEE Asia-Pacific Conference on Services Computing, 2006.

[7] http://ieee802.org/15/index.html

Chapter 3

Performance Analysis of the IEEE 802.15.4 MAC Layer

M.R. Palattella, A. Faridi, G. Boggia, P. Camarda, L.A. Grieco, M. Dohler, A. Lozano

CONTENTS

In this chapter, the IEEE 802.15.4 MAC layer is modeled using a per-node Markov chain model. Using this model, expressions for various performance metrics including delay, throughput, power consumption, and efficiency, are derived and such expressions are subsequently validated against the corresponding values obtained via simulation. The simplifying assumptions required by the Markov-chain analysis are studied and their impact on the performance metrics is quantified.

3.1 Introduction

As already alluded to in previous chapters of this book, ZigBee is arguably the most prominent alliance dedicated to low-power embedded systems. It is a facilitator of applications pertaining to home and building automation, smart metering, health care, among many others. Its link and access protocols rely on the specifications of IEEE 802.15.4 [1], whereas higher layers are subject to the profile definition of the ZigBee special interest group (SIG). The Internet engineering task force (IETF), however, has lately commenced standardizing networking protocols within their 6LoWPAN [2] and ROLL [3] which are assuming IEEE 802.15.4 link layer technology; ZigBee may thus adopt IETF's networking solution in the future. In short, ZigBee is gaining importance and the underlying IEEE standard ensures that technology is available from multiple vendors.

On the downside, true deployment success stories are fairly rare still which might be due to the fact that it operates in the highly interfered 2.4 GHz ISM band. Also, ZigBee is not alone and needs to compete with Wibree, a low-power solution based on Bluetooth; Wavenis, the only ultra low-power solution on the smart metering market today; Zwave, a short range solution backed by Intel and Cisco; IO-Homecontrol, an international alliance of worldwide leaders for building management solutions; Konnex/KNX, a European standard for home & building automation; Wireless HART, a SIG offering interoperable wireless communication standards for process measurement and control applications; just to mention a few.

The physical (PHY) layer, which is responsible for maintaining a reliable point-to-point link, is comprised of at least six different solutions. As such, the 2006 revision of the IEEE 802.15.4 standard defines four PHY layers:

- 868/915 MHz Direct Sequence Spread Spectrum (DSSS) with binary phase shift keying;
- 868/915 MHz DSSS with offset quadrature phase shift keying;

- 2450 MHz DSSS with offset quadrature phase shift keying;
- 868/915 MHz Parallel Sequence Spread Spectrum (PSSS), a combination of binary keying and amplitude shift keying.

The 2007 IEEE 802.15.4a version includes two additional PHY layers:

- 2450 MHz Chirp Spread Spectrum (CSS);
- Direct Sequence Ultra-Wideband (UWB) below 1 GHz, or within 3–5 or 6–10 GHz.

Beyond these PHYs at the three bands, there are IEEE 802.15.4c for the 314–316, 430–434 and 779–787 MHz bands in China and IEEE 802.15.4d for the 950–956 MHz band in Japan since these countries recognized that interference in the congested ISM bands is severely deteriorating performance. The above PHY solutions trade complexity with performance and energy efficiency but are all generally facilitating embedded operations at low power.

The medium access control (MAC) layer, which is responsible for maintaining a collision-free schedule among neighbors, is tailored to the low-power needs of embedded radios. There are generally two channel access methods, i.e., the non-beacon mode for low traffic and the beacon-enabled mode for medium and high traffic. The former is a traditional multiple access approach used in simple peer networks; it uses standard carrier sensing multiple access (CSMA) for conflict resolution and positive acknowledgments for successfully received packets. The latter is a flexible approach able to mimic the behavior of a large set of previously published wireless sensor network (WSN) MACs, such as framed MACs, contention-based MACs with common active periods, sampling protocols with low duty cycles, and hybrids thereof [4]; it follows a flexible superframe structure where the network coordinator transmits beacons at predetermined intervals. It successfully combats the main sources of energy drainage by minimizing idle listening, overhearing, collisions and protocol overheads — it may not be the optimum MAC for all applications but covers a large number of envisaged ZigBee applications sufficiently well.

The device classes that are supported by ZigBee are the full function device (FFD), which can be a simple node as well as a network coordinator, and the reduced function device (RFD), which cannot become a network coordinator and hence only talks to a network coordinator. A combination of FDD and RFD allows one to realize any networking topology, such as star, ring, mesh, etc.

Henceforth, we will assume that PHY and networking protocols are given. We will thus concentrate on formalizing the MAC behavior of IEEE 802.15.4 since it is a crucial step in a successful system deployment with multiple parties suffering from contention. In this chapter, we will only focus on the slotted CSMA/CA mechanism in the beacon-enabled mode. The center of our investigations is to understand the parameters and system assumptions of said MAC and to analyze its performance in terms of delay, throughput, power consumption and efficiency. With these tools at hand, a synthesis of parameters which optimizes a given metric, such as efficiency, becomes feasible. Such a synthesis, even though not explicitly conducted here for space reasons, is central to system designers as it allows one to use derive formulas to optimize the performance of the ZigBee network under given operating conditions. For instance, one could derive analytical expression for a suitable number of contending nodes to satisfy some trade-off between delay and energy efficiency. Such expressions are currently not available as the synthesis, i.e., inversion of equations characterizing performance, has been deemed too complex. This leaves a field engineer no other choice but to parametrize the rolled-out ZigBee network manually based on a visual inspection of performance graphs. The below outlined approach is hence a significant step forward in that such parametrization can henceforth be automated.

This chapter is structured as follows. In the following section, we will detail the IEEE 802.15.4 MAC structure and its key parameters. We will then review prior works that characterized the performance of said MAC, using a Markov chain model. We then move on to the characterization and analysis of the ZigBee MAC. Finally, conclusions are drawn and future research indications given.

3.2 IEEE 802.15.4 WPANs: An Overview

An IEEE 802.15.4 network [1], known also as Low Rate Wireless Personal Area Network (LR-WPAN), is composed of two different types of devices: FFDs and RFDs. An FFD can operate in three distinct modes serving as: a personal area network (PAN) coordinator, a coordinator, or a device. An FFD can exchange data with both RFDs and other FFDs, whereas an RFD can communicate only with an FFD. RFDs are usually employed in applications that are extremely simple, e.g., a light switch or a passive infrared sensor, where a very limited amount of data has to be sent to a single FFD. They can be implemented using minimal resources and memory capacity.

In an LR-WPAN, the PAN coordinator (i.e., the central controller) builds the network in its personal operating space. Communications from nodes to coordinator (uplink), from coordinator to nodes (downlink), or from node to node (ad hoc) are possible. Two networking topologies are supported: star and peer-to-peer (Figure 3.1). The star topologies are well suited to PANs covering small areas. In this case, the PAN coordinator controls the communication, acting as a network master. In the peer-to-peer topology, any device can communicate with any other device as long as they are in range of one another. With this kind of topology, more complex networks can be realized (e.g., mesh networks), supporting several types of applications, such as industrial control and monitoring, wireless sensor networks, asset and inventory tracking, and intelligent agriculture. In a peer-to-peer topology, multiple hops can route messages from any device to any other device in the network. A special case of a peer-to-peer network is the cluster tree in which most devices are FFDs. An RFD connects to such a network as a leaf device at the end of a branch because RFDs do not allow other devices to associate. Any of the FFDs can act as a coordinator, providing synchronization services to other devices or other coordinators. Only one of these coordinators can be the overall PAN coordinator, which may have greater computational resources than any other device in the PAN. This

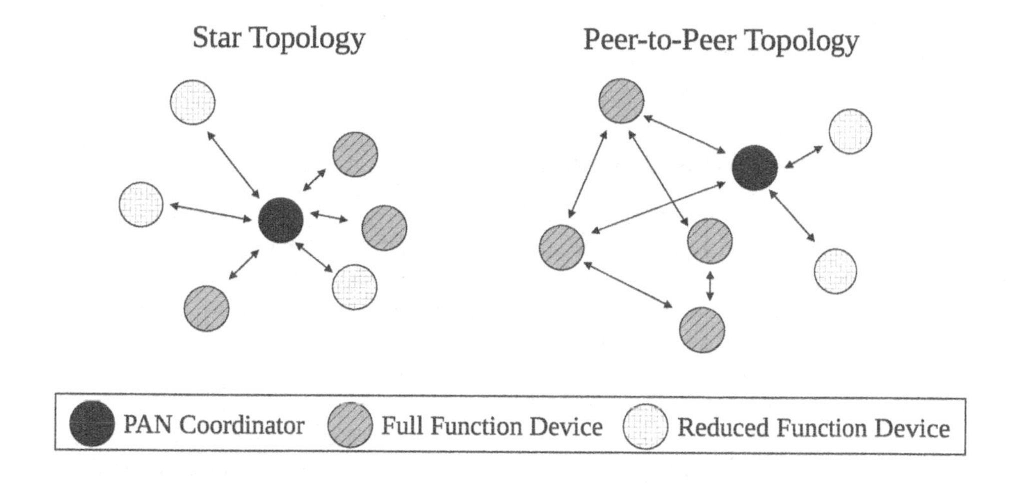

Figure 3.1: Example of IEEE 802.15.4 topologies.

requires a mechanism to decide the PAN coordinator and a contention resolution mechanism if two or more FFDs simultaneously attempt to establish themselves as a PAN coordinator.

LR-WPANs can operate in two distinct modes: *beacon-enabled* and *nonbeacon-enabled* modes. In the beacon-enabled mode, the time axis is structured as an endless sequence of superframes. Each one comprises an active part and an optional inactive part. The active part is made by a contention access period (CAP), during which a slotted CSMA/CA mechanism is used for channel access, and an optional contention free period (CFP). During the inactive part of the superframe, the devices do not interact with the PAN coordinator and could enter in a low-power state to save energy. In the nonbeacon-enabled mode, the superframe structure is not used, but the unslotted CSMA/CA mechanism is adopted. The use of beacon-enabled or nonbeacon-enabled modes depends on the application; for example, beacon transmissions are disadvantageous when no periodic or frequent messages are expected from the coordinator, and only sporadic traffic is transmitted by network devices.

It should be noted that the slotted CSMA/CA mechanism adopted with the beacon-enabled mode is different from the well-known IEEE 802.11 CSMA/CA scheme [5]. The main differences are: the time slotted behavior, the backoff algorithm, and the Clear Channel Assessment (CCA) procedure used to sense whether the channel is idle. In the slotted CSMA/CA algorithm, each operation (channel access, backoff count, CCA) can only begin at the boundary of time slots, called backoff periods (BPs). Moreover, unlike in the IEEE 802.11 CSMA/CA scheme, the backoff counter value of a node decreases regardless of the channel status. In fact, only when the backoff counter value reaches zero, the node performs two CCAs, during which it senses the channel to verify if it is idle. This allows great energy saving compared to IEEE 802.11 CSMA/CA scheme, given that during listening a significant amount of energy is spent.

3.2.1 Superframe Structure

Herein, more details about the superframe structure adopted in the beacon-enabled mode are given. The format of the superframe is defined by the PAN coordinator. As briefly described before, the superframe consists of an active period and an optional inactive period. All communications take place in the active period. In the inactive period, instead, nodes are allowed to power down and save energy. Each superframe is bounded by two beacon frames, as shown in Figure 5.2. The beacons are used to synchronize devices attached to the PAN, to identify the PAN, and to give the description of the superframe structure. They are also used for carrying service information for network maintenance, and to notify nodes about pending data in the downlink.

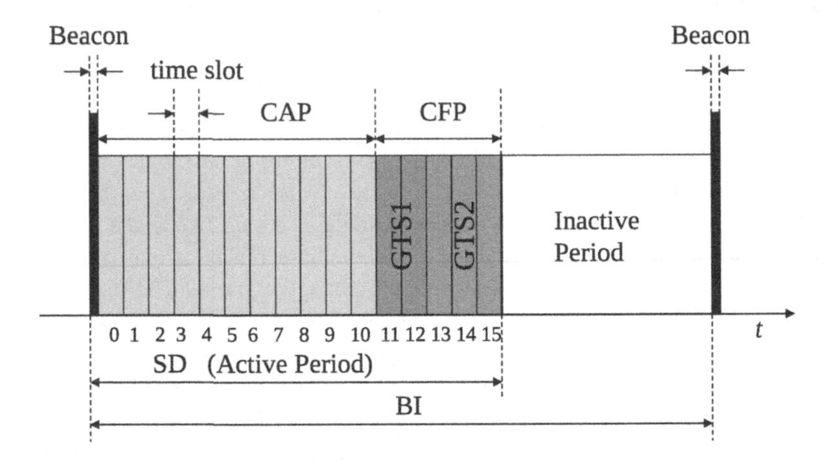

Figure 3.2: MAC superframe.

The length of the superframe, called the Beacon Interval (*BI*), and the length of its active part, called the Superframe Duration (*SD*), are determined by the beacon order (*BO*) and the superframe order (*SO*) as follows

$$BI = 2^{BO} \times aBaseSuperframeDuration \quad (3.1)$$

$$SD = 2^{SO} \times aBaseSuperframeDuration \quad (3.2)$$

The values of *BO* and *SO* are chosen by the coordinator, and have to fulfill the following inequality: $0 \leq SO \leq BO \leq 14$. Instead, the quantity *aBaseSuperframeDuration* denotes the minimum duration of the superframe (corresponding to $SO = 0$) and it is fixed to 960 modulation symbols. Note that in the 2.4 GHz ISM band, a modulation symbol period is equal to 16 μs. We hereafter refer to modulations symbols as simply symbols.

The active portion of a superframe is divided into 16 time slots, each with a duration of $2^{SO} \times aBaseSlotDuration$ symbols, where the constant *aBaseSlotDuration* is equal to 60 symbols. Moreover, as shown in Figure 5.2, the active portion consists of three parts: the beacon, a Contention Access Period (CAP) and a Contention Free Period (CFP). The beacon is sent by the PAN coordinator in the first time slot of the superframe. During the CAP, nodes access the channel using the slotted CSMA/CA. The optional CFP is activated upon request from the nodes to the PAN coordinator for allocating guaranteed time slots (GTS). Each GTS consists of some integer multiple of CFP slots and up to 7 GTS are allowed in a CFP.

3.2.2 The Slotted CSMA/CA Mechanism

The basic time unit of the IEEE 802.15.4 MAC protocol is the backoff period (BP), a time slot of length $t_{bp} = aUnitBackoffPeriod = 20$ symbols. In the slotted CSMA/CA algorithm, each operation (channel access, backoff count, CCA) can only start at the beginning of a BP.

Note that a BP is different and smaller than each of the 16 time slots that compose the active period of the superframe shown in Figure 5.2. For example, if $SO = 0$, the superframe slot duration is three times that of a BP. Therefore, a superframe slot duration is always a multiple of three BPs. Moreover, for every node, the first BP boundary in a superframe should be aligned with the first superframe slot boundary of the PAN coordinator.

Each node maintains three variables for each transmission attempt: *NB*, *CW*, and *BE*. *NB* indicates the backoff stage, or equivalently, the number of times the CSMA/CA backoff procedure has been repeated while attempting the current transmission. *CW* is the contention window length, which defines the number of BPs the channel has to be sensed idle before the transmission can start. *BE* is the backoff exponent, which is used to extract the random backoff value. The value of *BE* should fulfill the following inequality: $macMinBE \leq BE \leq macMaxBE$, where *macMinBE* and *macMaxBE* are constants (see Table 3.1).

The slotted CSMA/CA mechanism works as shown in Figure 3.3. Before every new transmission, *NB*, *CW* and *BE* are initialized to 0, 2 and *macMinBE*, respectively. The node waits for a random number of BPs specified by the backoff value, drawn uniformly in the range $[0, 2^{BE} - 1]$. Then, it performs the first CCA, i.e., it senses the channel and verifies whether it is idle. If the channel is idle, the first CCA succeeds and *CW* is decreased by one. The node then performs the second CCA and, if that is also successful, it can transmit the packet.

If either of the CCAs fail, both *NB* and *BE* are incremented by one, ensuring that *BE* is not more than *macMaxBE*, and *CW* is reset to 2. The node repeats the procedure for the new backoff stage by drawing a new backoff value, unless the value of *NB* has become greater than a constant $M = macMaxCSMABackoffs$. In that case, the CSMA/CA algorithm terminates with a *Channel Access Failure* status, and the concerned packet is discarded.

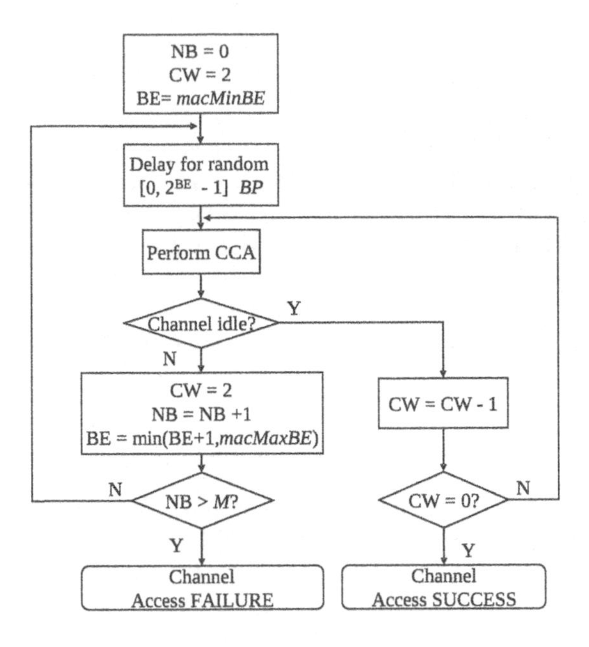

Figure 3.3: Flow chart of the channel access procedure.

A packet transmitted after a successful channel access procedure can be either received successfully or have a collision. The network can be operating in either acknowledged or unacknowledged transmission modes. Hereafter, we refer to these modes as ACK mode and no-ACK mode, respectively.

In ACK mode, a successful transmission is accompanied by the reception of a MAC acknowledgment (ACK), which has a fixed length L_{ack} of 11 bytes (i.e., 22 symbols in the ISM 2.4 GHz band). The ACK is fed back to the transmitter after a minimum time of $t^{-}_{ack} \leq aTurnaroundTime + t_{bp}$. The constant *aTurnaroundTime* represents time needed for switching the transceiver from one operative mode to another (transmission-to-reception or reception-to-transmission), and has a duration of 12 symbols. The ACK is expected by the sender node to be received before a maximum time t^{+}_{ack} (i.e., the *macAckWaitDuration*), that is equal to $t^{-}_{ack/max} + L_{ack}$ (i.e., 54 symbols). After this time, if the ACK frame is not correctly received, a collision is declared. In this case, the packet is retransmitted using a new transmission procedure with *NB*, *CW*, and *BE* set to their initial values. A packet can be retransmitted at most $R = aMaxFrameRetries$ times if required, before being discarded. The default values of the parameters used in the slotted CSMA/CA procedure, as indicated by the standard [1], are given in Table 3.1.

It should be noted here that, each time the node draws a random backoff value, it has to make sure that it has enough time for transmitting the packet within the current CAP. If it has enough time to finish both the backoff and the transmission (including CCAs and ACK reception), it shall proceed as indicated before. Otherwise, if it can

Table 3.1: Values of Slotted CSMA/CA Parameters

Parameter	Range	Default Value
macMinBE	0 - *macMaxBE*	3
macMaxBE	3 - 8	5
macMaxCSMABackoffs	0 - 5	4
aMaxFrameRetries	0 - 7	3

finish the backoff, but there is not enough time for completing the steps for transmission procedure, it shall defer the entire transmission by performing another backoff at the beginning of the next CAP and proceeding as usual. Finally, if there is not even enough time to finish the backoff procedure, it shall perform as many BPs of the backoff that fit in the current CAP, and then continue the remainder of the backoff in the next CAP.

3.3 Markov Chains for the Slotted CSMA/CA

In this section, we will briefly review the different Markov chain models available in the literature which model the behavior of the IEEE 802.15.4 MAC protocol. We will then focus in the next section on the model presented by Pollin et al. in [6], which is the model we will be using for the present work.

The behavior of the slotted CSMA/CA in an IEEE 802.15.4 network has been widely investigated in the literature using the same approach introduced by Bianchi [7] for the traditional IEEE 802.11 CSMA/CA. In all these works, the behavior of the network is analyzed through modeling a single node's behavior with a discrete-time Markov chain. The state of each node evolves through its corresponding Markov chain independently of other nodes' states except for when it is sensing the channel. As it was mentioned earlier, each node can attempt a transmission only after it has sensed the channel idle for two consecutive time slots (CCA_1 and CCA_2) after finishing its backoff. The probability that it senses the channel idle in these two time slots depends on whether other nodes are transmitting or not. In all the works discussed in this section, it is assumed that the probability of sensing the channel idle is independent of the backoff stage in which CCA_1 and CCA_2 are performed. We will base our model on the same assumption, but as we show in Section 3.6, these probabilities do depend on the backoff stage their corresponding CCA is performed in. We will also discuss the network metrics that are most affected by this assumption.

To avoid any confusion, in the following discussion we will consistently refer to the probability of sensing the channel busy when performing CCA_1 and CCA_2 as α and β respectively, independently of the notation used in the discussed work. Since the CCA states are the only states in the Markov chain where the dependence of the nodes come into play, calculating α and β accurately is the key to the correct analysis of the network. For this reason, in discussing the prior work, we will be mostly focused on the way these probabilities are calculated as the main criteria for evaluating the accuracy of the models.

Mišić et al. in [8] propose a Markov chain model to analyze the slotted CSMA/CA under saturated and unsaturated traffic conditions in a beacon-enabled network in ACK mode. The superframe structure and the retransmissions are not considered in their model. In [9], they extend the model of [8] in the unsaturated case, modeling also the superframe structure and retransmissions. For the saturated case in [8], they construct a per-node Markov chain. However, they only include one state for transmission, even though the transmission may take more than one BP. They also do not include the corresponding states for the time spent receiving and waiting for the ACK. Therefore, in the normalizing condition for the steady state probabilities of the chain, the BPs spent for transmission and ACK are not accounted for. Furthermore, even though when calculating α they do consider the ACK, they neglect its effect when calculating β. Finally, when calculating both α and β, they implicitly assume that the probability to start transmission is independent for all the nodes. But in fact, this probability is highly correlated for different nodes since before a transmission a node has to sense the channel idle for two BPs, and therefore, for two nodes to be transmitting simultaneously, they have to have sensed the channel exactly at the same time. Since in such a case both nodes observe the same channel state, the probability of two nodes transmitting at the same BP is highly dependent. For the unsaturated condition in both [8] and [9], they extend the Markov chain model of the saturated case, and therefore, all the aforementioned issues apply to these cases as well. In neither case the analytical results are validated with simulation.

It should be noted here that, in Bianchi's work on IEEE 802.11 in [7], the transmission states are also omitted from the chain. However, in that case this is justified and in fact necessary due to the structural difference between the IEEE 802.11 and the IEEE 802.15.4 CSMA/CA mechanisms. In IEEE 802.11, nodes are constantly sensing the channel while performing backoff, and freeze their backoff counter when there is a transmission on the channel. This implies that even though the backoff states in the Markov chain in [7] do not have a fixed duration, they all have the same *expected* duration. On the other hand, while a node is transmitting, its transmission is never interrupted and therefore, the transmission states cannot be included in the Markov chain model as they do not have the same expected duration as the backoff states. Thus, in Bianchi's model, all the states included in the chain have the same expected duration, and therefore, the steady state probabilities are well-defined. In IEEE 802.15.4, however, all transitions happen strictly at the boundaries of backoff periods and no sensing is done during the backoff. Instead, the sensing is done only after the backoff during CCA_1 and CCA_2. This means that the backoff and sensing states all have the same fixed duration of exactly one BP. Therefore, representing the entire transmission (which lasts a fixed multiple of a BP) with only one state in the chain renders the steady state probabilities ill-defined, because when they are calculated, states with different duration are given the same weight.

Pollin et al. in [6] suggest a simple but complete Markov chain model for the slotted CSMA/CA under saturated and unsaturated periodic traffic conditions in a

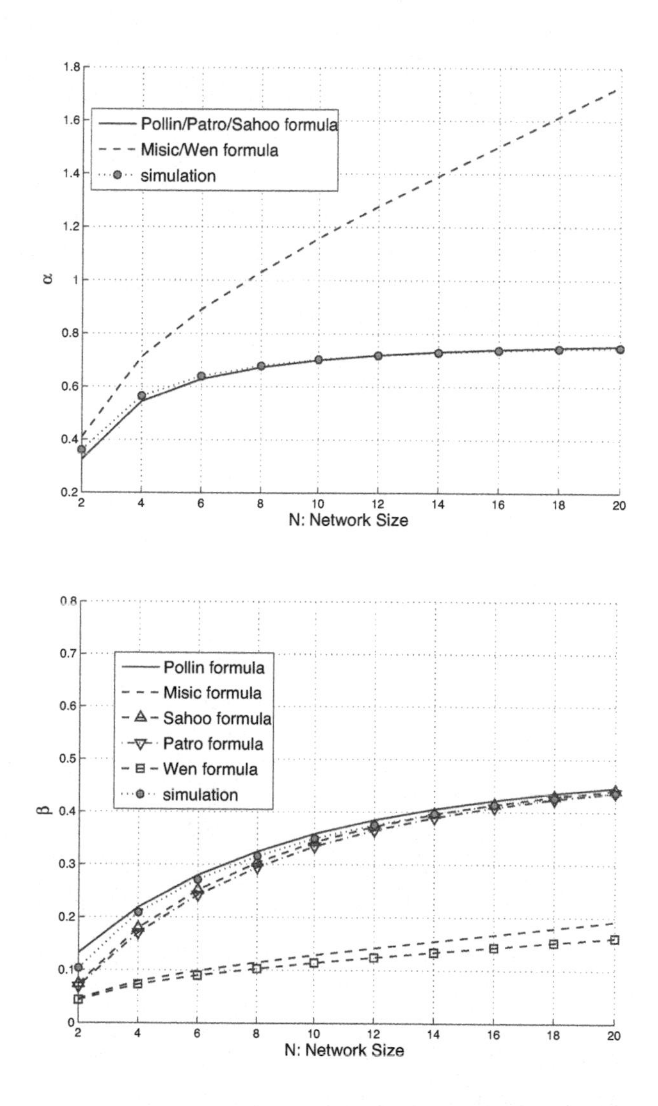

Figure 3.4: Comparison between the values of α and β obtained from formulas in [6] and [8]–[12], and the actual values obtained from simulation.

beacon-enabled network in both ACK and no-ACK modes. They do not model the superframe structure. Unlike [8] and [9], their Markov chain includes a number of transmission states equal to the packet length (in the number of BPs). They base their model on the main assumption that the probability to start sensing the channel is independent across the nodes and properly consider the dependence between the nodes when calculating α and β. They validate their analytical results with simulations. However, as we will explain in more detail in Section 3.4.2, in their model a group of states are accounted for twice.

Other works using similar approaches to Pollin and Mišić include [10] and [11]. However, similar issues as the ones just mentioned can be observed in their models and calculations as well. For example, in [10] only one transmission state is included in the chain, and in [11] the same Markov chain model as in [6] is used where similarly a set of states are accounted for twice. Furthermore, when writing the normalizing condition for the chain they allow the backoff exponent to increase above *macMaxBE*. The packet discard probability is not properly calculated either, as it does not include the cases where a packet fails the access procedure during a retransmission.

In more recent works, Wen et al. in [12] use a proper Markov chain model, but they do not consider the dependence in transmission probability of different nodes when calculating α and β. Finally, in [13], Jung et al. propose a new model based on the Mišić model in [9]. However, they too do not include all the transmission states in the Markov chain.

Figure 3.4 shows a comparison between the values of α and β obtained from the formulas in the aforementioned works and those obtained from simulation. Note that in all cases, we have plotted these parameters only for a network operating in no-ACK mode, with no retransmissions, no superframe structure, and under saturated traffic condition. We see that especially in the case of [8] and [12], the formulation for α and β offers an inaccurate approximation of the actual values of these parameters. This, as we mentioned before, is due to the fact that they ignore the high dependence between the probabilities of starting transmission for different nodes. All discussed models, except [8], [9], and [12], use the same formulation for α. This is not the case for β, in which case, we note that Pollin et al.'s model offers the best estimate for most network sizes.

The IEEE 802.15.4 MAC features modelled in each of the works in [6] and [8]–[12] are summarized in Table 3.2.

3.4 System Model and Notation

In the present work, we will be using the Markov chain model presented by Pollin et al. in [6], with some modifications, and will be mostly following the notation thereof, with minor changes when necessary. Therefore, in this section, we will first describe this modified Markov chain model and its corresponding formulation, and will then compare it to the one presented in [6]. To validate our results, we will use simulations

Table 3.2: Prior Work Summary

Model includes	Pollin [6]	Misic [8]	Misic [9]	Sahoo [10]	Patro [11]	Wen [12]	Jung [13]
Saturated traffic	Yes	Yes	No	No	Yes	Yes	No
Unsaturated traffic	Yes	Yes	Yes	Yes	No	Yes	Yes
Retransmissions	No	No	Yes	Yes	No	No	Yes
Superframe structure	No	No	Yes	No	No	No	Yes
Correct # of TX states	Yes	No	No	No	Yes	Yes	No
Correct # of CCA states	No	Yes	Yes	No	No	Yes	Yes

which reproduce the CSMA/CA mechanism in IEEE 802.15.4. More details about the simulation are to come in Section 3.5.

3.4.1 Markov Chain Model

In [6], the performances of a single hop LR-WPAN, made by N nodes and a PAN coordinator (i.e., the sink in a WSN) have been evaluated for uplink traffic. Both saturated and unsaturated periodic traffic conditions, and ACK and no-ACK modes have been considered. We have focused our attention on the saturated case (i.e., when each of the N nodes in the LR-WPAN, always has a packet available for transmission) in the no-ACK mode. The saturated case reflects a sensor network scenario in which an event is detected by many sensor nodes that want to transmit the gathered information, at the same time, to the sink node. We will not be considering the superframe structure in our model, i.e., we are assuming that the network is operating in a CAP with an infinite duration.

The behavior of a single node is modeled using a two-dimensional Markov chain, with states represented by $\{s(t), c(t)\}$ at a given backoff period t, as shown in Figure 3.5. Hereafter, we use the term time slot, or simply slot, to refer to a backoff period. All events happen at the beginning of a time slot. At a given time slot t, the stochastic process $s(t)$ represents the backoff stage when $s(t) \in \{0, \ldots, M\}$, and the transmission stage when $s(t) = -1$. When the node is in transmission (i.e., $s(t) = -1$), the stochastic process $c(t) \in \{0, \ldots, L-1\}$ represents the state of the transmission, i.e., the number of slots spent on the current transmission. L is the packet size, measured in the number of slots it takes for transmitting the packet, and includes the overhead introduced by the PHY and MAC headers. When the node is in backoff, $c(t) \in \{0, \ldots, W_i - 1\}$ represents the value of the backoff counter, where $W_i = 2^{\min\{macMinBE+i, macMaxBE\}}$ is the size of the backoff window at backoff stage $s(t) = i \in \{0, \ldots, M\}$. Finally, when the node is performing one of the CCAs, $c(t)$ represents the value of the CCA counter, with $c(t) = 0$ during CCA_1 and $c(t) = -1$ during CCA_2. Note that the state $\{s(t), c(t)\} = \{i, 0\}$, has to be seen as a CCA_1 state and not as a backoff state as in [6]. In fact, although the randomly picked backoff

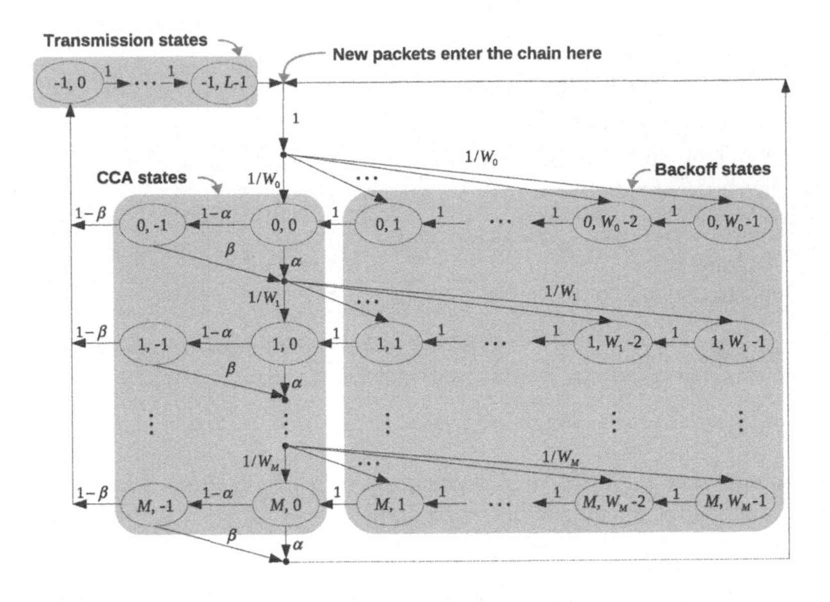

Figure 3.5: Markov model for slotted CSMA/CA.

window size at stage i can take any value in the set $\{0,\ldots,W_i-1\}$, the value zero indicates no waiting and immediate sensing. In other words, if the backoff counter of a node is equal to zero, it immediately starts sensing the channel (i.e., it performs CCA_1).

The parameter α in Figure 3.5 is the probability of assessing the channel busy during CCA_1, and β is the probability of assessing it busy during CCA_2, given that it was idle in CCA_1. In reality, the values of α and β for a given node depend not only on the backoff stage of that node, but also on that of other nodes in the system. However, for simplicity, we ignore this dependence and assume that the values of both α and β are the same for different backoff stages and are also independent of the backoff stage of other nodes. This assumption is used in all the prior work discussed in Section 3.3 and it is what allows us to model the network by only analyzing the individual Markov chain of a node in the network as shown in Figure 3.5. We will further scrutinize this assumption in Section 3.6. As it was mentioned earlier, in this per-node Markov chain model, the effect of other nodes on the behavior of a given node is captured only through the values of α and β, and therefore, these two parameters play a key role in the model.

Let $b_{i,k}$ be the steady state probability of being in state $\{i,k\}$, i.e., $b_{i,k} = \lim_{t\to\infty} P\{s(t)=i, c(t)=k\}$. These steady-state probabilities are related to one another through the following equations:

$$b_{i,k} = \frac{W_i - k}{W_i} b_{i,0} \qquad 0 \le i \le M, 0 \le k \le W_i - 1 \tag{3.3}$$

$$b_{i,0} = (1-y)^i b_{0,0} \qquad 1 \le i \le M \tag{3.4}$$

$$b_{i,-1} = (1-\alpha) b_{i,0} \qquad 0 \le i \le M \tag{3.5}$$

$$b_{-1,k} = y \sum_{j=0}^{M} b_{j,0} = y\phi \qquad 0 \le k \le L-1 \tag{3.6}$$

where $y = (1-\alpha)(1-\beta)$, and ϕ is the probability that a randomly picked slot is spent performing CCA_1 which is given by

$$\phi = \sum_{j=0}^{M} b_{j,0} = \frac{1-(1-y)^{M+1}}{y} b_{0,0}. \tag{3.7}$$

By determining the interactions between the N nodes on the medium, the expressions for probabilities α and β have been derived in [6]. In short, the probability α, of finding the channel busy during the first CCA is given by

$$\alpha = L\left[1-(1-\phi)^{N-1}\right] y. \tag{3.8}$$

In turn, β, the probability that there is a transmission in the medium when the considered node does its second sensing, is given by

$$\beta = \frac{1-(1-\phi)^N}{2-(1-\phi)^N}. \tag{3.9}$$

The values of ϕ, α and β can be determined by imposing the following normalizing condition:

$$\sum_{i=0}^{M} \sum_{k=0}^{W_i-1} b_{i,k} + \sum_{i=0}^{M} b_{i,-1} + \sum_{k=0}^{L-1} b_{-1,k} = 1 \tag{3.10}$$

which can be equivalently written as

$$\frac{b_{0,0}}{2} \left\{ [3 - 2\alpha + 2yL] \frac{1-(1-y)^{M+1}}{y} + 2^d W_0 \frac{(1-y)^{d+1} - (1-y)^{M+1}}{y} + W_0 \frac{1-(2-2y)^{d+1}}{2y-1} \right\} = 1 \tag{3.11}$$

where $W_0 = 2^{macMinBE}$ and $d = macMaxBE - macMinBE$.

As can be seen in Figure 3.6, (3.8) and (3.9) offer a good approximation to the values of α and β for large network sizes. But for example at $N = 2$, the formulas introduce about 10% and 30% approximation errors for α and β, respectively. As we will see in the next sections, the impact of this error can be negligible when calculating some performance metrics of the network, but for certain other metrics, it has a noticeably negative impact.

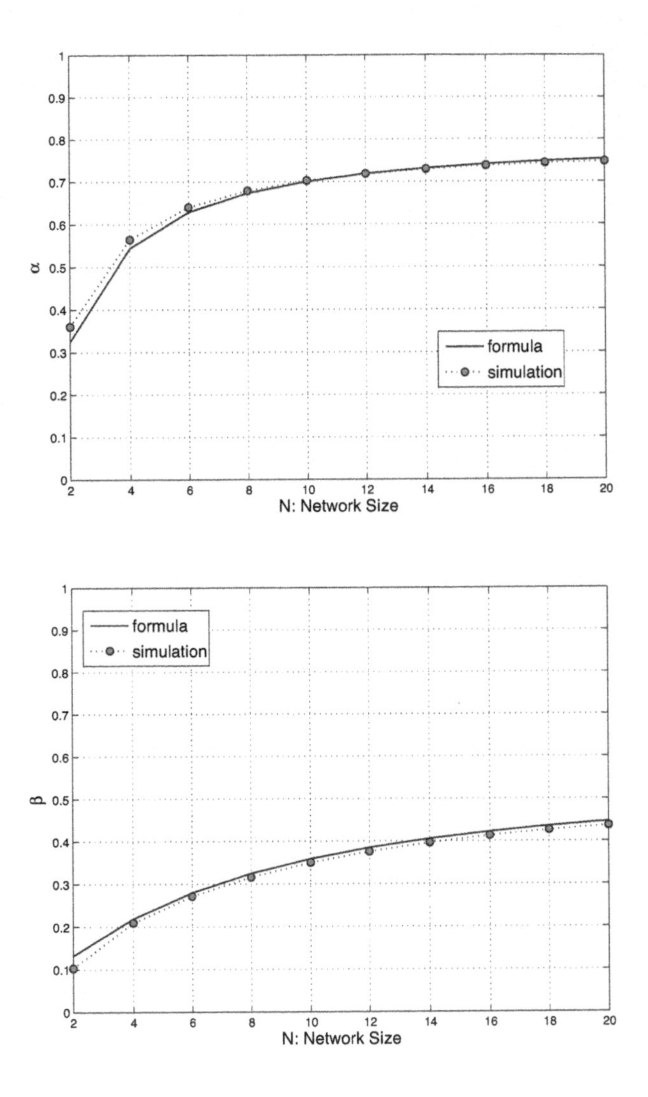

Figure 3.6: Comparison between the values of α and β obtained from (3.8) and (3.9), and those obtained from simulation.

3.4.2 Discussion of the Pollin Model

As was mentioned earlier, the model described in the previous section closely follows that of the Pollin model. However, there were some necessary modifications made to that model which we list here.

The main assumptions: The main assumptions that enable us to use the described Markov chain model are the following:

1. The probability to start sensing the channel, ϕ, is independent across nodes;
2. The probability to sense the channel busy during CCA_1 and CCA_2 does not depend on the backoff stage where the corresponding CCA is performed. In other words, $\alpha_i = \alpha$ and $\beta_i = \beta$ for $i \in \{0,\dots,M\}$;
3. The probability that a node is in a given backoff stage, is independent of that of other nodes;
4. The probability of sensing the channel busy during a CCA does not depend on the random backoff value drawn in the backoff stage preceding the CCA.

In [6], A1 is the only stated assumption. In what follows, especially in Section 3.6, we will discuss the significance of these assumptions.

The normalizing condition of the Markov chain steady state probabilities: In [6] the following equation is given as the normalizing condition for the steady state probabilities:

$$\sum_{i=0}^{M}\sum_{k=0}^{W_i-1} b_{i,k} + \sum_{i=0}^{M} b_{i,-1} + \sum_{i=0}^{M} b_{i,-2} + \sum_{k=0}^{L-1} b_{-1,k} = 1 \qquad (3.12)$$

where $b_{i,-1}$ and $b_{i,-2}$ in their notation correspond to the first and second CCA, respectively. On the other hand, states $b_{i,0}$ also correspond to the first CCA, because, as we mentioned earlier, when the backoff counter reaches zero, the node will immediately start sensing. Even though the Markov chain depicted in [6] does not contain any state $b_{i,0}$, in the above normalizing condition, both of the states $b_{i,0}$ and $b_{i,-1}$ are counted and therefore, the states $b_{i,0}$, as per our notation, are counted twice. This normalizing condition together with (3.7), (3.8), and (3.9) constitute a system of equations that is used in [6] to numerically obtain α, β, and ϕ. Therefore, the aforementioned overcount in the normalizing condition affects the values obtained for these variables.

3.5 Results

In this section, we use the Markov chain model to calculate different important network metrics, such as the average delay experienced by successfully transmitted packets, the average power consumed by a node, and the efficiency in terms of the number of bits a node can transmit per unit of energy.

In order to validate the analytical results, we simulate the behavior of the slotted CSMA/CA using an event-driven simulator written in MATLAB.® We reproduce the network under the same conditions of the analytical model, i.e., we analyze the channel access mechanism in the no-ACK mode, and disregarding the superframe structure. It should be noted that for the simulation we do not make any assumption on the dependence of the nodes or backoff stages, and therefore, the simulation truly reflects the behavior of the network under the aforementioned conditions, and not the behavior of the Markov chain model.

In what follows, all the metrics have been defined as functions of α, β and ϕ. In [6], ϕ is calculated by numerically solving their corresponding (3.8) – (3.10). In our work, when plotting the curves from the analytical formulation to validate their accuracy, we use the value of ϕ found by simulation, and then calculate α, β, and the network metrics from ϕ. These curves are indicated in the figures' legends by "formula".

For the MAC parameters, we use the default values defined by the standard (see Table 3.1) and fix the packet length for all nodes to $L = 7$ time slots. The simulation is run for a duration of $T = 10^8$ slots.

3.5.1 Average Delay

In the no-ACK mode, the average delay for a successfully transmitted packet, i.e., the number of slots it takes from the moment it reaches the head of the line to the moment it arrives at its destination, is given by

$$\bar{D} = \bar{n}_{B_{suc}} + \bar{n}_{C_{suc}} + L \tag{3.13}$$

where $\bar{n}_{B_{suc}}$ and $\bar{n}_{C_{suc}}$ are the mean number of slots spent performing backoff and CCA, respectively, before a successful transmission. The average delay calculated above, can be equivalently viewed as the packet service time in a saturated IEEE 802.15.4 network of queues.

Note that based on assumptions A2 and A3, the number of slots spent in backoff or CCA before transmission for packets that are successfully transmitted and those that end up having a collision should be the same, i.e., $\bar{n}_{B_{suc}} = \bar{n}_{B_{col}} = \bar{n}_{B_{tx}}$ and $\bar{n}_{C_{suc}} = \bar{n}_{C_{col}} = \bar{n}_{C_{tx}}$. For this reason, in Section 3.5.3, we obtain the expressions for the values of $\bar{n}_{B_{tx}}$ and $\bar{n}_{C_{tx}}$, which we use to calculate the delay as described above.

As can be seen in Figure 3.7, the value of delay calculated from (3.13) and that obtained from simulation have a constant difference of about two slots for all values of network size N. This difference is due to assumption A2. A similar difference is observed in the values of $\bar{n}_{B_{tx}}$ and $\bar{n}_{C_{tx}}$ as we will see in Section 3.5.3, where we will discuss this issue in more detail.

3.5.2 Average Power Consumption

For every packet that is transmitted or discarded, the node uses different power levels depending on whether it is in backoff, sensing, or transmission. Therefore, the average power consumption per node (in joules/s) is given by

$$\bar{P} = \frac{\bar{n}_B W_{id} + \bar{n}_C W_{rx} + L(1 - p_f)W_{tx}}{\bar{n}_B + \bar{n}_C + L(1 - p_f)} \tag{3.14}$$

where $\bar{n}_B$ and $\bar{n}_C$ are the average number of slots spent in backoff and CCA per attempt, respectively. These parameters are derived in Section 3.5.3. In turn, p_f is the packet discard probability due to access procedure failure, and it is given by

$$p_f = (1-y)^{M+1}. \tag{3.15}$$

The power consumed by the radio transceiver in mode x, is given by $W_x = I_x \times V_{DD}$. Here, I_x is the amount of current consumption for operating mode x, and V_{DD} is the supply voltage. For illustrative purposes, we use the parameter values specified for Chipcon 802.15.4-compliant RF transceiver CC2430 [14]. The CC2430 provides four different power modes, PM0 to PM3. PM0-TX and PM0-RX are used for transmission and reception modes, respectively and we have supposed that the device uses PM2 as idle mode. The current consumptions of CC2430 power modes and the supply voltage, V_{DD}, are given in Table 3.3.

Table 3.3: CC2430 Specifications

Parameter	Value	Unit	Power Mode
I_{tx}	26.9	mA	PM0-TX
I_{rx}	26.7	mA	PM0-RX
I_{id}	0.5	μA	PM2
V_{DD}	3	V	–

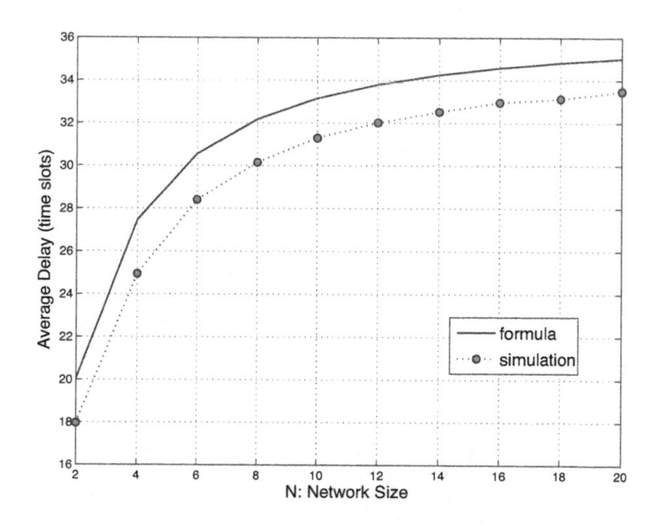

Figure 3.7: Comparison between formula and simulation values for $\bar{D}$.

As can be seen in Figure 3.8, the value for the average power obtained from (3.14) matches well its value from simulation. We see here that the average power consumption per single node decreases with larger N. This is because the packet discard probability, p_f, increases with N and therefore, each node is able to transmit fewer packets as the network size grows. As most of the power consumption happens during the actual transmission of the packet, this means that on average less power is consumed.

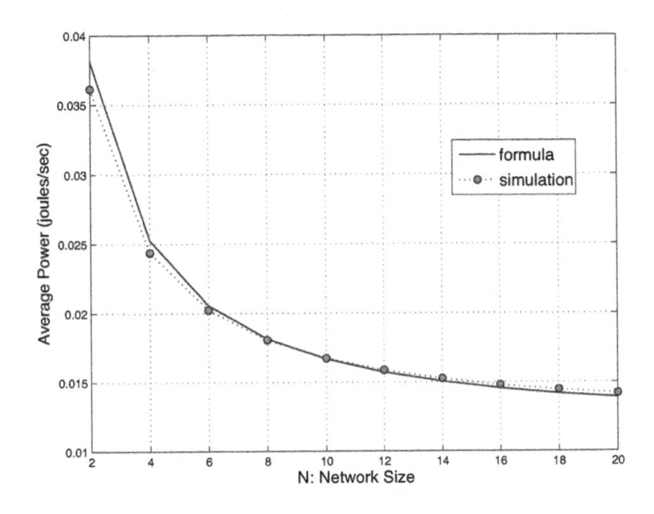

Figure 3.8: Comparison between formula and simulation values for $\bar{P}$.

3.5.3 Mean Number of Backoffs and CCAs

As we saw earlier, in order to calculate the average delay and the average power consumption, we need to know how many slots on average a node spends performing backoff and CCA before transmitting or discarding every packet. In this section, we will derive these parameters using the Markov chain model described earlier.

3.5.3.1 *Mean Number of Backoffs*

The average number of backoffs a transmitted packet goes through, $\bar{n}_{B_{tx}}$, is different from that of a discarded packet, $\bar{n}_{B_f}$. This is because a discarded packet always goes through all the backoff stages of the chain before exiting it, whereas a transmitted packet might exit the chain from any backoff stage.

The mean number of slots spent in backoff stage i, every time it is entered, is given by $(W_i - 1)/2$. This is because the backoff value is drawn according to a discrete uniform distribution in $[0, W_i - 1]$. Therefore, $\bar{n}_{B_{tx}}$, the mean number of backoffs before transmission, is given by

$$\bar{n}_{B_{tx}} = \sum_{i=0}^{M} \left(\sum_{k=0}^{i} \frac{W_k - 1}{2} \right) \frac{p_{S_i}}{1 - p_f} \tag{3.16}$$

where p_{S_i} is the probability that the channel access procedure ends successfully (i.e., the packet is sent) in backoff stage i and it is given by

$$p_{S_i} = y(1-y)^i, \qquad 0 \le i \le M. \tag{3.17}$$

After a bit of algebra, we obtain

$$\begin{aligned} \bar{n}_{B_{tx}} = \frac{1}{1-p_f} \Bigg\{ & W_0 y \frac{1-(2-2y)^{d+1}}{2y-1} - \frac{1-y}{2y} - \frac{W_0+1}{2} + 2^{d-1} W_0 \left(3 + \frac{1-y}{y}\right)(1-y)^{d+1} \\ & - \left[W_0 2^{d-1}(3-d+M) - \frac{W_0+M+2}{2} + \left(\frac{W_0 2^d - 1}{2}\right)\left(\frac{1-y}{y}\right)\right](1-y)^{M+1} \Bigg\}. \end{aligned} \tag{3.18}$$

Figure 3.9(a) shows the value of $\bar{n}_{B_{tx}}$ obtained from (3.16) and that obtained from simulation. As we see here, there is an almost constant difference of less than two slots between the analytical and simulation results. This is again due to assumption A2. For this reason, in the same figure we have also plotted the value of $\bar{n}_{B_{tx}}$ calculated using the actual values of α_i and β_i for every backoff stage i obtained from simulation. In this case, the value of $\bar{n}_{B_{tx}}$ is still given by (3.16), but p_{S_i} has to be calculated using the following expression

$$p_{S_i} = \begin{cases} y_0, & i = 0, \\ y_i \prod_{k=0}^{i-1} (1 - y_k), & 1 \le i \le M \end{cases} \tag{3.19}$$

where $y_i = (1-\alpha_i)(1-\beta_i)$. Also, the probability p_f has to be calculated as follows

$$p_f = \prod_{k=0}^{M} (1 - y_k). \tag{3.20}$$

In turn, a packet is discarded when the node goes through every backoff stage in the chain and fails to access the channel in all the $M+1$ consecutive backoff stages. Therefore, $\bar{n}_{B_f}$, the mean number of slots spent in backoff before an access procedure failure, is given by

$$\bar{n}_{B_f} = \sum_{k=0}^{M} \left(\frac{W_k - 1}{2}\right) \frac{p_{S_M}}{p_f} = \sum_{k=0}^{M} \frac{W_k - 1}{2} = W_0 2^d \left(1 + \frac{M-d}{2}\right) - \frac{W_0 + M + 1}{2}. \tag{3.21}$$

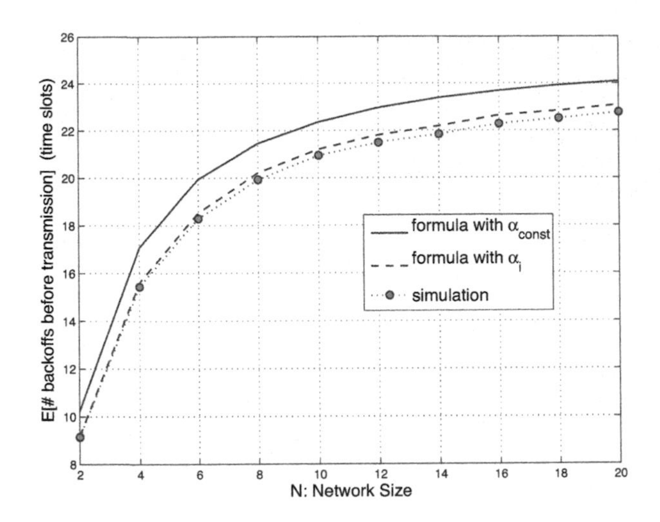

(a) $\bar{n}_{B_{tx}}$ from simulation compared to its value from formula applied to the constant α, and to α_i.

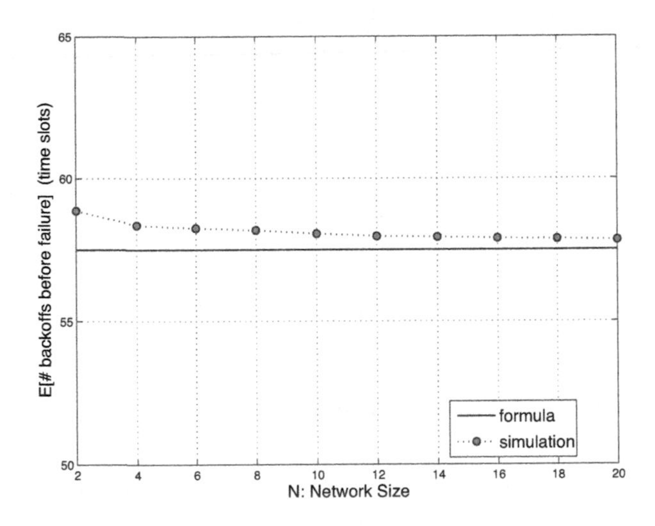

(b) $\bar{n}_{B_f}$ from simulation and formula applied to constant α.

Figure 3.9: Comparison between formulas and simulation values for $\bar{n}_{B_{tx}}$ and $\bar{n}_{B_f}$.

As seen in (3.21), $\bar{n}_{B_f}$ is not a function of α and β and therefore, the expression is valid independently of assumption A2. Moreover, $\bar{n}_{B_f}$ is not a function of N, because every discarded packet goes through all the backoff stages, and hence, the number of slots spent in backoff for a discarded packet only depends on the random backoff value drawn at every backoff stage. Assuming A4, for a given backoff stage i, an average of $(W_i - 1)/2$ slots are spent in backoff before performing the CCA. Figure 3.9(b) shows the value of $\bar{n}_{B_f}$ from simulation and formula. We see here that the value obtained from (3.21) is very close but always slightly lower than the one obtained from simulation. This is because α is not completely independent of the random backoff value drawn. In fact, the packets that end up being discarded are those that are less "fortunate" and draw a larger backoff value.

Finally, for a generic packet, the mean number of slots that a node spends in backoff before transmitting or discarding the packet, is given by

$$\bar{n}_B = \bar{n}_{B_{tx}}(1 - p_f) + \bar{n}_{B_f} p_f. \tag{3.22}$$

3.5.3.2 Mean Number of CCAs

For a packet to be transmitted, a node needs to succeed in the access procedure in some backoff stage. If the access procedure succeeds in stage i, it means that it had two successful CCAs in that stage, and at least one failed CCA in stages 0 to $i-1$. In other words, it must have had k successful CCA$_1$'s and failed CCA$_2$'s, for some $k \leq i$, and $i-k$ failed CCA$_1$'s. This event happens with probability

$$p_{f_{i,k}} = \binom{i}{k} [(1-\alpha)\beta]^k \cdot \alpha^{i-k}. \tag{3.23}$$

In this case, there will be a total of $i+k$ CCAs performed during the failed accesses plus an additional two successful CCAs at stage i where the access procedure succeeds. Thus, considering that the successful access at stage i happens with probability y, the mean number of CCAs before a successful access procedure is given by

$$\begin{aligned} \bar{n}_{C_{tx}} &= y \sum_{i=0}^{M} \sum_{k=0}^{i} (i+k+2) \frac{p_{f_{i,k}}}{1-p_f} \\ &= 2 + [2(1-y) - \alpha] \left[\frac{1}{y} - (M+1) \frac{(1-y)^M}{1-p_f} \right]. \end{aligned} \tag{3.24}$$

In the case of an access procedure failure, there are $M+1$ failed attempts and no successful consecutive CCAs. Therefore, the mean number of CCAs due to an access failure is given by

$$\bar{n}_{C_f} = \sum_{k=0}^{M+1} (M+1+k) \frac{p_{f_{M+1,k}}}{p_f} = (M+1)\left(2 - \frac{\alpha}{1-y}\right). \tag{3.25}$$

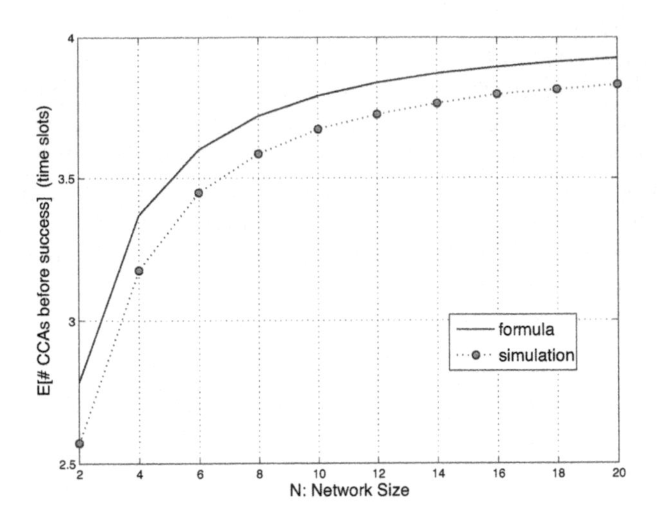

(a) $\bar{n}_{C_{tx}}$ from simulation and formula.

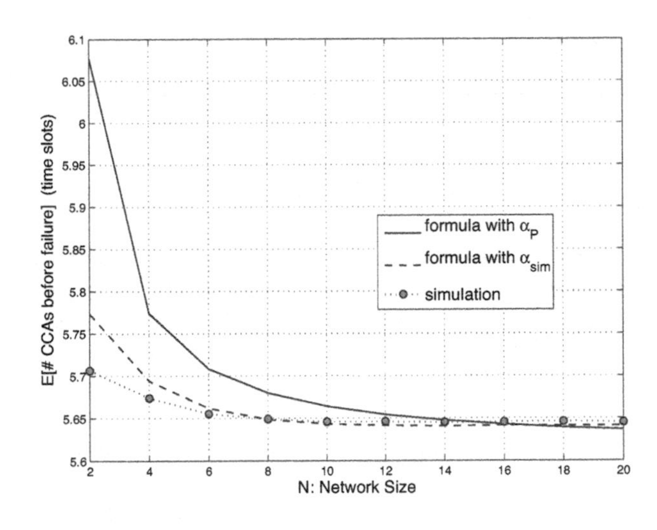

(b) $\bar{n}_{C_f}$ from simulation compared to its value obtained from formula applied to α_P (α calculated from formula proposed by Pollin et. al), and to α_{sim} (α from simulation).

Figure 3.10: Comparison between formula and simulation values for $\bar{n}_{C_{tx}}$ and $\bar{n}_{C_f}$.

Figure 3.10(a), shows the value of $\bar{n}_{C_{tx}}$ obtained from (3.24) and from simulation. As in the case of $\bar{n}_{B_{tx}}$, we see here that there is a difference (although smaller) between simulation and formula. We conjecture that this difference is also due to assumption A2. However, we will not validate this conjecture here, since calculating $\bar{n}_{C_{tx}}$ as a function of α_i and β_i is a lot more complicated than calculating it for a constant α and β. This is because to calculate $\bar{n}_{C_{tx}}$ as a function of α_i and β_i, we have to consider each different possible success sequence separately. In other words, when we assume that α and β are constant, we only need to consider the number of CCA_1's and CCA_2's performed before a transmission. But when α and β are a function of the backoff stage, we also need to consider what exactly happens in each backoff stage for all possible cases.

The value of $\bar{n}_{C_f}$ obtained from analysis and simulation is compared in Figure 3.10(b). In this figure, we have plotted two different curves for the value of $\bar{n}_{C_f}$ from formula. Indicated by "formula with α_P" is what we have been thus far indicating by only "formula", and it is basically found by applying (3.25) to α and β as derived from (3.8) and (3.9). Instead, the curve indicated by "formula with α_{sim}" is plotted by applying the same equation to the value of α and β directly obtained from simulation. We see that in the latter case, we get significantly better results. This means that the difference between the simulation and formula is not due to any inaccuracy in (3.25), but due to the inaccuracy in calculating α and β using (3.8) and (3.9) for small values of N. This inaccuracy was also evidenced in Figure 3.6.

Finally, the mean number of CCAs for a generic packet is given by

$$\bar{n}_C = \bar{n}_{C_{tx}}(1 - p_f) + \bar{n}_{C_f} p_f. \tag{3.26}$$

3.5.4 Efficiency

We define the efficiency, η, as the ratio between the per-node throughput and the power consumption, in bits/joule, that is given by

$$\eta = \frac{A \times S}{P}. \tag{3.27}$$

Here P is the average power consumption given by (3.14); S is the per-node throughput, defined as in [6] as the proportion of time that a node spends in successful transmission, given by $S = Ly\phi(1-\phi)^{N-1}$; and finally, A is a constant to convert the throughput to bits/s and is given by $A = B_{bp}/t_{bp}$, where $B_{bp} = 80$ is the number of bits transmitted in one backoff period, and $t_{bp} = 0.32$ ms is the duration of a backoff period.

Figure 3.11 shows the comparison between simulation and analysis for both throughput, S, and efficiency, η. As we can see in both cases, the analytical formulas offer a good approximation to the actual value of the corresponding network metrics.

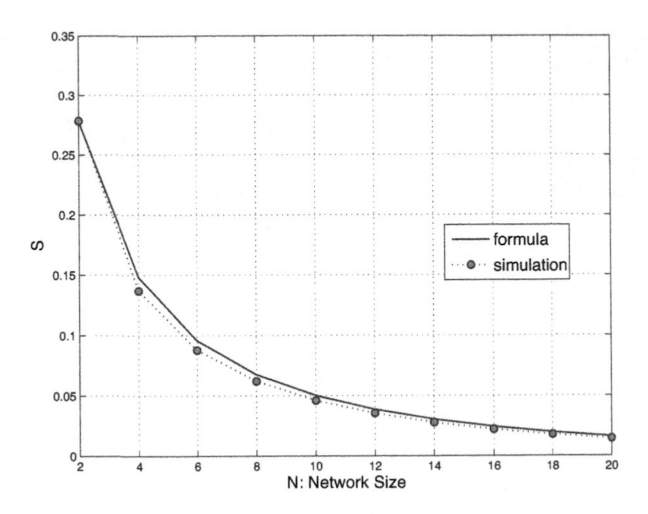

(a) Per-node throughput

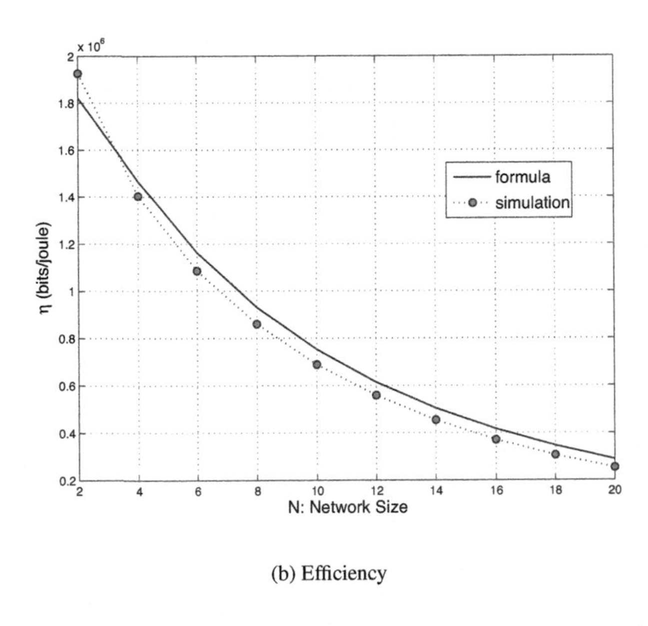

(b) Efficiency

Figure 3.11: Comparison between formula and simulation values for throughput and efficiency.

3.6 Analysis of the Model Assumptions

3.6.1 Dependence of α and β on the Backoff Stage

As was mentioned earlier, in the Markov chain model of Figure 3.5, it was assumed for simplicity that the probability of sensing the channel busy during CCA_1 and CCA_2 at a given time t was independent of the backoff stage the node is in (assumption A2). This allowed us to use a constant value for both α and β for all backoff stages, which in turn greatly simplified the derivations and analysis which followed. But as we saw in the previous section, when calculating the value of certain parameters, such as $\bar{n}_{B_{tx}}$, $\bar{n}_{C_{tx}}$, and therefore the average delay $\bar{D}$; this assumption does not result in a very good approximation of those parameters.

Figure 3.12 shows the values of α_i and β_i for each backoff stage i. As we see here, α is in fact very dependent on the backoff stage. This difference is particularly noticeable between the first backoff stage ($i = 0$) and the rest of the backoff stages. This can be explained by observing that a node that is in the first backoff stage on average draws a smaller backoff value than other nodes that are competing with it. Additionally, the joint probability of two nodes being in the first backoff stage is particularly small (as we will see in Section 3.6.2). This means that a node that is in the first backoff stage is given an opportunity with not much competition compared to when it is in any other stage. Therefore, it is very likely for it to find the channel idle. We further see that most of this dependence is absorbed by the first CCA as when it comes to β, different backoff stages experience a very similar probability of finding the channel busy.

3.6.2 Dependence of the Backoff Stage of a Node on That of Other Nodes

Figure 3.13 shows the dependence of the backoff stage of a node on that of other nodes. The dependence metric $\gamma(N)$ is defined as the maximum relative difference between the joint probability and the product of the marginal probabilities of two nodes N_1 and N_2 being in backoff stages i and j respectively, when there are N nodes in the network. In other words,

$$\gamma(N) = \max_{i,j} \left\{ \frac{f_{S_1,S_2|N}(i,j) - f_{S_1|N}(i) f_{S_2|N}(j)}{f_{S_1,S_2|N}(i,j)} \right\} \tag{3.28}$$

where S_i is the random variable indicating the backoff stage of node i at a randomly chosen time slot in the steady state, and $f_X(x)$ indicates the probability mass function (pmf) of the random variable X at a given point x.

As we see in Figure 3.13, for small number of nodes, there is a strong dependence between the backoff stages of different nodes. However, as the network grows, this dependence becomes less and less noticeable.

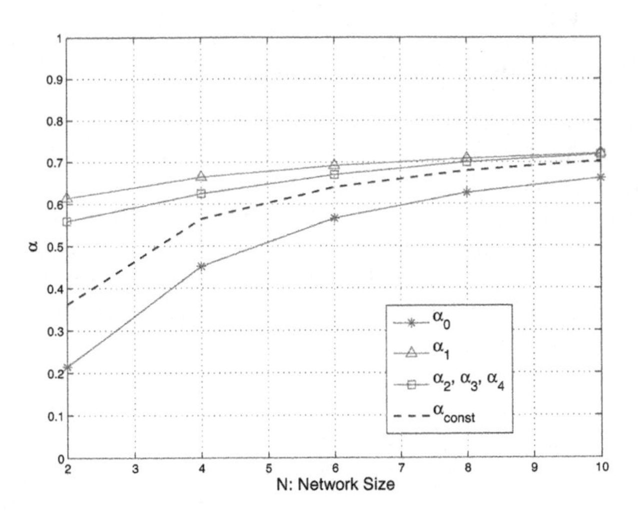

(a) α for different backoff stages

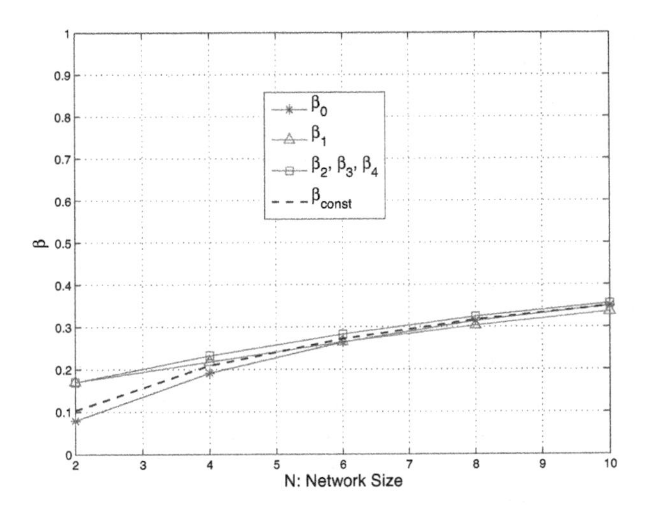

(b) β for different backoff stages

Figure 3.12: Dependence of α and β on the backoff stage.

Figure 3.14(a) shows the joint pmf for the backoff stages of two nodes for the two cases of $N = 2$ and $N = 20$. We see that for $N = 2$, with high probability, at least one node is in the first backoff stage, while the other is in one of the other backoff stages. In this case, the probability of both nodes being in higher backoff

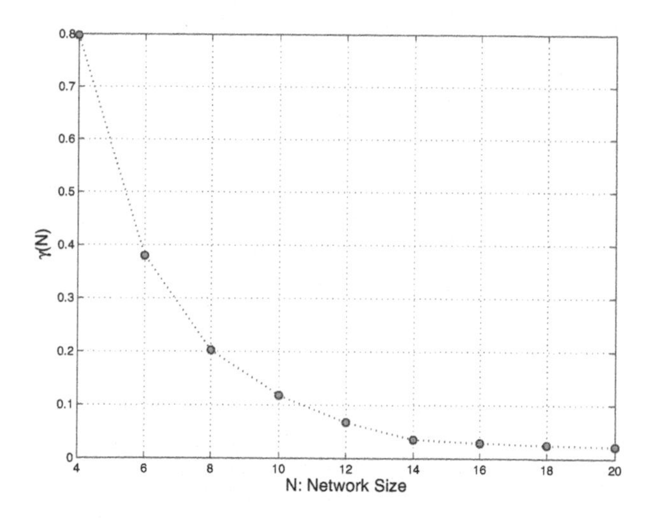

Figure 3.13: Dependence metric, $\gamma(N)$.

stages simultaneously is extremely small. This indicates a high chance of successful transmission for $N = 2$ (and in general for small numbers of nodes). The situation is quite different in a network with a larger number of nodes. We see in Figure 3.14(b) that for $N = 20$, the nodes spend more time in the higher backoff stages of the chain, indicating a greater chance of failure. However, we observe that in both cases, even though the probability of a node being in the first backoff stage is not small (each packet has to go through the first backoff stage once in its lifetime irrespective of the value of N), the joint probability of two nodes being in the first backoff stage is relatively very small. As we mentioned earlier, this explains why α_0 is different from α_i for $i > 0$.

3.7 Conclusion and Outlook

ZigBee, with IEEE 802.15.4 as its link and access technology, will undoubtedly play a central part in the growth of the Internet of Things (IoT), the wireless, and heterogeneous extension of the wireline Internet. This cannot only be attributed to its large international support by major vendors but also to its sufficiently good performance. This performance behavior has been the focus of this chapter, where we concentrated solely on the MAC assuming PHY and networking protocols given.

After having reviewed key parameters contributing to the MAC's performance behavior and prior publications in this field, we have moved on to its characterization and then analysis. We have exposed the correct mathematical formulation of

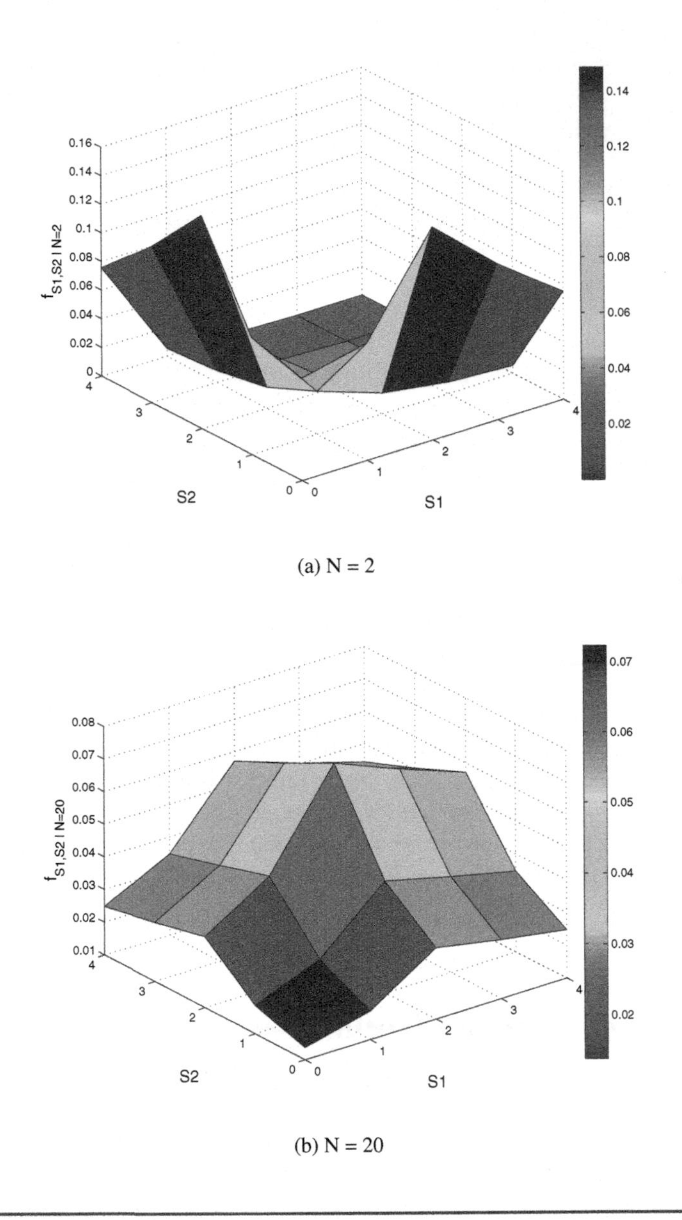

(a) N = 2

(b) N = 20

Figure 3.14: Joint pmf of the backoff stages of two nodes $f_{S_1,S_2|N}(i,j)$.

the throughput, delay and energy efficiency behavior as a function of the number of contenting nodes, the packet length, the depth of the backoff stage, etc. This formulation can then be used to tune the system parameters in order to optimize a given metric, e.g., efficiency.

In the process of developing the aforementioned formulation, we have also been able to pinpoint the reasons why the results obtained from event simulations sometimes differ from those gathered from a Markov-chain-based analysis. This identifies the limitations of this popular approach, thereby facilitating its adequate utilization.

At the same time, several important issues have not been touched upon, be it because no analytical expressions are available yet or due to space limit. Far from being exhaustive, some of these are summarized below and subsequently discussed in greater details:

- impact of finite buffer lengths;
- non-saturated traffic conditions;
- impact of capture effect due to shadowing;
- relayed traffic in more complicated topologies;
- parametric mapping to known WSN MAC families;
- extension to emerging IEEE 802.15.4e, .15.4g and .15.4f MACs.

Finite buffers lead to buffer overflows and hence to additional packet losses which have so far not been catered for. These losses are non-linear functions of the packet arrival rate and the buffer length. To include this in the current analysis, some additional transitions need to be introduced into the Markov state diagram.

Current analysis only caters to saturated traffic conditions, i.e. every node has always a packet to transmit. To extend the analysis rigorously to any possible traffic arrival conditions, additional states and transitions need to be introduced. However, some simplifying heuristics can be invoked using a binomial expansion of active nodes and the results of the above exposed theory.

Capture is referred to the effect when a node can receive a packet even though more than one transmission happens simultaneously. This is facilitated by shadowing at link level where an ongoing transmission is not perturbed simply because the receiver is shadowed by larger objects from any other ongoing transmission. A rigorous analysis of the capture effect is quite involved but, again, involving some simplifying heuristics may help finding a suitable solution.

Relayed traffic is — in a sense — redundant traffic since it involves the retransmission of already received information. It clearly impacts end-to-end delay and throughput, as well as the system's energy efficiency. Relayed packets are also usually treated differently from newly generated packets. All this is currently not reflected in the analysis where a rigorous approach is again seen to be prohibitively complex.

As already alluded to in the introduction, MACs for WSNs typically follow the following taxonomy: framed MACs for delay-constrained high traffic loads;

contention-based MACs with common active periods for medium traffic loads; duty-cycled sampling protocols for low traffic loads; and hybrids thereof [4]. These have proved to be optimum under respective traffic conditions. A formal derivation of IEEE 802.15.4 MAC parameters realizing these MACs is still an open problem and certainly worth investigating.

Finally, novel MAC families are emerging at the time of the writing of this book. These are mainly the IEEE 802.15.4e, .15.4g and .15.4f MACs, where the first is for delay-constrained embedded system solutions, the second for active RFID systems, and the last for smart utility networks. Applying the derived analysis and synthesis to these emerging standards is an open issue deserving attention.

References

[1] IEEE std. 802.15.4. *Part. 15.4: Wireless Medium Access Control (MAC) and Physical Layer (PHY) Specifications for Low-Rate Wireless Personal Area Networks (LR-WPANs).* IEEE standard for Information Technology, IEEE-SA Standards Board, Sept. 2006.

[2] IETF WG. IPv6 over Low power Wireless Personal Area Networks. Available online: http://tools.ietf.org/wg/6lowpan/.

[3] IETF WG. Routing Over Low Power and Lossy Networks. Available online: http://tools.ietf.org/wg/roll/.

[4] A. Bachir, M. Dohler, T. Watteyne, and K.K. Leung. MAC Essentials for Wireless Sensor Networks. *IEEE Communications Surveys and Tutorials*, to appear.

[5] IEEE std. 802.11. *Information Technology - Telecommun. and Information Exchange between Systems. Local and Metropolitan Area Networks. Specific Requirements. Part 11: Wireless LAN MAC and PHY Specifications.* ANSI/IEEE Std. 802.11, ISO/IEC 8802-11, 1st edition, 1999.

[6] S. Pollin, M. Ergen, S. C. Ergen, B. Bougard, L. Van der Perre, I. Moermann, A. Bahai, P. Varaiya, and F. Catthoor. Performance Analysis of Slotted Carrier Sense IEEE 802.15.4 Medium Access Layer. *IEEE Transactions on Wireless Communications*, 7(9), Sept. 2008.

[7] G. Bianchi. Performance Analysis of the IEEE 802.11 Distributed Coordination Function. *IEEE Journal on Selected Areas in Communications*, 18(3):535–547, 2000.

[8] J. Mišić, S. Shafi, and V. B. Mišić. The Impact of MAC Parameters on the Performance of 802.15.4 PAN. *Ad Hoc Networks*, 3(5):509–528, Sept. 2005.

[9] J. Mišić and V. B. Mišić. Access Delay for Nodes with Finite Buffers in IEEE 802.15.4 Beacon Enabled PAN with Uplink Transmissions. *Computer Communications*, 28(10), Jun. 2005.

[10] P.K. Sahoo and J. Sheu. Modeling IEEE 802.15.4 based Wireless Sensor Network with packet retry limits. In *Proc. of the 5th ACM Symposium on Performance Evaluation of Wireless Ad Hoc, Sensor, and Ubiquitous Networks*, Canada, Oct. 2008.

[11] R. K. Patro, M. Raina, V. Ganapathy, M. Shamaiah, and C. Thejaswi. Analysis and improvement of contention access protocol in IEEE 802.15.4 star network. In *Proc. of IEEE Internatonal Conference on Mobile Adhoc and Sensor Systems (MASS'07)*, 2007.

[12] H. Wen, C. Lin, Z.J. Chen, H. Yin, T. He, and E. Dutkiewicz. An Improved Markov Model for IEEE 802.15. 4 Slotted CSMA/CA Mechanism. *Journal of Computer Science and Technology*, 24(3):495–504, 2009.

[13] C.Y. Jung, H.Y.Hwang, D.K.Sung, and G.U.Hwang. Enhanced Markov Chain Model and Throughput Analysis of the Slotted CSMA/CA for IEEE 802.15.4 Under Unsaturated Traffic Conditions. *IEEE Transactions on Vehicular Technology*, 58(1), Jan. 2009.

[14] Chipcon Products from Texas Instruments. *CC2430 Preliminary Data Sheet (rev. 2.1) SWRS036F*, Jun. 2007.

Chapter 4

Advanced MAC Protocols for ZigBee Networks

Baozhi Chen, Dario Pompili

CONTENTS

ZigBee is a short-range wireless technology for low data rate, low power, low cost wireless systems operating in unlicensed spectrum bands. To improve the performance of ZigBee networks, a number of Medium Access Control (MAC) protocols that use one or more channels[1] [16] have been proposed. Most protocols that use only *one channel* focus on designing sleep/wake-up coordination schemes to reduce energy consumption or channel access latency, as reviewed in Sect. 4.2.1. Though efficient in energy saving, they may experience much interference, or the network throughput is limited, due to the use of only one channel.

The limitations with these one-channel protocols can be overcome using multiple channels. Interference can be greatly reduced by using channels of different frequencies. As a result, simultaneous transmissions in the neighborhood are possible and throughput can be improved. A number of *multi-channel* MAC protocols are devised to coordinate the nodes for communications over multiple channels. As summarized in Sects. 4.2.2 and 4.2.3, they mainly concentrate on the design of channel assignment/coordination and medium access mechanisms with *one* or *more transceivers*. However, since channel quality is not taken into account, these protocols may end up using channels experiencing severe interference or fading. Therefore, the performance of these MAC protocols is not optimized as link characteristics such as quality, link reliability, and coherence time[2] are not captured.

In this chapter, we propose a Multi-channel Quality-based MAC (MQ-MAC) protocol, which exploits the measured Link Quality Information (LQI) to capture channel quality variations in order to allow simultaneous transmissions over good quality channels. Our MAC protocol uses only one transceiver and is able to track channel quality variations and to select the best quality channel for packet transmission. To the best of our knowledge, this is the first ZigBee-compliant multi-channel MAC that exploits measured LQI information for optimal performance. Based on this MQ-MAC, we further propose a cross-layer solution that jointly considers the interaction of MAC, routing, and scheduling. The proposed solution was implemented in TinyOS and tested on our TelosB (with IEEE 802.15.4) sensor network testbed. Our experiments showed that our solution offers better performance than existing protocols in terms of end-to-end delay, reliability, and energy consumption.

4.1 Introduction

Over the past decades, there has been tremendous growth in wireless communications and networks all over the world. More and more wireless devices are being deployed to communicate with each other using the shared wireless medium. These devices may interfere with each other when they access the wireless medium at the same time. Therefore, it is necessary to coordinate the transmissions of these devices

[1] In the ZigBee standard, different channels refer to channels with different frequency bands.

[2] The coherence time is a measure of the minimum time required for the magnitude change of the channel to become uncorrelated from its previous value.

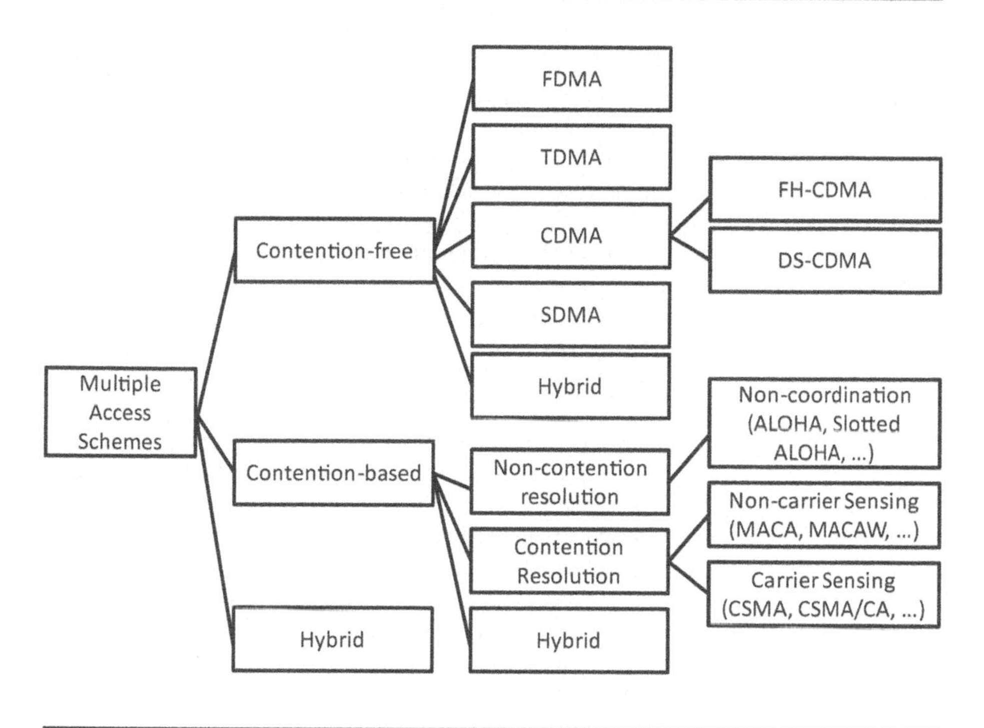

Figure 4.1: Classification of multiple access schemes.

in order to reduce the interference, and medium access control is one of the most important ways to achieve this goal.

Numerous medium access control or multiple access schemes, most of which are based on the basic schemes listed in Table 4.1, have been proposed to improve the performance of wireless networks. A way to study these schemes is to categorize them into different classes using different criterions. One widely adopted classification method is based on the way they resolve the contention when accessing the channel [38, 57]. *Contention* occurs when two close nodes both attempt to access the communication channel at the same time, which leads to message *collisions*. As shown in Fig. 4.1, these schemes can be classified into three classes: *contention-free* schemes, *contention-based* schemes and *hybrid* schemes. Contention-free schemes are designed so that nodes can access the channel without causing collisions. They are mainly used in a centralized network architecture. On the other hand, contention-based schemes are used to access the channel in a random way and hence may lead to collisions, while the hybrid schemes are a combination of the schemes from different classes.

Based on the type of diversity that the multiple access scheme exploits, contention-free schemes can be divided into the following sub-classes (see Table 4.1 for full names of the acronyms): FDMA, TDMA, CDMA, SDMA, and the hybrid schemes (a combination of these sub-class schemes). On the other hand, contention-based schemes are divided into sub-classes of non-contention resolution schemes,

Table 4.1: Basic Multiple Access Schemes

Scheme	Full Name & Main Features
FDMA [38]	Frequency Division Multiple Access: different users use different frequency bands for communications.
TDMA [38]	Time Division Multiple Access: time is divided into slots and in each slot only one user is allowed to access the channel for communications.
CDMA [38]	Code Division Multiple Access: use spread-spectrum techniques to let multiple users share the same channel. *Spread-spectrum techniques* are methods by which a signal with a particular bandwidth is deliberately spread in the frequency domain, resulting in a signal with a wider bandwidth.
SDMA [38]	Space Division Multiple Access: spatially separated users are allowed to share the channel at the same time using technologies such as smart antenna techniques.
FH-CDMA [50]	Frequency Hopping CDMA: one of the spread-spectrum techniques where signals are transmitted by switching among frequency channels (following a pseudorandom hopping pattern).
DS-CDMA [50]	Direct Sequence CDMA: one of the spread-spectrum techniques where baseband signals are spread by multiplying a pseudorandom sequence.
ALOHA [23]	A user accesses a channel as soon as a message is ready to be transmitted.
Slotted ALOHA [23]	It is an improvement to ALOHA. Time is divided into slots and a message is only transmitted at the beginning of the time slot.
MACA [20]	Multiple Access with Collision Avoidance: use *RTS-CTS mechanism* to inform other nodes to keep silent. That is, a node with data to transmit first sends a Request-To-Send frame (RTS). Upon receiving this RTS, the destination node replies with a Clear-To-Send frame (CTS). Any other node receiving the RTS or CTS should refrain from sending data for a given time.
MACAW [4]	Multiple Access with Collision Avoidance for Wireless: it is an improvement to MACA. In addition to the RTS-CTS mechanism, the receiver replies with an ACK after successful data reception.
CSMA [23]	Carrier Sense Multiple Access: before transmitting, the sender checks if the channel is busy. It will hold off the transmission until the channel is idle.
CSMA/CA [23]	Carrier Sense Multiple Access With Collision Avoidance: It is a modification of CSMA. In addition to carrier sensing, the RTS-CTS mechanism is used to avoid collision.

contention resolution schemes and hybrid schemes. Contention resolution schemes try to avoid collision by checking channel status (e.g., CSMA), using control overhead (e.g., MACA, MACAW), or both (e.g., CSMA/CA), while non-collision resolution schemes such as ALOHA and Slotted ALOHA send out packets without using any way to avoid collision. CDMA can further be divided into FH-CDMA and DS-CDMA depending on what spread spectrum techniques the protocols use.

Another way to classify the multiple access schemes is by the number of frequency channels they use. That is, they can be divided into two classes: *one-channel* schemes that use only one channel for communications and *multi-channel* schemes that use multiple (≥ 2) channels for communications. Depending on the number of transceivers in use, the multi-channel schemes are further divided into two sub-classes: those with only one transceiver and those with multiple (≥ 2) transceivers. Performance of the second sub-class is generally better than that of the first sub-class due to the ability of simultaneously receiving and transmitting packets, and the ability of receiving multiple packets from different transmitting nodes.

In this chapter, we focus on the MAC protocols that are designed for ZigBee networks. ZigBee is a low data rate, low energy consumption, low cost wireless networking technology based on the IEEE 802.15.4 standard [16] for wireless personal area networks (WPANs) [42] or Wireless Sensor Networks (WSNs) [2]. This technology is expected to provide low cost and low power connectivity for (possibly disposable) nodes whose lifetime is expected to range from months to years but do not require data transfer rates as high as those enabled by Bluetooth. ZigBee has become the de facto standard for WSNs and has been applied for a broad range of applications such as healthcare monitoring, intelligent agriculture, building automation, home area network, industrial automation, security, smart metering, and transportation.

IEEE 802.15.4 defines basic communication functionalities of the physical (PHY) and MAC layers. The PHY functionalities include activation and deactivation of the radio transceiver, energy detection (ED), link quality indication (LQI), channel selection, clear channel assessment (CCA), and transmitting as well as receiving packets, while the MAC functionalities include beacon management, channel access, guaranteed time slot (GTS) service, frame validation, acknowledged frame delivery, association and disassociation.

Based on these functionalities, a number of MAC protocols have been proposed for ZigBee wireless networks. Many of these MAC protocols rely on only one channel for communication. Most of these protocols, e.g., Sensor-MAC (S-MAC) [58], Berkeley-MAC (B-MAC) [36], aim at reducing energy consumption by using different sleep/wake-up coordination schemes. These protocols achieve limited throughput due to the inefficient use of the 16 channels available to ZigBee. Because current ZigBee devices only have one half-duplex wireless interface and the use of multiple channels is complicated as it requires some form of coordination among nodes, for the sake of simplicity these MAC protocols use only one channel for communication. Although pre-optimized offline schemes could be employed to achieve optimal performance for static configurations of the network, channel variations may lead to conditions different from those used to compute the optimum. In such cases, performance may not be optimal any more.

To overcome these limitations, researchers proposed multi-channel MAC protocols to increase throughput and to reduce signal interference and, ultimately, packet collisions. For example, Multi-channel MAC (McMAC) [44] transmits on multiple channels using a pseudo-random hopping pattern, while Slotted Seeded Channel Hopping (SSCH) [3] employs a scheduling reservation mechanism to handle multi-channel communications. However, these protocols do not take channel quality measurements into account. For this reason, they cannot capture channel variations in order to avoid using channels experiencing severe interference or fading. Therefore, the performance of these MAC protocols is not optimized. There are also some proposals using the Receive Signal Strength Indicator (RSSI) to optimize multi-channel communication. However, RSSI only characterizes received signal energy on a channel, which cannot capture link characteristics such as quality, reliability, and coherence time. For example, the RSSI value for a link with interference may be high but the link quality may still be poor, or the RSSI value may be low but the link quality is good enough. Therefore, in an interference sensitive environment such as hospitals (see Sect. 4.3.1), using RSSI value for interference reduction may not be appropriate.

In this chapter, we propose a Multi-channel Quality-based MAC (MQ-MAC) [6] protocol, which exploits the measured Link Quality Information (LQI) to capture channel quality variations in order to minimize interference, maximize channel utilization, and allow simultaneous transmissions. The best channel to transmit is selected by the receiver based on corresponding measured link quality as channels are asymmetric and forwarding channel quality can only be measured at the receiver. By using the functionalities provided by the ZigBee standards, our MAC protocol is able to monitor channel quality variations and select the best quality channel for packet transmission with only one transceiver. To the best of our knowledge, this is the first ZigBee-compliant multi-channel MAC with one transceiver that exploits measured LQI information for optimal performance. Our MQ-MAC is extended to a cross-layer solution that jointly considers the interaction of MAC, routing, and scheduling. The routing module collects channel quality information from the MAC module and the queue utilization information from neighboring nodes. The scheduling module estimates packet deadlines using information from MQ-MAC and routing, while MQ-MAC selects the best channel to forward packets.

Cross-layer wireless communication solutions allow for an efficient use of the scarce resources such as bandwidth and battery energy. However, although we advocate integrating highly specialized communication functionalities to improve ZigBee wireless network performance and to avoid duplication of functionalities by means of cross-layer design, it is important to consider the ease of design by following a modular design approach [37]. This will also allow improving and upgrading particular functionalities without the need to re-design the entire communication system. For these reasons, in our work we rely on the above-mentioned design guidelines and propose a cross-layer communication solution incorporating MAC, routing, and scheduling functionalities. Our cross-layer solution is based on current ZigBee standards and TelosB sensors, which means that it can be implemented and deployed without the need for a new ZigBee standard or hardware platform.

Within MQ-MAC, a dynamic probing scheme is designed to monitor the communication quality of all available channels. Moreover, an enhanced RTS-CTS mechanism is proposed that makes use of a Common Control Channel (CCCH) to facilitate the negotiation between the transmitter and receiver and select the best channel. Our enhanced RTS-CTS mechanism can also efficiently handle the Exposed and Hidden Terminal Problems[3]. In particular, the routing algorithm selects the route with the maximum *Route Quality Indicator* (RQI), the minmax LQI value of the links (with multiple channels) along a route. The RQI values are distributed using a Bellman-Ford algorithm. Furthermore, the scheduling module selects the packet that is expected to expire first according to link quality and the number of hops to the destination.

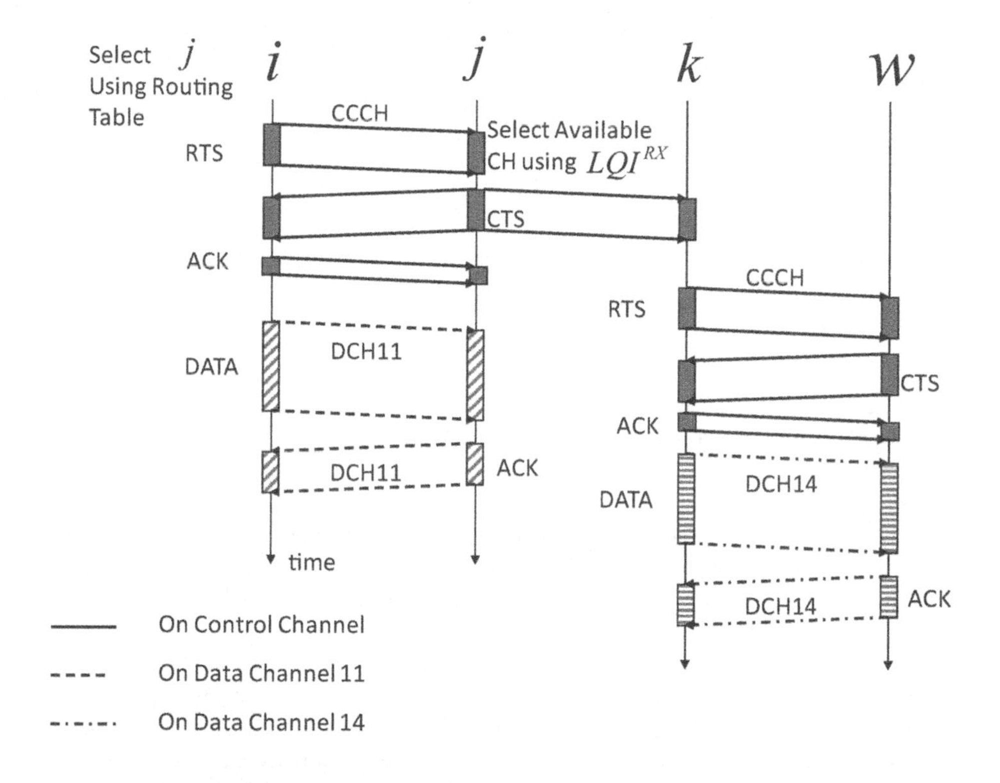

Figure 4.2: Proposed MQ-MAC protocol.

[3]As discussed in [4], the exposed node problem occurs when a node cannot send packets due to an active neighbor transmitter. In the hidden terminal problem [46], packet collision happens at the intended receiver if there is a transmission from a hidden terminal. Here a hidden terminal is a node that cannot sense the ongoing transmission but is able to introduce enough interference to corrupt the reception. More discussion is found in Sect. 4.3.2 and Fig. 4.4.

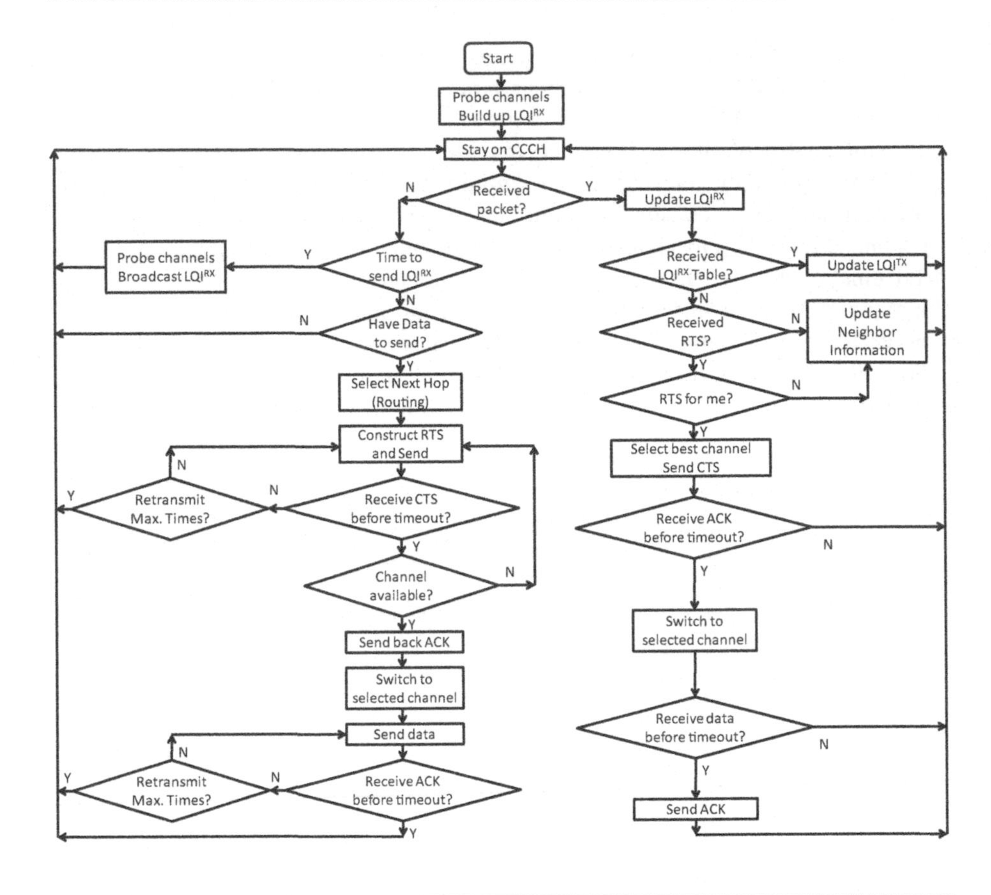

Figure 4.3: Flowchart of our MQ-MAC protocol.

The proposed solution was implemented in TinyOS and tested on our TelosB sensor network testbed. Our experiments showed that our solution offers better performance than existing protocols in terms of end-to-end (e2e) delay, reliability, and energy consumption.

The remainder of this chapter is organized as follows. In Sect. 4.2, we review MAC protocols and scheduling schemes in ZigBee wireless sensor networks. The proposed cross-layer solution is presented and discussed in Sect. 4.3. The performance of our solution is then evaluated and compared using as example a healthcare monitoring network in Sect. 4.4; finally, conclusions are drawn in Sect. 4.5.

4.2 Related Work

In this section, we first give a brief chronological review and qualitative comparison of the two classes of MAC protocols for ZigBee networks: one-channel protocols and

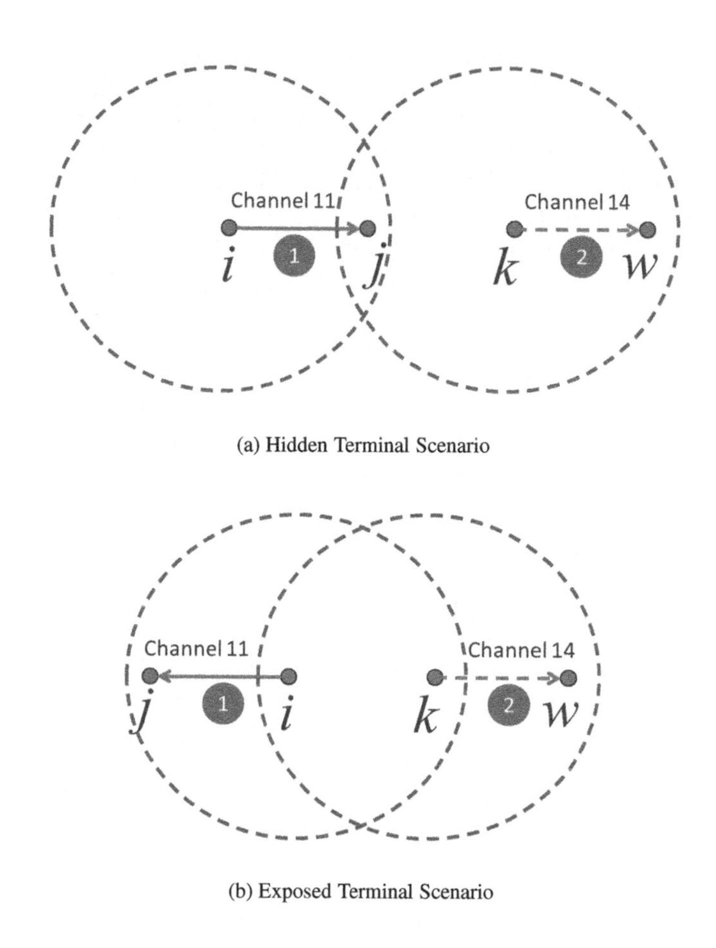

Figure 4.4: Hidden/exposed terminal scenario.

multi-channel protocols. Our MQ-MAC protocol is designed to be a practical solution for existing low-cost ZigBee devices, which have only one transceiver. Therefore, it falls into the sub-class of multi-channel protocols with one transceiver. After describing these MAC protocols, we review the packet scheduling algorithms with an emphasis on cross-layer solutions incorporating the scheduling functionality.

4.2.1 One-channel MAC Protocols

Most of the one-channel protocols focus on reducing energy consumption or latency by using different sleep/wake-up coordination schemes. In this subsection, we summarize existing one-channel MAC protocols in chronological order.

S-MAC [58] is a MAC protocol designed for WSNs. It tries to reduce the waste of energy from collision, overhearing, and idle listening using the following three techniques: periodic sleeping, setting radio to sleep during transmission of other nodes, and applying *message passing* to reduce latency (i.e., divide the long message into small fragments and then transmit them in a burst). It is shown that S-MAC consumes 2/3 to 6/7 less energy than 802.11-like MAC protocols.

Although idle listening time is reduced in S-MAC, the active time under variable message rate is not optimal. A solution for this problem is provided in Timeout-MAC (T-MAC) [48], which reduces idle listening by transmitting all messages in bursts of variable length and by sleeping between bursts. The length of the active time under variable load is dynamically determined to achieve optimality. Simulations and experiments have shown that T-MAC outperforms S-MAC in energy consumption when the message rate fluctuates.

In [29], it is shown that both S-MAC and T-MAC suffer from a *data forwarding interruption problem*, where nodes on a multihop path to the sink are not all notified of data delivery in progress, resulting in significant sleep delay. The authors then designed a MAC protocol for data gathering called DMAC to solve this problem by staggering the active/sleep schedule of the nodes in the data gathering tree according to its depth in the tree to allow continuous packet forwarding flow. Data prediction is employed to solve the problem when each single source has low traffic rate but the aggregated rate at an intermediate node is larger than the rate that the basic duty cycle can handle. Simulation results show that DMAC achieves both energy savings and low latency over S-MAC when used with data gathering trees in wireless sensor networks.

While the above protocols are focused on ad-hoc sensor networks, Wireless Sensor MAC (WiseMAC) [11] gives a downlink solution to the infrastructure network by exploiting the access point's unconstrained energy supply. WiseMAC uses non-persistent CSMA[4] with preamble sampling to minimize idle listening by exploiting the knowledge of the sensor nodes sampling schedules. This technique provides very low energy consumption when the channel is idle. Also, it was shown that WiseMAC can provide significantly lower energy consumption for the same delay.

S-MAC, T-MAC and WiseMAC are classical MAC protocols that perform channel access arbitration (e.g., by RTS-CTS mechanism) and are tuned for good performance over a set of workloads that are thought to be representative of the domain. Applications and services must rely on internal policies of the protocols to adjust their operation as node and network conditions change, and such changes are opaque to the applications. In contrast, the lightweight and reconfigurable B-MAC [36] contains only a small core of media access functionalities. It employs an adaptive preamble sampling scheme to reduce duty cycle and to minimize idle listening. By using an effective clear channel assessment, it is shown to achieve better packet delivery rates, throughput, latency, and energy consumption than S-MAC and T-MAC.

S-MAC and T-MAC are hybrid protocols with CSMA and TDMA since they maintain the synchronized time slots. Zebra-MAC (Z-MAC) [39] is also a hybrid of CSMA and TDMA, where CSMA is used as the baseline MAC scheme and a

[4]In nonpersistent CSMA, when the medium becomes idle, the sender transmits immediately. Otherwise, the sender waits for a random amount of time, and checks the medium, repeating the process.

TDMA scheme is used to enhance contention resolution. A time slot assignment is performed at deployment and a node assigned to a time slot has higher priority over other nodes. This priority scheme reduces contention probability although it is locally unfair when compared to S-MAC, T-MAC and B-MAC. Over a long period, the high overhead introduced by the time slot assignment is eventually compensated by improved throughput and energy efficiency. The performance results show that Z-MAC has an energy consumption advantage over B-MAC under medium to high contention, while it shows competitive, but slightly lower, performance than B-MAC under low contention.

In WSNs, as packets move closer toward the sink, traffic intensity, collision, congestion, packet loss, and energy drain will increase significantly. This is called *funneling effect*, and contention-based approaches such as S-MAC, T-MAC and B-MAC are not capable to mitigate it because of the large built-up losses in nodes closer to the sink. To handle this funneling effect, funneling-MAC [1] is proposed. It is a localized, sink-oriented MAC based on a CSMA/CA scheme being implemented network-wide, with a localized TDMA algorithm overlaid in the funneling region (i.e., within a small number of hops from the sink). It does not have the scalability problems associated with the network-wide deployment of TDMA. Experimental results demonstrate that it mitigates the funneling effect, and significantly outperforms other competing protocols such as B-MAC, and more recent hybrid TDMA/CSMA MAC protocols such as Z-MAC.

The long preamble sampling used by B-MAC and T-MAC is a simple and efficient way to enable low power communication. However this long preamble introduces excess latency at each hop, is suboptimal in terms of energy consumption, and suffers from excess energy consumption at nontarget receivers. These problems can be solved by using a short preamble, as proposed in X-MAC [5]. In X-MAC, address information of the target is embedded in the preamble so that non-target receivers can quickly go back to sleep. Further, a strobed preamble is used to allow the target receiver to interrupt the long preamble. It is demonstrated through implementation and evaluation that this approach significantly reduces energy usage and per-hop latency, while offering additional advantages such as flexible adaptation to both burst and periodic sensor data sources.

Periodic synchronization messages are used in S-MAC, T-MAC and DMAC to schedule duty cycling and packet transmissions. Such message exchanges consume significant energy even when no traffic is present. Convergent MAC (CMAC) [28] is proposed to reduce energy wastage like this by utilizing three mechanisms: aggressive RTS, anycast, and convergent packet forwarding. By real experiments and ns-2 [33] simulations, the authors show that CMAC outperforms other duty cycle scheduling protocols in all aspects while providing comparable throughput and latency performance as fully awake CSMA protocol.

Although efficient in energy saving or latency performance, the above protocols achieve limited throughput due to the inefficient use of the 16 channels available. There are some other one-channel protocols targeting for throughput improvement and Concurrent-MAC (C-MAC) [41] (not the one in [28]) is one of them. By disabling clear channel assessment function and employing a block-based communication mode, this C-MAC can allow a concurrent wireless channel, outperforming the state-of-art CSMA protocol with respect to system throughput, delay and energy consumption.

Although throughput is improved in C-MAC, the interference between the nodes still exists due to the use of only one channel. On the other hand, with the same hardware complexity, multi-channel protocols have the potential to further improve the throughput as interference can be greatly reduced by using orthogonal channels.

4.2.2 Multi-channel MAC Protocols with One Transceiver

As mentioned above, multi-channel MAC protocols have been proposed to increase throughput and to reduce signal interference. Many of these protocols have been categorized and compared in [54, 52, 32]. Here, we give a brief chronological, up-to-date and more complete review of the multi-channel protocols with one transceiver that were proposed recently as follows.

Receiver-Initiated Channel-Hopping with Dual Polling (RICH-DP) MAC [47] is a receiver-initiated collision-avoidance protocol that does not require carrier sensing or the unique code assignment for collision-free reception. All nodes in a network follow a common channel hopping sequence. Nodes that are not sending or receiving data listen on the common channel. To send data, nodes engage in a receiver-initiated dialogue over the current channel hop. Nodes that succeed in a collision-avoidance handshake remain in the same channel hop while the rest of the nodes continue to follow the common channel hopping sequence.

Unlike RICH-DP, which switches control channels according to a hopping sequence, Multi-channel MAC (MMAC) [45] uses a default channel for traffic indication with the power saving mechanism. The multi-channel hidden terminal problem is solved using temporal synchronization. The main idea is to divide time into fixed-time intervals using beacons, and have a small window at the start of each interval to indicate traffic and negotiate channels. Network throughput is improved significantly, especially in high network congestion conditions.

Compared to the above two protocols, Ad hoc Multichannel Negotiation Protocol (AMNP) [8] adopts a different way to exchange control information. Instead, it relies on the RTS-CTS mechanism on a common control channel for channel negotiation. A transmitting node selects a free channel and informs the receiver with RTS, and the receiver will send back a CTS message for channel selection. Higher throughput can be provided compared to its single channel counterpart.

Like RICH-DP, Slotted Seeded Channel Hopping (SSCH) protocol [3] also employs a channel hopping scheme. But the channel hopping scheme is used in a different way. Channel hopping is not only for control but also for data transmissions in SSCH, while it is only used for control channels in RICH-DP. Scheduling packets are employed to arrange channel hopping schedule so that communications do not interfere with each other, while novel synchronization techniques are introduced to distribute traffic across channels. To be effective, SSCH must adapt its schedule continuously so that frequently communicating nodes overlap in channels frequently.

The four protocols summarized so far are designed on IEEE 802.11 standards so their performance is not optimal for WSNs, although revisions can be made for ZigBee networks. To fix this problem, four optional frequency assignment schemes, exclusive frequency assignment, even selection, eavesdropping and implicit-consensus, are proposed for a protocol called Multifrequency Media access control for wireless Sensor Networks (MMSN) [60]. MMSN is designed on the constraint of small MAC layer packet size so it is more suitable for WSN applications than SSCH, which embeds the scheduling information in IEEE 802.11 Long Control Frame Header. Lightweight frequency assignment schemes are used to reduce the overhead instead of using the RTS-CTS negotiation. The non-scalable "one control channel + multiple data channels" design as in AMNP is avoided. An optimal non-uniform backoff algorithm is derived and its lightweight approximation is implemented in MMSN, which significantly reduces potential conflicts among neighboring nodes.

A component-based channel assignment protocol called *component level channel assignment* is proposed in [49], where nodes belonging to a component that is formed by nodes belonging to intersecting flows are assigned a single channel. It is shown to have less complexity than MMAC and SSCH. Moreover, MMAC and SSCH have practical limitations such as synchronization, scheduling and switching delay, while these do not exist for the proposed component-based protocol. It is shown that such a strategy can result in considerable performance gains.

Local Coordination-based Multichannel MAC (LCM-MAC) [31] uses a similar approach as AMNP to perform coordinated channel negotiations and channel switching. The only difference is that the transmitter sends a RES packet after receiving the CTS in LCM-MAC to inform its neighbors while a similar packet is not used in AMNP. On the other hand, LCM-MAC does not require network-wide synchronization as SSCH and MMAC. The effectiveness of this protocol over MMAC [45] is demonstrated via extensive ns-2 simulations.

McMAC [44] is proposed to avoid control channel congestion, which may happen for AMNP and LCM-MAC, so that it can scale to use a large number of channels efficiently. The MAC address of each node is used as a seed to generate its hopping sequence. Therefore, the hopping pattern of a receiver can be predicted so different node pairs are able to rendezvous concurrently on multiple channels and communicate with each other. Compared to SSCH, the hopping pattern is chosen at random and careful pair-wise scheduling is not needed. Also, network-wide synchronization is not needed as in SSCH.

A two-dimensional negotiation is introduced in TDMA based Multi-channel MAC (TMMAC) [59] for multi-channel communications: lightweight explicit time negotiation and explicit frequency negotiation. In this protocol, the advantages of both multiple channels and TDMA are exploited, and aggressive energy savings are achieved by allowing inactive nodes to doze. The negotiation window size is dynamically adjusted based on different traffic patterns, and this is different from MMAC. The performance evaluation shows that TMMAC achieves up to 113% higher communication throughput while consuming 74% less per packet energy over MMAC.

So far existing multichannel MAC protocols require either synchronization or a dedicated control channel. Otherwise the network will be partitioned due to the use of different channels. Busy Tone Multi-Channel (BTMC) MAC [12] is proposed as a complementary solution. It uses hash functions to distribute the control overhead over all the channels. In addition, it uses busy tones to reserve the channels and solve the hidden terminal and the channel rendezvous problems in multiple channels.

A protocol called Y-MAC [21] works in a very similar way as McMAC, but it requires accurate synchronization that McMAC does not need. In Y-MAC, if a node receives a unicast message on the base channel, it hops to the next channel to receive the following message according to the hopping sequence generation algorithm. Nodes sending to this receiver hop to the same channel and contend for the transmission. Experimental results show that Y-MAC achieves low duty cycle under light traffic conditions, while maintaining low energy consumption under high traffic conditions.

Le et al. [25] use a distributed heuristic scheme to partition nodes among channels in a way that keeps cross-channel communication to a minimum. A simple feedback control strategy was designed to ensure stability and avoid congestion. It is probably the first multi-channel MAC protocol that is based on control theory. This not only results in higher bandwidth, but also adaptively alleviates network congestion and avoids channels with high interference.

Network nodes with all the above protocols rely only on themselves to gather control information. As a contrast, Cooperative Asynchronous Multichannel MAC (CAM-MAC) [30] is a cooperative MAC protocol that allows nodes to collaborate and share control information with each other such that nodes can make more informed communication decisions. Transmitter-receiver pairs are notified of channel conflicts and deaf terminals to prevent collisions and retransmissions by neighboring nodes. The performance improvement of it is validated by experiments on hardware.

4.2.3 Multi-channel MAC Protocols with Multiple Transceivers

Protocols reviewed in the previous section are based on one transceiver in order to keep the cost and complexity of the hardware platform low. On the other hand, there are also some proposals to use multiple transceivers for further performance improvement (in chronological order).

Dynamic Channel Assignment (DCA) [56] protocol utilizes two packet interfaces, one as a control channel interface and the other as data forwarding channel interface. RTS/CTS packets are exchanged on the control channel to select a channel for data transmission. Because a dedicated channel is used for control information exchange, when the number of channels is small a considerable wastage of resources occurs.

While using multiple transceivers makes it possible to send or receive multiple packets simultaneously, it may lead to increased energy consumption. Therefore Power-Saving Multi-channel MAC (PSM-MMAC) [53] is proposed for energy saving by reducing the collision probability and the nodes' waiting time in the "awake" state. It estimates the number of active links, selects channels and power states according to queue lengths and channel conditions, followed by the optimization of the medium access probability in p-persistent CSMA[5]. Simulation results show that this solution improves both throughput and energy efficiency.

Kyasanur and Vaidya proposed protocols for multi-channel communications with multiple transceivers by classifying available interfaces into "fixed" and "switchable" interfaces [24]. Fixed interfaces stay on specified "fixed channels" for long intervals of time, while switchable interfaces can be switched more frequently, as necessary, among the non-fixed channels. By distributing fixed interfaces of different nodes on different channels, all channels can be utilized, while the switchable interface can be used to maintain connectivity. Simulation results have shown that network capacity can be significantly improved by using multiple channels, even if only two interfaces per node are available.

The Extended Receiver Directed Transmission protocol (xRDT) [31] solves the multichannel hidden terminal and deafness problems by using an additional busy tone interface and few additional protocol operations. It is shown to provide a far superior performance with respect to control channel based protocol (DCA) [56], but only using busy tones instead of control channels and thus solving the issue of determining the right bandwidth to allocate for the control channel.

The protocols above do not take channel variations into account so they may not work well in channels under severe fading. To handle this problem, opportunistic transmission schemes are used in Opportunistic Multiradio MAC (OMMAC) [7] as a way to utilize the physical layer feedback from multiple sources to improve performance through MAC, packet scheduling, and rate adaptation. Channel fading statistics, and geographical difference of users, variable maximum data rates can be supported on different transmission links over the same channel. Transmissions are scheduled on a per-channel basis, i.e., selecting the best transmission pair for each channel, rather than on a per-packet basis, which selects the best channel for transmission, and multicast RTS, virtual multi-CTS, and channel-based packet scheduling are used to collect receiver-measured channel conditions. Simulations in ns-2 show that OMMAC significantly improves the network throughput in both single and multi-hop networks.

In DCA, the hidden node problem may happen if a node chooses a channel used by an active two-hop neighbor. Dynamic Channel Selection with Snooping (DCSS) is proposed in [40] to handle such a problem by snooping. Nodes snoop data channels during idle times and then select an idle data channel based on both the snooping results. Since snooping results indicate realistic channel usages within carrier sensing range, nodes can select a truly idle data channel and avoid the hidden node problem. Simulation results verify that the proposed channel selection approach can effectively avoid the multi-channel hidden node problem and improve the network wide performance.

As discussed, most of the above proposals are designed to avoid contention or collision without considering channel fading. Although OMMAC [7] considered channel fading, it is evaluated only by simulations so it is not clear how well it

[5] In p-persistent CSMA, if the medium is sensed busy, a node continues to listen until it becomes idle. If the medium is idle, the node transmits with probability p, and delays for the worst case propagation delay of one packet with probability $1 - p$.

performs in real environments as a real propagation environment is much more complicated. These protocols do not take the actual channel quality measurements into account and hence cannot avoid using channels experiencing severe interference or fading. Therefore, the performance of these MAC protocols is not optimized. There are also some proposals such as [55, 51] using the RSSI value to optimize multichannel communication. However, RSSI only characterizes received signal energy on a channel, which cannot capture link characteristics such as quality, reliability, and coherence time.

4.2.4 Packet Scheduling Algorithms

In data networks, packet scheduling has been used to ensure Quality of Service (QoS) as a way to allocate bandwidth and control packet delays. Packets are usually put into queues when they arrive to a node, and a scheduling algorithm is run to decide which packet should be served first based on their QoS requirements. An overview of scheduling algorithms for wireless networking is provided in [13], where a number of desirable features have been summarized, and many classes of schedulers have been compared on the basis of these features. Here we will give a brief introduction of the well-known packet scheduling algorithms, which include First-Come-First-Serve (FCFS), Priority Queuing (PQ), Fair Queuing (FQ), and Weighted Round Robin (WRR), Generalized Processor Sharing (GPS) [34, 35], and Earliest Deadline First (EDF) [9], followed by a review on cross-layer solutions incorporating the scheduling functionality.

Before the scheduling algorithms, we would like to introduce some metrics to measure their performance. These metrics, which will be used in our introduction, include: 1) *fairness*, which is used to determine whether users or applications are receiving a fair share of system resources; 2) *queuing delay* or *latency*, which measures the time a packet waits in the queue until it is served; 3) *packet loss ratio*, defined as the ratio of the number of packets that are dropped to the total number of arrival packets; 4) *jitter*, defined as the absolute value of the difference between the arrival times of two adjacent packets minus their departure times (IETF RFC 2598).

As the name suggests, FCFS serves the packets in the ascending order of their arrival time. It is the simplest scheduling algorithm. Buffered packets are not reordered so FCFS cannot serve one class of traffic differently from other classes of traffic. During congestion, FCFS benefits the User Datagram Protocol (UDP) flows[6] over Transmission Control Protocol (TCP) flows since the transmission rate for TCP flows is reduced while the transmission rate for UDP flows remains constant. Furthermore, a bursty flow may take up the whole buffer of a FCFS queue, causing all other flows be denied.

Unlike FCFS, the priority queuing algorithm can offer services to different classes of traffic with different requirements. In PQ, each packet is assigned a priority and placed into a hierarchy of queues based on priority. When there are no more

[6] A flow is defined as the traffic of the same type with the same QoS requirement sent along a path connecting a source-destination pair.

packets in the highest queue, the next lower priority queue is served. The problem with this method is that lower priority packets may get little attention. A misbehaving high-priority flow may add significant delay and jitter experienced by other high-priority flows sharing the same queue.

PQ is not a fair algorithm since high priority traffic is always served first. In contrast, in FQ each packet is assigned a type (flow) and placed into the queue for that type. All the queues are served in the round-robin manner: a packet from one queue, a packet from the next and so on. FQ provides a more uniform service to all packet types than priority queuing. As a result, an extremely bursty or misbehaving traffic flow does not degrade the QoS delivered to other flows due to the fair service. But flows with different bandwidth requirements may not be served well since all the queues are treated equally. And real-time services cannot be easily supported, either.

As we have seen, PQ and FQ are still not satisfying for traffic with different requirements. One way to improve the service is to combine them together, which leads to the WRR algorithm. In WRR, packets are assigned a class and placed into the queue for that class of service. Packets are accessed in the round-robin style, but classes can be given different weights. This allows flows with different bandwidth requirements to be served well without denying services to the low-priority traffic. The primary limitation of WRR is that it provides the correct assigned percentage of bandwidth to different traffic class only if all packets in all the queues are of the same size or when the mean packet size is known in advance.

This limitation of WRR can be overcome by using another algorithm: GPS (also called Weighted Fair Queuing (WFQ) [10]). GPS schedules packets of each flow with guaranteed minimum bandwidth according to specified weights that give fair bandwidth sharing among the flows. It is similar to WRR but it does not require packets be of the same size or mean packet size be known in advance as WRR does. It has been shown that end-to-end delay requirements can be mapped into a bandwidth allocation problem by appropriate admission control in a network with deterministic end-to-end delay bounds [35]. However, it has also been shown that the close coupling between delay and rate under GPS in deterministic delay bounds leads to sub-optimal performance and decreased network utilizations [14]. Moreover, GPS comes at a cost of greatly increased complexity to implement the scheduling discipline. WRR can achieve similar service differentiation and is a simpler scheme to implement, with similar performance.

Instead of scheduling based on the weights assigned to the queues, an EDF scheduler assigns "deadlines" to packets arriving at the scheduler and then serves the packets in the ascending order of their assigned deadlines. Specifically, every time a packet arrives at one of the queues it is assigned a deadline equal to its arrival time plus the maximum tolerable queuing delay of the packets. Every packet needs to be sorted according to its deadline upon its arrival at the node, and the packet with the least deadline is served first.

It has been shown that optimal performance can be obtained with EDF policy for a single switch [9], and certain EDF techniques can achieve better performance than those using GPS [14]. Yet, the implementation of an EDF server is more complicated than that of a GPS one, and proper techniques have to be devised to make the cost of such a server affordable in practice.

To give optimal network performance, many researchers have proposed cross-layer design solutions that incorporate traffic scheduling mechanisms. Here, we give some examples without aiming at being exhaustive.

In [27], a cross-layer design for multiuser scheduling at the data link and physical layers is derived to simultaneously enable QoS requirements and efficient bandwidth utilization. The cross-layer scheduler is implemented using a low-complexity approach. It also provides service isolation and scalability, decouples delay from dynamically-scheduled bandwidth, and is backward compatible with existing separate-layer designs.

In [26], a general class of cross-layer transmission schemes is proposed to address the problem of optimizing the packet transmission schedule in a multihop network with end-to-end delay constraints. A recursive non-homogeneous Markovian analysis framework is proposed to study the effect of the lifetime-distance factor on packet loss probability in a general multihop environment, with different configurations of peer-node channel contention. It is demonstrated that the proper balance between distance and lifetime in a transmission schedule can significantly improve the network performance, even with an imperfect schedule implementation.

In [22], the problem of scheduling messages with probabilistic deadline constraints is presented and solved, with consideration of the practical erroneous channel condition and the receiver energy consumption. An Integer Linear Program (ILP) for the scheduling problem is formulated, and efficient heuristic scheduling algorithms that minimize the energy consumption while providing the required guarantees are proposed. The proposed heuristic algorithms are shown to achieve energy savings comparable to those obtained using the linear programming methodology under practical channel conditions.

4.3 Proposed Cross-layer Multi-channel Communication Solution

In this chapter, we propose a cross-layer multi-channel solution, which includes three modules: Multi-channel Quality-based MAC (MQ-MAC), Channel Quality Based Routing (CQBR) and Delay Guaranteed Scheduling (DGS). With this cross-layer design, the scarce network resources can be used efficiently to improve network performance while meeting the QoS requirements of many applications.

4.3.1 *Motivations for the Cross-layer Multi-channel Solution*

Wireless technology has experienced explosive growth over the past decade. Wireless wide area networks (WWAN), including mobile phone networks, and wireless local area networks (WLAN) are being deployed almost everywhere at an incredible speed, resulting in increased spectrum use by a variety of heterogeneous devices, standards, and applications. This holds especially true for the unlicensed Industrial, Scientific and Medical (ISM) bands that host a number of heterogeneous networks. For example, many devices of different standards for WLANs such as Bluetooth, ZigBee, IEEE 802.11b/g all operate in the same 2.4 GHz ISM (Industrial-Scientific-Medical) frequency band.

Because radio waves of the same frequency emitted from these devices interfere with each other, coexistence of them has become an important issue in order to ensure that wireless services can maintain their desired performance requirements. For instance, in a critical environment such as medical emergency scenarios, it is extremely important to avoid the failure of the medical devices due to radio frequency interference.

With an ever-increasing use of electronics in medical devices of all kinds as well as many wireless communication devices in medical environments, some unforeseen problems are coming: the interactions between the products emitting the electromagnetic (EM) energy and sensitive medical devices. Even the devices themselves can emit EM energy which can react with other devices or products. It has been reported that medical devices may fail to operate correctly due to the existence of electromagnetic interference[43].

To guarantee wireless services in such environments, it is necessary to design a system that can handle such interference. Compared to single-channel communications, multi-channel communications and networking have more potential to guarantee the services due to the availability of multiple non-interfering frequency bands. Therefore, a multi-channel MAC protocol is adopted in our proposal to minimize the interference. A MAC protocol with partial cognitive radio capability will be devised with with an *extended RTS/CTS mechanism* to handle the hidden/exposed terminal problem.

Our MQ-MAC is a multi-channel MAC protocol that uses only one transceiver. Currently most ZigBee devices are designed to be of low cost with limited resources. Using more than one transceiver would result in more cost and hence it is not desirable. We aim to design a practical solution that can be implemented and deployed on these low-cost devices therefore a MAC with one transceiver is a proper choice.

In wireless sensor networks, there are different types of sensing data and these data may have different QoS requirements for wireless sensor networks. For example, voice or video applications have QoS requirements that are quite different from those of email applications in terms of end-to-end delay and reliability. In this chapter, we aim to guarantee the e2e delays of different types of data. As seen in Sect. 4.2.4, an EDF scheduler is clearly an appropriate choice for this requirement.

Considering the above requirements, to further improve the efficiency of using the scarce resources such as bandwidth and battery energy, we adopt a cross-layer approach that jointly considers the interaction of MAC, routing, and scheduling. With this cross-layer design, nodes can select the best channel for packet forwarding along the best quality route in a timely manner and network performance is improved in terms of end-to-end delay, reliability, and energy consumption.

Table 4.2: LQI^{RX} (LQI^{TX}) Table at Receiver (Transmitter)

Transmitter (Receiver)	DCH11	DCH12	···	DCH26
i_1	$LQI_{i_1}^{(1)}$	$LQI_{i_1}^{(2)}$	···	$LQI_{i_1}^{(16)}$
i_2	$LQI_{i_2}^{(1)}$	$LQI_{i_2}^{(2)}$	···	$LQI_{i_2}^{(16)}$
.	.	.	.	.

4.3.2 Proposed MQ-MAC Protocol

Our Multi-channel MAC is designed to have partial cognitive radio capability. In a cognitive radio system, the cognitive process typically starts with spectrum sensing, followed by channel identification and spectrum management [15]. In our case spectrum sensing is achieved by probing and sensing qualities of the 16 channels. The measured channel quality serves as a metric to characterize channel activities and variations such as fading. A channel with the best quality, i.e., the channel experiencing the least interference or fading, is chosen for packet forwarding. Moreover, as the channel quality is observed at the receiver, the channel is selected by the *receiver* in our MAC. By using the measured LQI information on multiple channels, our MAC gives a solution to capture channel variations, maximize channel utilization, reduce interference, and allow simultaneous transmissions.

The spectrum (channel) sensing protocol to probe and track the LQI values of all the channels is designed as follows: A dedicated Common Control CHannel (CCCH) is used for all the nodes to exchange control information. During initialization, a node sends probes and listens on all available data channels. This can be done by competing on CCCH with probes. The winner sends probes over the available channels according to a previously or dynamically agreed schedule. Upon reception of these probes, the probe receivers build their LQI^{RX} tables (Table 4.2), which records the LQI values of the channels. LQI^{RX} is also updated upon the reception of a packet on one channel. This LQI^{RX} table is broadcast periodically on CCCH for the transmitters to build up their LQI^{TX} table (Table 4.2), which will be used by the transmitter for its egress (outgoing) channel information. LQI^{TX} is used by the transmitter for routing, as discussed in Sect. 4.3.3.

The schematic of our MQ-MAC is shown in Fig. 4.2, and a detailed flowchart including the transmitting and receiving part is given in Fig. 4.3. After LQI^{RX} is built, nodes can forward data to each other with MQ-MAC. When i has a data packet for j, it first sends out a RTS packet with the unpreferred channels on CCCH, and l, the data packet's length. Here l is used by the receiver and its neighbors to estimate the time that the channel is used. Upon receiving the RTS, j selects the best available channel as,

$$CH_{ij}^{*} = \arg \max_{\substack{CH \in \mathcal{ICH}(j) \\ CH \notin \Lambda(i,j)}} \{LQI_i^{RX}\}, \tag{4.1}$$

where LQI_i^{RX} is the set of ingress LQIs from i in table LQI^{RX}, $\mathcal{ICH}(j)$ is the idle channel set (estimated by overhearing RTS/CTS messages) in j's neighborhood, and $\Lambda(i,j)$ is the set of the unavailable channels just received from i and the adjacent channels of those active channels near j. Channels in $\Lambda(i,j)$ are not selected so as to avoid interference at the transmitter and that from adjacent channel[19]. CTS is then broadcast with CH_{ij}^* and l on CCCH. If no channel can be selected, CTS is broadcast with channel number 0, telling i to backoff.

After receiving CTS, i sends back an ACK message with CH_{ij}^* if this channel is not reserved or active in the neighborhood. On hearing this message, i's neighbors will mark CH_{ij}^* as reserved. Otherwise, i sends back a new RTS to ask j to re-select another best channel. This negotiation is repeated until they settle on an available channel for both sides.

Finally, i tunes its channel to CH_{ij}^* and starts the transmission of the data packet to j. An ACK message will be sent back to i for acknowledgement. If all these steps are successful, i.e., all messages are received, i and j will tune their channels back to CCCH. Otherwise, retransmission will be made after a timeout expires until the maximum number of retransmissions is reached. The channel will be tuned back to CCCH if the maximum number of retransmissions has been reached.

Remark: *1) Although the quality of egress channels can be obtained in LQI^{TX}, this information may not be up-to-date. Hence, relying on the receiver to use CTS to reserve the channel is a better way. 2) ACK is used to reserve the channel at the transmitter while the new RTS is for channel renegotiation.*

Hidden terminal problem is handled in Fig. 4.4(a). Suppose at first i is sending to j on channel 11 and k needs to send a packet to w. In one-channel MAC protocols, this may result in one of the three cases: neither i nor k can send; collision happens at j; k will backoff.

In our MQ-MAC, according to the above design, RTS is broadcast with unpreferred channel 11, asking w to select a channel via CTS. w chooses channel 14 according to equation (4.1). With the help of CTS and ACK, k can use channel 14 to forward data to w. With different channels, i and j, and simultaneously k and w, can communicate without blocking each other.

Even another delicate situation can be handled. Suppose that, right after, i sends RTS to j, k also sends RTS to w, and then k receives j's CTS before w's CTS, both trying to reserve channel 11. In such case, as j reserved channel 11 first, k has to give up channel 11 and sends out another RTS to renegotiate with w. Finally, i and j settle on channel 11 while k and w settle on channel 14. Simultaneous communications via different channels are still possible. Hence, hidden terminal problem is handled.

On the other hand, *exposed terminal problem* is also handled in Fig. 4.4(b). In one-channel MAC protocol, when this happens, normally either k will not transmit or k still uses the same channel. Yet interference between both communication pairs still exists due to usage of the same channel. In MQ-MAC, by overhearing i's RTS and ACK, k is able to select a different channel for transmitting and interference is greatly reduced. Therefore, exposed terminal problem is better handled.

To show the advantage of our cognitive radio approach, we compare our solution with the standard one channel CSMA/CA MAC (denoted by OC-MAC). The

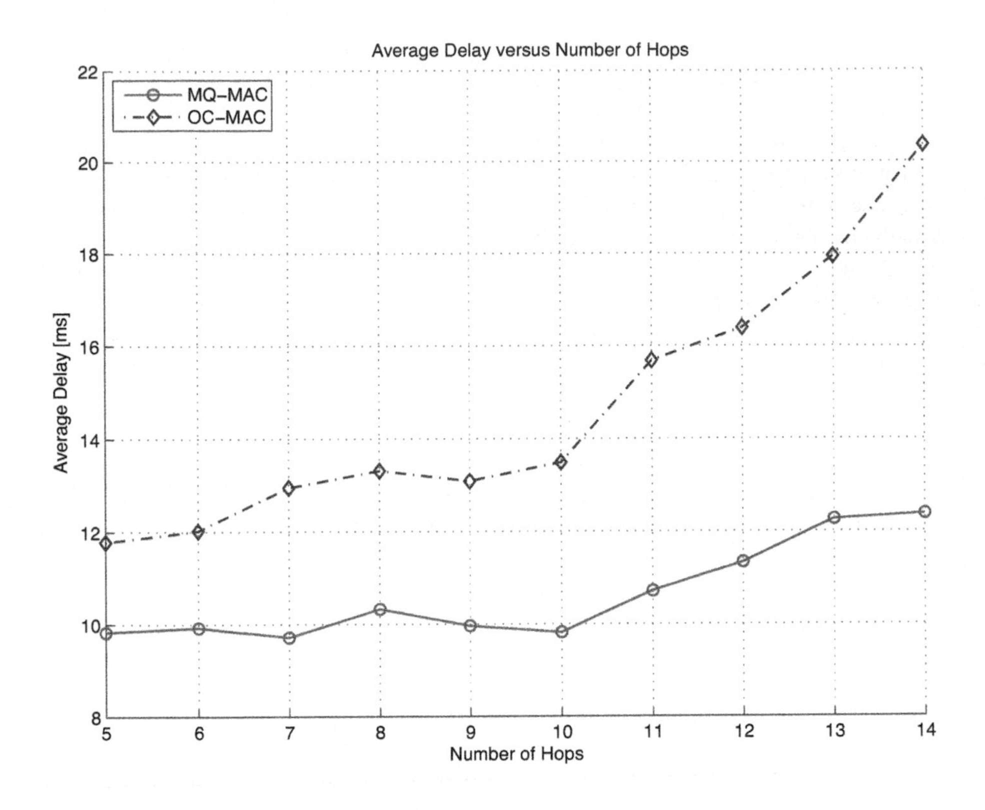

Figure 4.5: e2e delay: one-channel MAC (OC-MAC) versus MQ-MAC.

performance comparison is shown in Fig. 4.5 and Fig. 4.6, where *e2e delay* and *e2e reliability* are defined as in Sect. 4.4.2.

For a multi-hop scenario, they are tested on a network which routes on shortest paths, i.e., paths with the least number of hops to the sink. Our results show that exploiting channel diversity brings a lot of improvement. The communication protocols are implemented on TinyOS, the embedded operating system designed for networked sensors. These protocols are loaded into the TelosB sensors and tested in our testbed. Experiments were carried on in the aisles of the floor (around $50 \times 20\text{m}^2$) in our building with a sink node located at one corner as shown in Fig. 4.11.

From the results above, it is clear that our approach that exploits the channel diversity with measured channel quality achieves higher packet delivery ratio than those using only one channel.

4.3.3 Channel Quality Based Routing

Results above have shown that our MQ-MAC protocol using multi-channel and measured channel quality information can achieve better MAC layer performance. We

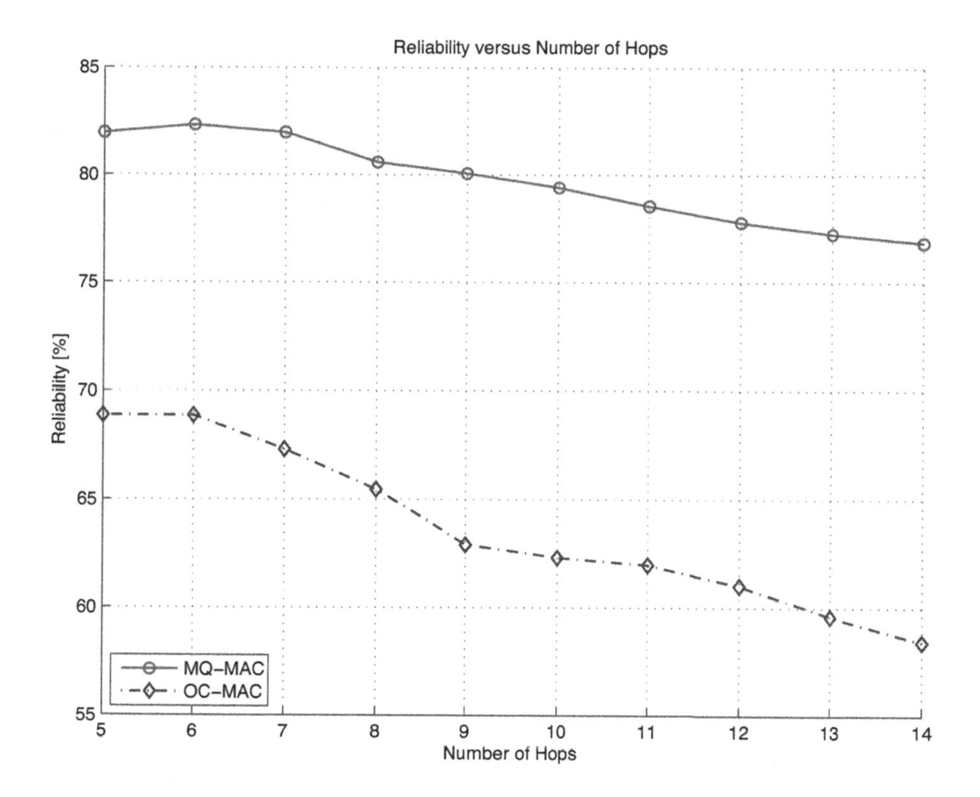

Figure 4.6: e2e reliability: one-channel MAC (OC-MAC) versus MQ-MAC.

are interested to extend it into routing and see how much improvement we can obtain. Motivated by this, we propose a Channel Quality Based Routing (CQBR) algorithm that uses multi-channels and measured channel quality information for routing. Moreover, the MQ-MAC scheme proposed in Sect. 4.3.2 is also incorporated into the routing algorithm.

We aim at achieving full utilization of the channels and robustness against interference. The objectives of our CQBR protocol are:

1) Use measured channel quality to select a route with the best quality;

2) Select the optimal channel in terms of link quality;

3) Allow simultaneous transmissions on different channels in the neighborhood.

Define *Route Quality Indicator* RQI_{SD}, a metric to measure the route quality from source S to D, as

$$RQI_{SD} = \min_{(i,j)\in\mathcal{R}_{SD}} \max_{c\in ACH_{ij}} LQI_{ij}^{c}, \tag{4.2}$$

where (i, j) is the link from i to j, $\mathcal{R}_{SD}$ is the set of links along the route from S to D, ACH_{ij} is the set of available channels for link (i, j), and LQI_{ij}^{c} is the LQI from i to j via channel c.

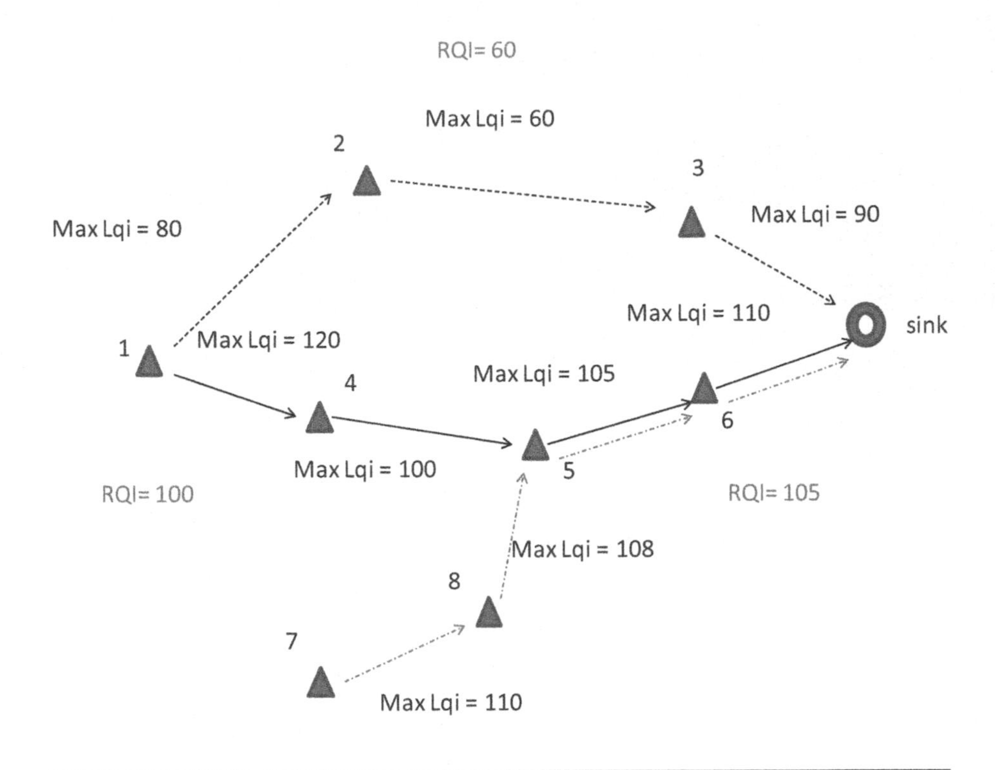

Figure 4.7: A CQBR example.

By combining routing and MQ-MAC, we obtain a cross-layer protocol CQBR, which selects the next hop decided by the RQI estimated at the *transmitter*, and the best LQI channel at the *receiver*. When a node needs to route traffic to the sink, it selects the best next hop using RQI in the routing table (Table 4.3), i.e., data traffic is forwarded to the neighbor that has the best RQI. With the MQ-MAC protocol, the selected next hop commands the transmitter to tune to the optimal channel for data forwarding.

Assume that i needs to route a packet to some destination D, CQBR works by selecting the optimal available next hop with the best RQI as,

$$j^* = \arg \max_{j \in \mathcal{N}(i)} RQI_{ijD}, \tag{4.3}$$

where $\mathcal{N}(i)$ is the set of i's non-busy neighbors (estimated by looking into the overheard RTS/CTS packets with the data packet length), and RQI_{ijD} is the RQI value of the route from i to sink D via j. Then, it utilizes MQ-MAC to select the best channel and forward the packet to next hop.

To illustrate the routing protocol, an example is given in Fig. 4.7. The maximum LQI values of each communication pair are given in the figure, and the RQI values for route from 1 to sink via 2, from 1 to sink via 4, and from 7 to sink are 60, 100, and

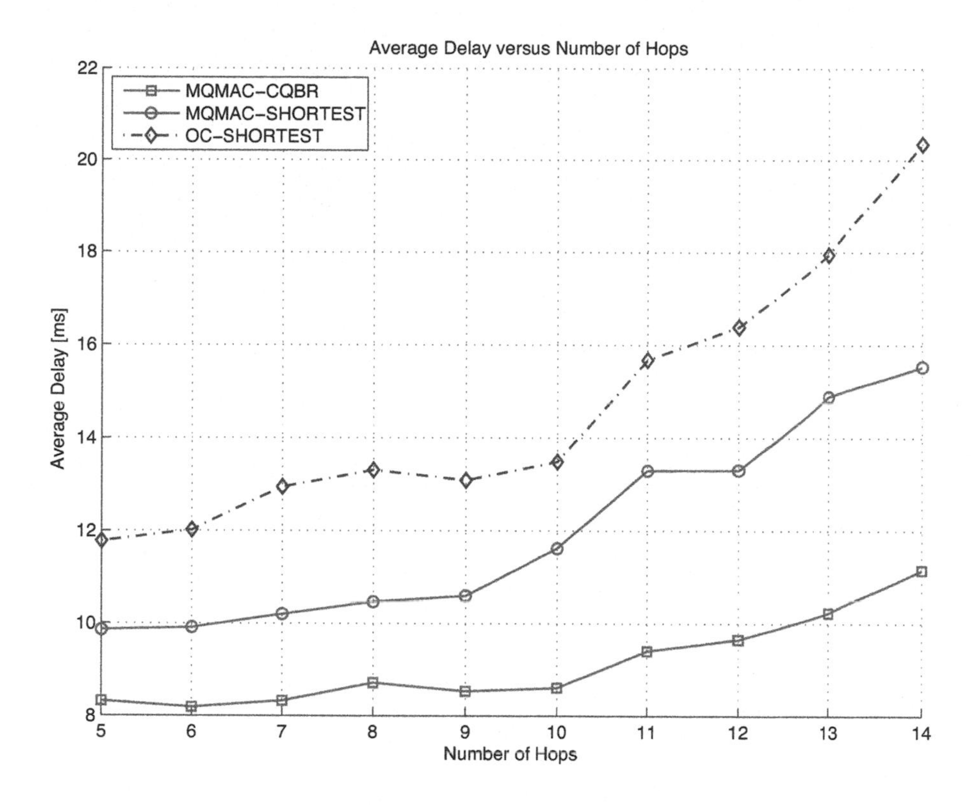

Figure 4.8: e2e delay comparison.

105 respectively according to Eq. (4.2). When node 1 has data to send to the sink, it selects the route via 4 since the RQI of this route is greater than that of the route via 2. By using a route with high RQI, the node can always ensure that there is no bad LQI channel along the route.

Routing tables, LQI^{TX} and LQI^{RX}, are created and maintained as follows:

- **Initialization:** Channels are scanned and LQI^{RX} is created at each node. This table is then broadcast for the neighbors to create their LQI^{TX} tables. Routing table is created from LQI^{TX} with only the entries to neighbors.
- LQI^{RX} is updated upon reception of packets.
- Periodically, LQI^{RX} and routing table are broadcast to neighbors.
- Update LQI^{TX} with LQI^{RX}s from the neighbors.
- Update routing table with LQI^{TX} and routing information from neighbors.

The flowchart of our MQ-MAC in Fig. 4.3 already includes the steps to create and update LQI^{RX} and LQI^{TX} so we only need to add in the creation of a routing table

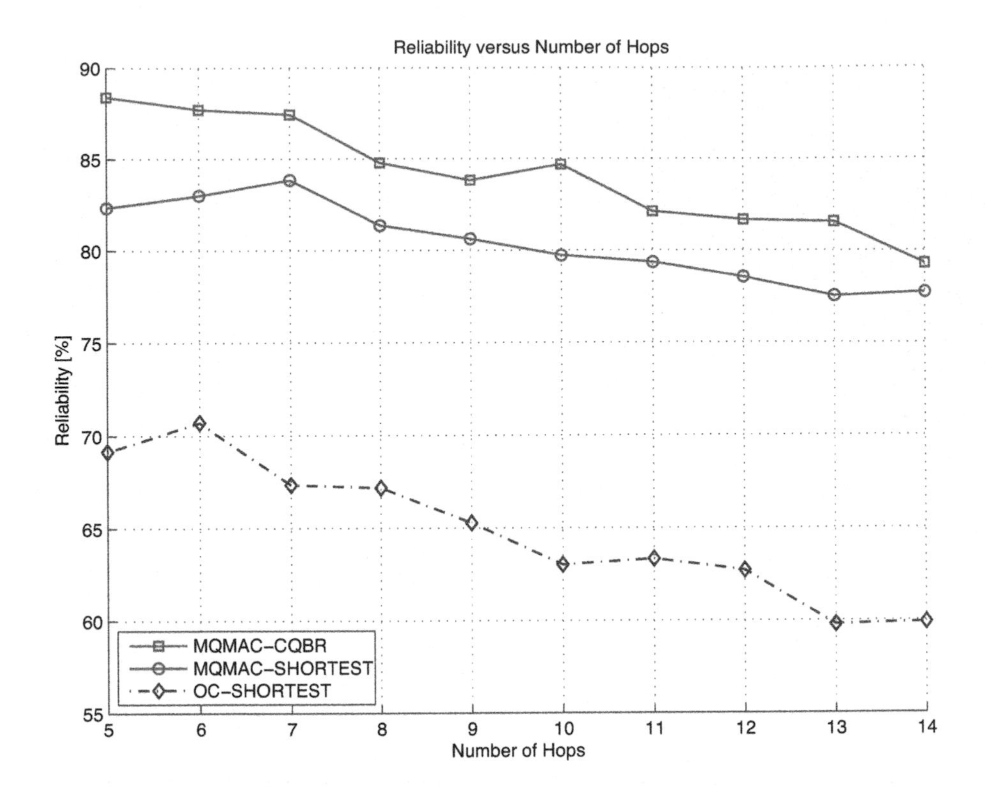

Figure 4.9: e2e reliability comparison.

during initialization, broadcast of routing information after broadcasting LQI^{RX} and update of routes at the receiving part.

The e2e performance of our MQ-MAC with CQBR routing (MQMAC-CQBR) in terms of delay and reliability is compared with the following protocols:

1) Protocol stack with one-channel CSMA/CA MAC protocol and routing on the path with the shortest number of hops (OC-SHORTEST);

2) Protocol stack with our Multi-Channel MAC and routing on the path with the shortest number of hops (MQMAC-SHORTEST).

The evaluation results in Fig. 4.8 and 4.9 show that our solution gives better performance in terms of e2e delay and reliability than the other two.

4.3.4 Cross-layer Solution: CQBR + MQ-MAC + Scheduling

The above comparisons show that the cross-layer routing design in Sect. 4.3.3 gives performance improvement over conventional protocols using one channel. Yet protocols so far have no mechanism to guarantee the delay requirements of the traffic. From Sect. 4.2.4, an EDF scheduling scheme is an effective way to bound the packet

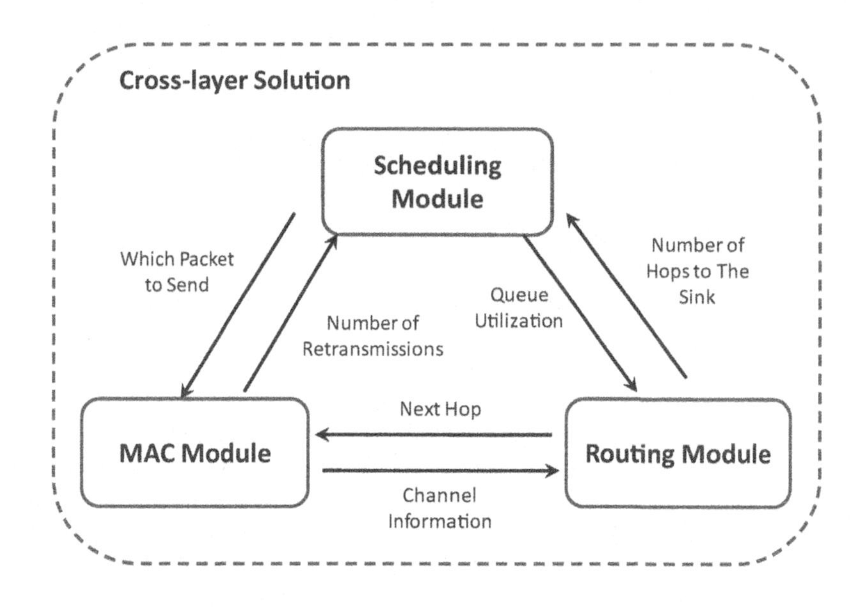

Figure 4.10: Interactions between the network modules.

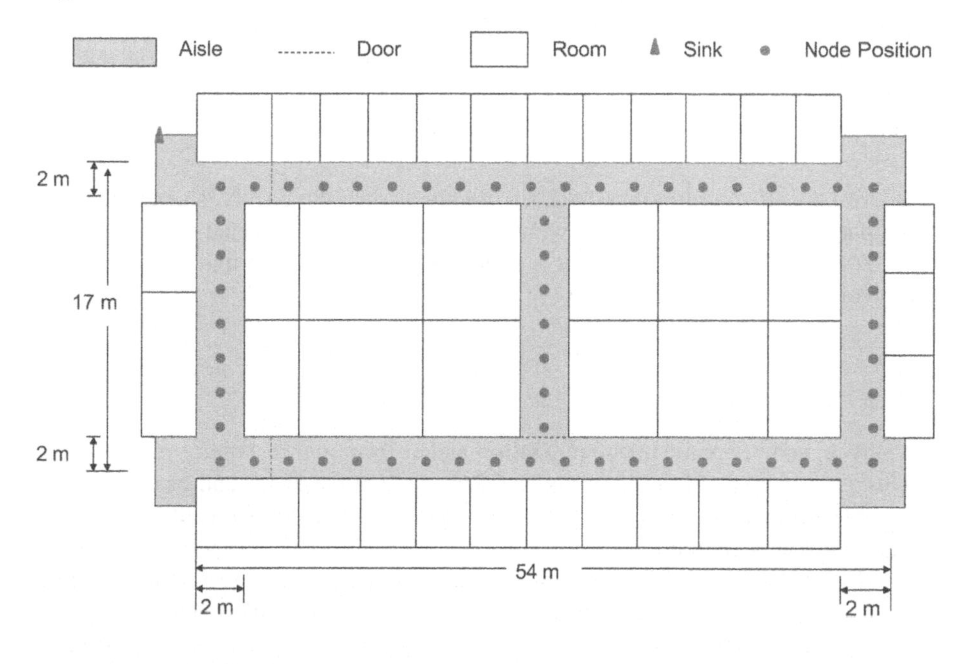

Figure 4.11: Floorplan of the experiment environment.

delay. The scheduler is designed to be delay guaranteed, and hence we denote it by DGS.

Table 4.3: Routing Table at Node *i*

Destination	Next Hop	Route Quality	Hops to Sink
d_1	j_1	$RQI_{ij_1d_1}$	n_1
d_2	j_2	$RQI_{ij_2d_2}$	n_2
.	.	.	.

Interactions between the DGS scheduling module, MAC module, and routing module are shown in Fig. 4.10. Each module operates on the input from the other two modules and feed back information in the reverse directions so that the other modules can adjust their operations. The *MAC module* collects the channel quality information and passes it to the routing module, which uses this information and the neighbors' *queue utilization* information, a metric L representing the percentage of queueing space in use in the scheduler, to decide the route to the sink. The *routing module* estimates the number of hops N_{sink} to the sink, passing it to the scheduling module, which also utilizes N_{tx}, the average number of transmissions made to successfully send a packet, from the MAC module to select a packet to transmit. Given a packet's maximum e2e delay requirement t_{e2e}, its maximum tolerable delay t_d at this node can be estimated by

$$t_d = \frac{t_{e2e} - (t_{cur} - t_0)}{N_{sink} \cdot N_{tx}}, \tag{4.4}$$

where t_{cur} is the current time and t_0 is the time when the packet is generated. Then the packet that expires first, i.e., with minimum t_d, will be selected for transmission.

A queue utilization constraint is added to equation (4.3) to find a balanced route to avoid overloading the nodes by

$$j^* = \arg \max_{\substack{j \in \mathcal{N}(i) \\ L_j < \tau_j}} RQI_{ijD}, \tag{4.5}$$

where L_j and τ_j are the queue utilization and utilization threshold of j, respectively. This L_j is embedded in the periodic LQI^{RX} update messages and broadcast to the neighbors, while an entry is added to the LQI^{TX} table to save this utilization information for the neighbors.

With the close interaction of these modules, the traffic is served with different priority based on a packet's expiration time. Furthermore, the queue utilization is used to balance the traffic of the routes and thus improve the network performance.

It is obvious that the operation of each module depends on the information from the other two modules, which leads to a cross-layer framework.

4.4 Performance Evaluation

To fully evaluate the whole solution introduced from Sect. 4.3.2 to Sect. 4.3.4, we implemented the whole solution on TinyOS and loaded them into the TelosB sensors and tested in our testbed.

4.4.1 Experiment Scenario: Healthcare Monitoring

We base the experiments on a healthcare monitoring application. The scenario is assumed to be a healthcare monitoring network with wireless sensors for multiple patients. Wireless healthcare monitoring in home or hospital is envisioned to be one important application of wireless sensor networks or ZigBee networks, where low-power, short-range, wireless sensor devices are deployed on the patients to collect critical patient-specific information, including temperature, pulse-oximetry, blood glucose levels, electrocardiogram (EKG) readings, blood pressure levels and respiratory carbon dioxide (CO_2) level. Moreover, these different types of information have different delay requirements as shown in Table 4.4. It is critical that this information of multiple patients should be collected and transmitted *reliably in real time* to medical devices or terminals accessed by doctors to enable efficient patient monitoring from any location. Consequently, a wireless healthcare monitoring system should be able to handle these different delay requirements of a patient's vital signs. On the other hand, as mentioned in Sect. 4.3.1, it is desirable to have a healthcare system that is able to detect and handle the EM interference.

Given the above requirements and the limited networking and processing capabilities of the sensors, it is necessary to find a solution to use the limited network resources efficiently to satisfy the requirements of such a critical environment. This can be possible with the help of cross-layer network optimizations so a cross-layer design is a natural choice.

Our cross-layer solution offers the capability to track and avoid radio interference while guaranteeing the e2e requirements of the data. Therefore it is an appropriate choice for the wireless healthcare monitoring scenario like this so we would like to evaluate its performance in such a scenario. Our proposal can also be used in many other applications such as multimedia WPAN networking, real-time industrial process monitoring and control, real-time environment and habitat monitoring with different types of data, especially in the environment with interference.

The experiment scenario is assumed to be a multi-patient wireless heathcare monitoring network located in the hospital or healthcare center with medical devices or WLAN networks. Sensors are deployed on a number of patients to monitor their vital signs with real delay requirements as in Table 4.4. These vital signs are forwarded to a sink, which is assumed to collect and process the patients' data.

Experiments were carried on in the aisles of the floor (around $50 \times 20\text{m}^2$) in our building with a sink node located at one corner. There are IEEE 802.11b/g wireless routers deployed on this floor so there is interference from WLAN communications. N_B nodes are evenly distributed in the aisles, as shown in Fig. 4.11. Every TelosB node emulates the activities of monitoring a patient's vital signs, generating three

Table 4.4: Typical Bit Rate and Delay Requirements of Healthcare Data (IEEE Standard 11073 [17])

Data Source		Bit Rate [bps]	Delay [s]	Sampling Rate [Hz]
Electrocardiogram (EKG)		1 - 8k	<10	63 - 500
Blood Pressure [mmHg]	Arterial Line	1k	10 - 30	63
	CVP (Central Venous Catheter)	1k	> 120	63
	Non-Invasive Cuff	0.05	30 - 120	0.025
Cardiac Output [L/min]		1k	< 10	63
Pulse Oximeter SpO_2 Saturation [%]		1k	< 10	63
Patient ID Band		0.05	> 120	0.0002
Inter-Cranial Brain Pressure [mmHg]		16	10 - 30	1
CO_2 Concentration (for respiration monitoring) [ppm]		1k	30 - 120	63
Temperature [°C]		0.3	> 120	0.02

types of data according to Table 4.4, and forwards them to the sink. These three types of data are EKG, CO_2, and CVP blood pressure, representing delays of *low, medium, high*. Queue utilization threshold τ is taken to be 90% here.

4.4.2 Experiment Results

As MQ-MAC exploits channel diversity with quality information to avoid interference, we are mainly interested in comparing it with approaches with only *one channel* and those without link quality information. Specifically, it is compared with the following two protocols:

1. One-channel protocol without LQI information, which only knows if a link to another node is available or not and routes on the path with the *least* number of hops (we denote this protocol by "LEAST" for convenience);

2. One-channel protocol with probing (PROBE). It probes for the best channel first, then stays on this channel for communications, and routes with the maximum RQI, defined as the minimum LQI value among the links along a route (i.e., $\min_{(i,j)\in R_{SD}} LQI_{(i,j)}$).

We are interested in measuring and comparing the end-to-end performance including delay, reliability, and the fairness associated with delay and reliability. In this paper, we adopt Jain's fairness [18] , which is defined as $f(x_1,x_2,\ldots,x_n) = \frac{(\sum_i x_i)^2}{n(\sum_i x_i^2)}$ for a set of values $x_1,x_2,\ldots,x_n$, and compare these protocols. It ranges from $1/n$ (worst case, i.e., most unfair) to 1 (best case, i.e., fair for all these n values). These metrics are defined as follows.

- **E2e delay:** Average difference between the time when one data packet is generated in the source and the time when it is received by the destination.

- **E2e reliability:** Average number of data packets received from one source over the total number of data packets sent by this source.
- **E2e energy efficiency:** The energy consumed per successful received payload data per node from source to destination.
- **Delay fairness:** It is defined as Jain's fairness for the e2e delay among all the sources. It is a measure of how fair the network serves different source-destination pairs in terms of e2e delay.
- **Reliability fairness:** It is defined as Jain's fairness for the e2e reliability among all the sources. It measures how fair the network serves different source-destination pairs in terms of e2e reliability. Note that since the data rate for CC2420 transceiver (used by TelosB) at 2450 MHz is constant (250 Kbps), e2e throughput is proportional to the e2e reliability. Hence throughput fairness is the same as reliability fairness.

The curves for our solution in the following figures are denoted by "X-MQMAC-CQBR-DGS", where X represents the patient's data type (EKG/CO_2/CVP). As for e2e delay, our experiment (Fig. 4.12) shows that: MQMAC-CQBR-DGS < PROBE

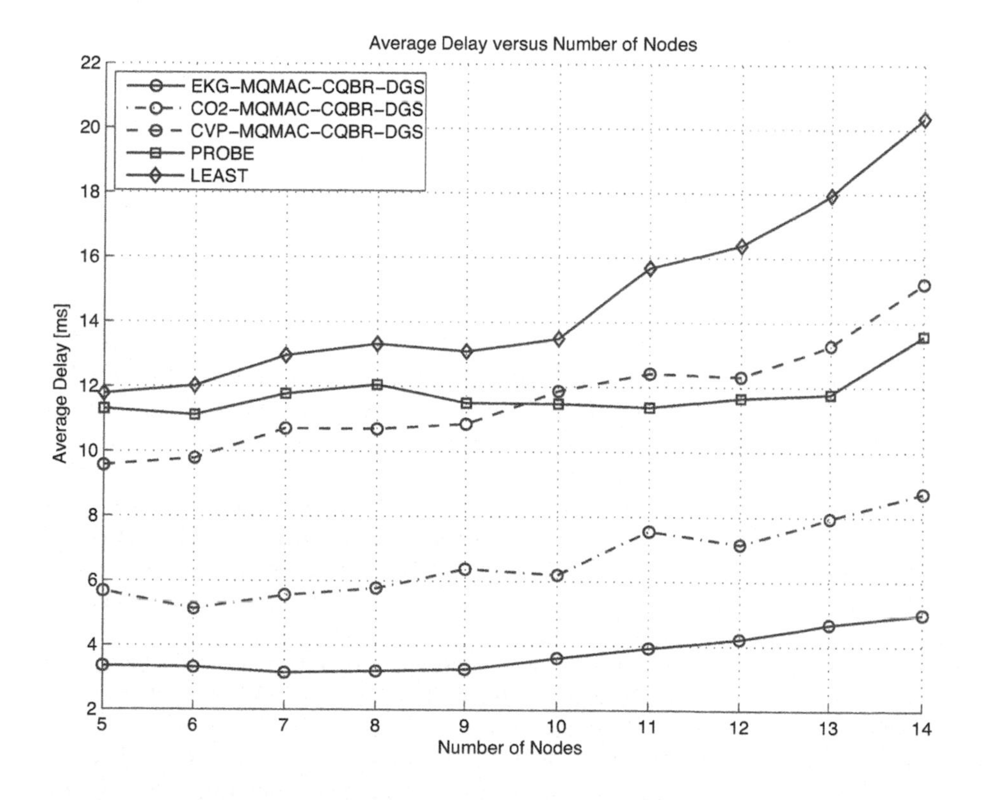

Figure 4.12: e2e delay comparison.

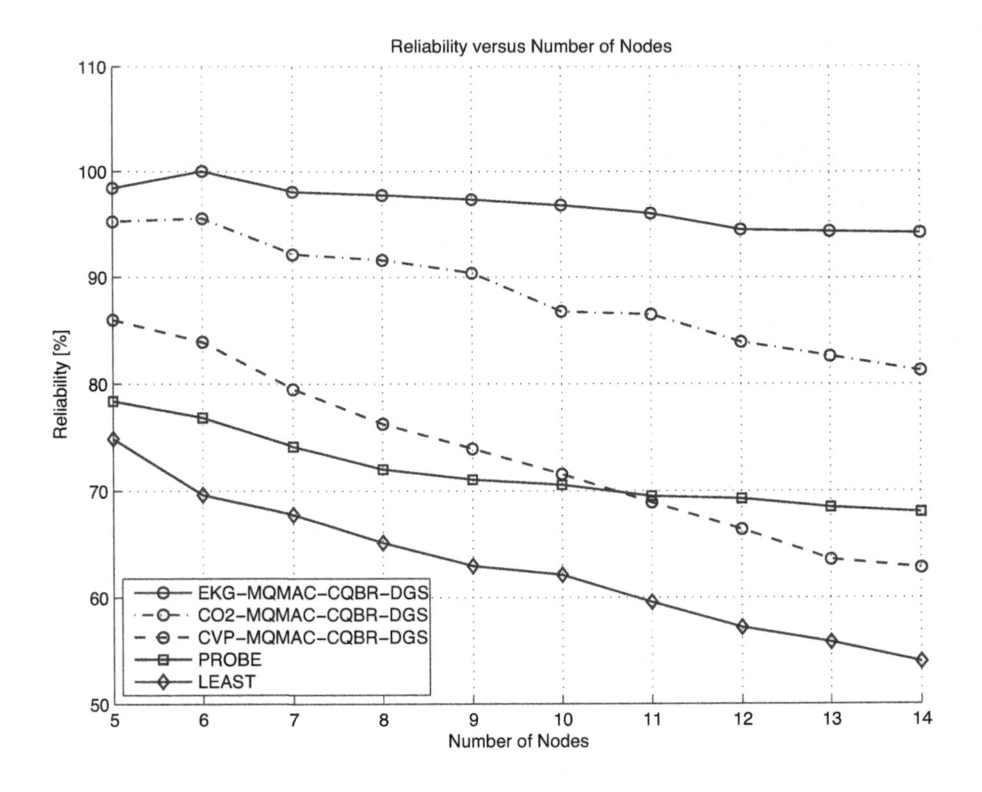

Figure 4.13: e2e reliability comparison.

< LEAST. On average, MQMAC-CQBR-DGS has the least e2e delay, LEAST has the greatest delay, while PROBE lies in the middle. The reason why MQMAC-CQBR-DGS is better than PROBE is that it has a delay guaranteed scheduler, and Packet Error Rate (PER) is reduced by exploiting channel diversity to allow simultaneous transmissions and avoid interference, and by selecting the best route with best link quality, while PROBE only selects the best route with good link quality without the option of choosing better channels. On the other hand, PROBE is better than LEAST as the latter chooses the route with the shortest number of hops, leading to the severest interference from neighbor nodes while PROBE tends to select better links for packet forwarding.

Our proposal provides effective priority services based on the vital sign's e2e delay requirements. As for e2e delay, we have EKG $<$ CO_2 $<$ CVP. Note that when N_B is small, the delay of the CVP data is less than that of PROBE since it can choose better channels while PROBE cannot. When N_B becomes bigger, the lowest priority offered to the CVP data introduces much delay that cannot be offset by using

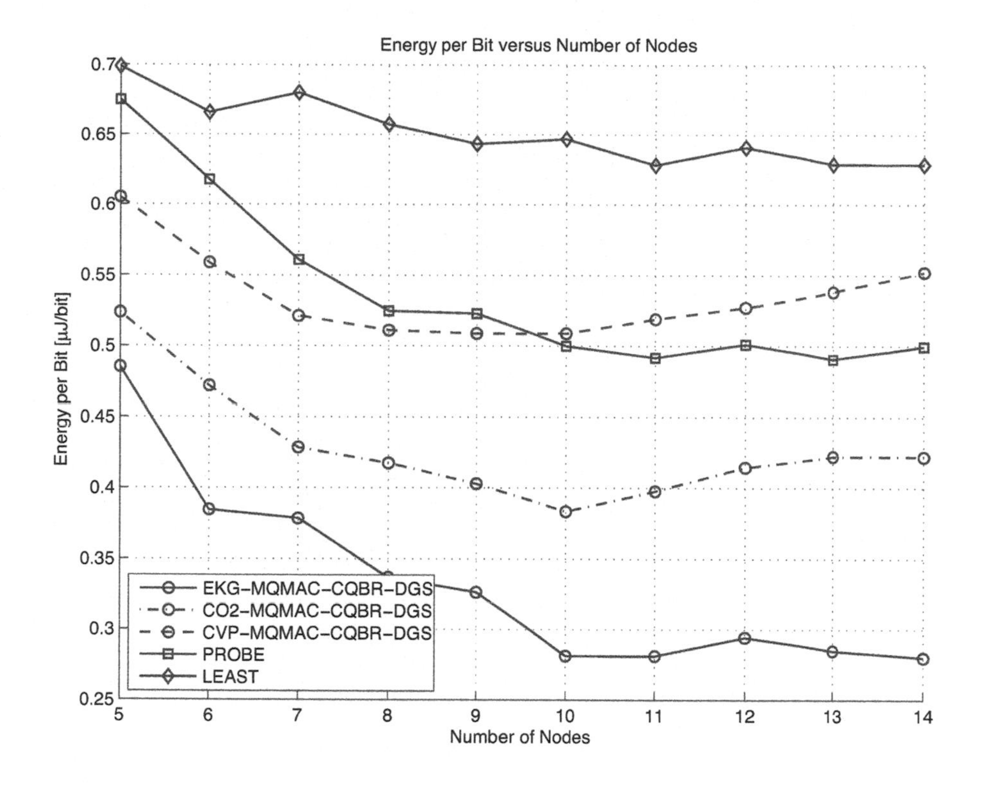

Figure 4.14: Energy consumption comparison.

good LQI channels so the delay is greater than that with the non-priority PROBE. This is also observed in terms of other interested performance measures in the paper.

As far as the e2e reliability is concerned (Fig. 4.13), as N_B increases, all these protocols show decreasing reliability since there is more interference and a higher number of hops to the sink, resulting in increased PER. Due to similar reasons as above, MQMAC-CQBR-DGS has the highest reliability while LEAST has the worst reliability, and data with more stringent delay requirement get better reliability, too.

Energy efficiency is plotted in Fig. 4.14, which shows that MQMAC-CQBR-DGS consumes the least amount of energy, while LEAST consumes the most energy and PROBE lies in between.

As depicted in Fig. 4.15, MQMAC-CQBR-DGS has better e2e delay fairness than the other two competing protocols. It has more distributive routes because when a certain link or channel becomes bad, or when the queues in certain nodes are overloaded, it can select another one with higher RQI or LQI or less queue

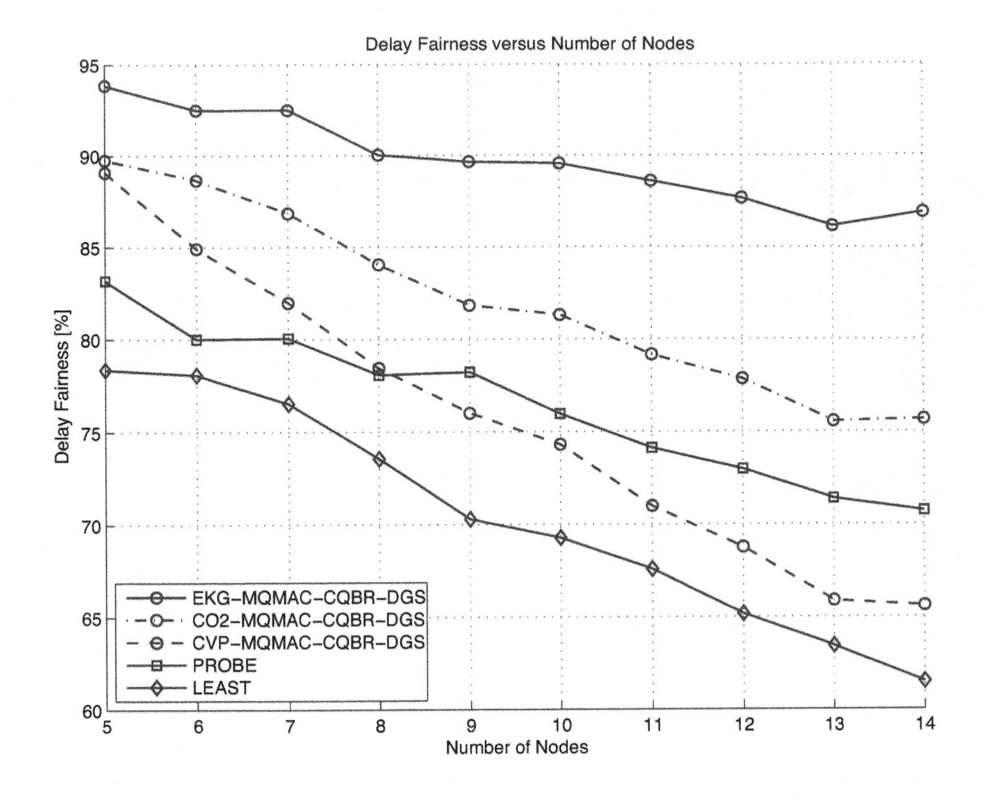

Figure 4.15: Delay fairness comparison.

utilization to forward packets. On the other hand, LEAST does not offer such flexibility so traffic cannot get through so quickly, resulting in decreased fairness for distant nodes. PROBE only has the option to dynamically select routes but not the channel so its delay fairness lies in between the other two protocols. Data with more stringent delay requirements obtain better fairness because they are forwarded in a timely manner.

A similar case happens when reliability fairness is taken into account as shown in Fig. 4.16. With options to dynamically choose routes (based on RQI and queue utilization) and channels (based on channel quality), MQMAC-CQBR-DGS gives the best reliability fairness, while PROBE has the medium performance with only one option for dynamic route selection and LEAST gives the worst fairness due to the inflexibility without both options. A similar reason above gives better fairness to vital signs with more stringent delay requirement.

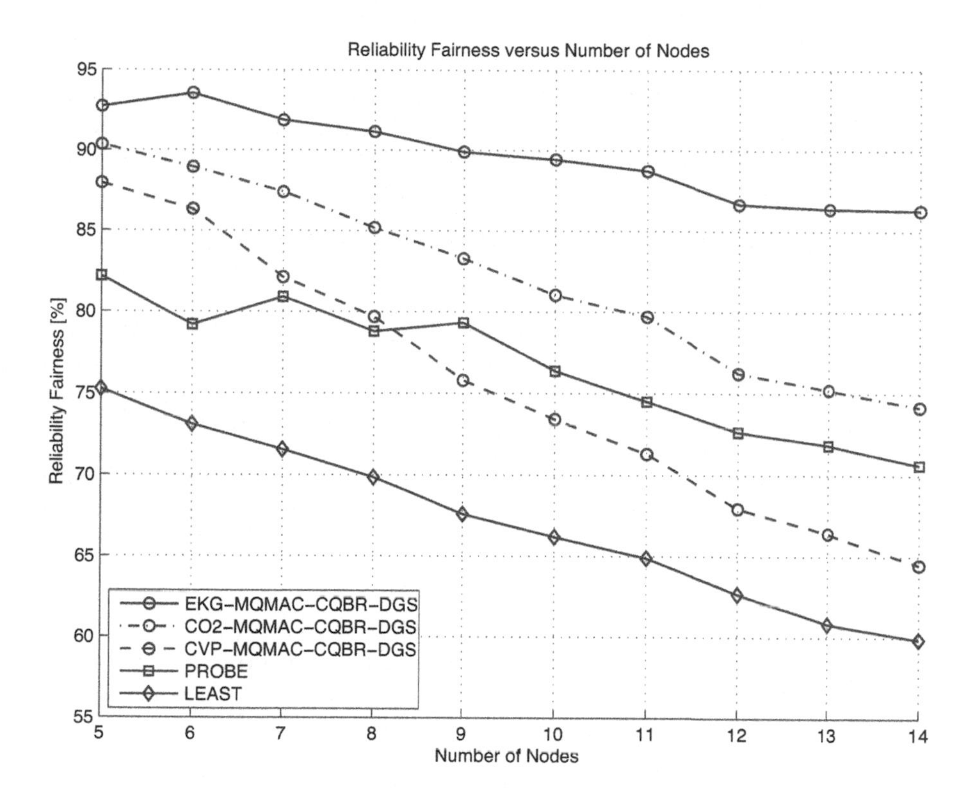

Figure 4.16: Reliability fairness comparison.

4.5 Conclusions and Future Work

In this chapter, we summarized existing MAC protocols for ZigBee networks. We also proposed, implemented, and evaluated a novel multi-channel MAC protocol that is based on link quality measurement. The proposed MQ-MAC has partial cognitive radio capability that is able to avoid interfering active users, monitor channel quality variations and select the best quality channel for packet transmission. It is further extended to a cross-layer approach that jointly considers the interaction of MAC, routing, and scheduling. With this cross-layer design, e2e delay requirement can be guaranteed and traffic of the routes can be balanced. Our experiments on a healthcare scenario show that our solution offers better performance than conventional protocols in terms of end-to-end delay, reliability, and energy consumption in an environment with interference from IEEE 802.11b/g networks. Future work can be the extension of MQ-MAC or the cross-layer solution to other ZigBee network applications with different requirements.

References

[1] G.-S. Ahn, S. G. Hong, E. Miluzzo, A. T. Campbell, and F. Cuomo. Funneling-MAC: a localized, sink-oriented MAC for boosting fidelity in sensor networks. In *Proc. of International Conference on Embedded Networked Sensor Systems (Sensys)*, Boulder, Colorado, USA, Nov. 2006.

[2] I. F. Akyildiz, W. Su, Y. Sankarasubramaniam, and E. Cayirci. Wireless sensor networks: a survey. *Computer Networks*, 38:393–422, 2002.

[3] P. Bahl, R. Chandra, and J. Dunagan. SSCH: Slotted Seeded Channel Hopping for Capacity Improvement in IEEE 802.11 Ad-Hoc Wireless Networks. In *Proc. of ACM Annual International Conference on Mobile Computing and Networking (MobiCom)*, Philadelphia, PA, USA, Sept. 2004.

[4] V. Bharghavan, A. Demers, S. Shenker, and L. Zhang. MACAW: a media access protocol for wireless LAN's. In *Proceedings of ACM Conference on Communications Architectures, Protocols and Applications (SIGCOMM)*, London, United Kingdom, Aug. 1994.

[5] M. Buettner, G. V. Yee, E. Anderson, and R. Han. X-MAC: A Short Preamble MAC Protocol for Duty-Cycled Wireless Sensor Networks. In *Proc. of ACM International Conference on Embedded Networked Sensor Systems (Sensys)*, Boulder, Colorado, USA, Nov. 2006.

[6] B. Chen, D. Pompili, and I. Marsic. Continuous Vital Sign Monitoring via Wireless Sensor Network. In *Malignant Spaghetti: Wireless Technologies in NYU-Poly Hospital Healthcare Workshop*, New York, NY, Nov. 2008.

[7] F. Chen, H. Zhai, and Y. Fang. An Opportunistic MAC in Multichannel Multiradio Wireless Ad Hoc Networks. In *Proc. IEEE Wireless Communications and Networking Conference (WCNC)*, Las Vegas, NV, USA, Mar. 2008.

[8] J. Chen and Y.-D. Chen. AMNP: ad hoc multichannel negotiation protocol for multihop mobile wireless networks. In *Proc. of IEEE International Conference on Communications (ICC)*, New York, NY, USA, June 2004.

[9] F. Chiussi and V. Sivaraman. Achieving high utilization in guaranteed services networks using early-deadline-first scheduling. In *Proc. of IEEE Sixth International Workshop on Quality of Serice (IWQoS)*, Napa, CA, May 1998.

[10] A. Demers, S. Keshav, and S. Shenker. Analysis and simulation of a fair queueing algorithm. In *Proc. of ACM SIGCOMM*, Austin, Texas, USA, pages 1–12, 1989.

[11] A. El-Hoiydi and J.-D. Decotignie. WiseMAC: an ultra low power MAC protocol for the downlink of infrastructure wireless sensor networks. In *Proc. of International Symposium on Computers and Communications (ISCC)*, Alexandria, Egypt, June 2004.

[12] M. Elhawary and Z. J. Haas. Busy Tone Multi Channel (BTMC): A New Multi Channel MAC protocol for Ad Hoc Networks. In *Proc. of IEEE International Conference on Pervasive Computing and Communications (PerCom)*, Hong Kong, China, Mar. 2008.

[13] H. Fattah and C. Leung. An overview of scheduling algorithms in wireless multimedia networks. *IEEE Wireless Communications*, 9(5):76–83, 2002.

[14] L. Georgiadis, R. Guérin, V. Peris, and K. N. Sivarajan. Efficient network qos provisioning based on per node traffic shaping. *IEEE/ACM Transactions on Networking*, 4(4):482–501, Aug. 1996.

[15] S. Haykin. Cognitive radio: brain-empowered wireless communications. *IEEE Journal on Selected Areas in Communications (JSAC)*, 23(2):201–220, 2005.

[16] IEEE Computer Society. *IEEE standard 802.15.4 - 2006: Wireless Medium Access Control (MAC) and Physical Layer (PHY) Specifications for Low-Rate Wireless Personal Area Networks (WPANs)*, Sept. 2006.

[17] IEEE Engineering in Medicine and Biology Society. *ISO/IEEE Standard 11073: Health Informatics - PoC Medical Device Communication - Part 00101: Guide–Guidelines for the Use of RF Wireless Technology*, Sept. 2008.

[18] R. Jain, D. Chiu, and W. Hawe. A Quantitative Measure Of Fairness And Discrimination For Resource Allocation In Shared Computer Systems. Technical Report TR-301, DEC Research, Sept. 1984.

[19] G. E. Jonsrud. *Application Note AN041 (Rev. 1.0) – CC2420 Coexistence*. Texas Instruments, June 2006.

[20] P. Karn. MACA: A New Channel Access Method for Packet Radio. In *Proc. of ARRL Computer Networking Conference*, London, UK, Apr. 1990.

[21] Y. Kim, H. Shin, and H. Cha. Y-MAC: An Energy-Efficient Multi-channel MAC Protocol for Dense Wireless Sensor Networks. In *Proc. of International Conference on Information Processing in Sensor Networks (IPSN)*, St. Louis, MO, USA, Apr. 2008.

[22] G. S. A. Kumar, G. Manimarana, and Z. Wang. Energy-aware scheduling with probabilistic deadline constraints in wireless networks. *Ad Hoc Networks (Elsevier)*, 7(7):1400–1413, 2009.

[23] J. Kurose and K. Ross. *Computer Networking: A Top-Down Approach*. Addison Wesley, 4 edition, 2007.

[24] P. Kyasanur and N. H. Vaidya. Routing and link-layer protocols for multi-channel multi-interface ad hoc wireless networks. *SIGMOBILE Mob. Comput. Commun. Rev.*, 10(1):31–43, 2006.

[25] H. K. Le, D. Henriksson, and T. Abdelzaher. A Practical Multi-Channel Media Access Control Protocol for Wireless Sensor Networks. In *Proc. of International Conference on Information Processing in Sensor Networks (IPSN)*, St. Louis, MO, USA, Apr. 2008.

[26] B. Liang and M. Dong. Packet Prioritization in Multihop Latency Aware Scheduling for Delay Constrained Communication. *IEEE Journal on Selected Areas in Communications (JSAC)*, 25(4):819–830, 2007.

[27] Q. Liu, S. Zhou, and G. B. Giannakis. Cross-layer scheduling with prescribed qos guarantees in adaptive wireless networks. *IEEE Journal on Selected Areas in Communications (JSAC)*, 23(5):1056–1066, 2005.

[28] S. Liu, K.-W. Fan, and P. Sinha. CMAC: An Energy Efficient MAC Layer Protocol Using Convergent Packet Forwarding for Wireless Sensor Networks. In *Proc. of IEEE Communications Society Conference on Sensor, Mesh and Ad Hoc Communications and Networks (SECON)*, San Diego, CA, June 2007.

[29] G. Lu, B. Krishnamachari, and C. Raghavendra. An adaptive energy-efficient and low-latency MAC for data gathering in wireless sensor networks. In *Proc. of IEEE International Parallel and Distributed Processing Symposium (IPDPS)*, Santa Fe, New Mexico, USA, Apr. 2004.

[30] T. Luo, M. Motani, and V. Srinivasan. Cooperative Asynchronous Multichannel MAC: Design, Analysis, and Implementation. *IEEE Transactions on Mobile Computing*, 8(3):338–352, 2009.

[31] R. Maheshwari, H. Gupta, and S. R. Das. Multichannel MAC Protocols for Wireless Networks. In *Proc. 3rd Annual IEEE Communications Society on Sensor and Ad Hoc Communications and Networks (SECON)*, Reston, VA, USA, Sept. 2006.

[32] J. Mo, H.-S. W. So, and J. Walrand. Comparison of Multichannel MAC Protocols. *IEEE Transactions on Mobile Computing*, 7(1):50–65, Jan. 2008.

[33] The Network Simulator - ns-2. http://www.isi.edu/nsnam/ns/.

[34] A. K. Parekh and R. G. Gallager. A generalized processor sharing approach to flow control in integrated services networks: the single-node case. *IEEE/ACM Transactions on Networking*, 1(3):344–357, June 1993.

[35] A. K. Parekh and R. G. Gallagher. A generalized processor sharing approach to flow control in integrated services networks: the multiple node case. *IEEE/ACM Transactions on Networking*, 2(2):137–150, Apr. 1994.

[36] J. Polastre, J. Hill, and D. Culler. Versatile Low Power Media Access for Wireless Sensor Networks. In *Proc. of ACM Conference on Embedded Networked Sensor Systems (SenSys)*, Baltimore, MD, Nov. 2004.

[37] D. Pompili, M. C. Vuran, and T. Melodia. Cross-layer Design in Wireless Sensor Networks. In N. P. Mahalik, editor, *Book on Sensor Network and Configuration: Fundamentals, Techniques, Platforms, and Experiments*. Springer-Verlag, Germany, 2006.

[38] T. S. Rappaport. *Wireless Communications: Principles and Practice*. Prentice Hall, Upper Saddle River, NJ, 1999.

[39] I. Rhee, A. Warrier, M. Aia, and J. Min. ZMAC: a Hybrid MAC for Wireless Sensor Networks. In *Proc. of ACM International Conference on Embedded Networked Sensor Systems (Sensys)*, San Diego, CA, USA, Nov. 2005.

[40] M. Seo, Y. Kim, and J. Ma. Multi-channel MAC protocol for multi-hop wireless networks: Handling multi-channel hidden node problem using snooping. In *Proc. of IEEE Military Communications Conference (MILCOM)*, San Diego, CA, USA, Nov. 2008.

[41] M. Sha, G. Xing, G. Zhou, S. Liu, and X. Wang. C-MAC: Model-driven Concurrent Medium Access Control for Wireless Sensor Networks. In *Proc. of IEEE Conference on Computer Communications (INFOCOM)*, Rio de Janeiro, Brazil, Apr. 2009.

[42] T. M. Siep, I. C. Gifford, R. C. Braley, and R. F. Heile. Paving the way for personal area network standards: an overview of the IEEE P802.15 Working Group for Wireless Personal Area Networks. *IEEE Personal Communications*, 7(1):37–43, 2000.

[43] J. Silberberg. Performance degradation of electronic medical devices due to electromagnetic interference. *Compliance Eng*, 10(5):25–39, 1993.

[44] H. W. So, J. Walrand, and J. Mo. McMAC: A Parallel Rendezvous Multi-Channel MAC Protocol. In *Proc. of IEEE Wireless Communications and Networking Conference (WCNC)*, Hong Kong, China, Mar. 2007.

[45] J. So and N. H. Vaidya. Multi-channel MAC for Ad Hoc Networks: handling multi-channel hidden terminals using a single transceiver. In *Proc. of ACM international symposium on Mobile ad hoc networking and computing (MobiHoc)*, Tokyo, Japan, May 2004.

[46] F. Tobagi and L. Kleinrock. Packet Switching in Radio Channels: Part II–The Hidden Terminal Problem in Carrier Sense Multiple-Access and the Busy-Tone Solution. *IEEE Transactions on Communications*, 23(12):1417–1433, 1975.

[47] A. Tzamaloukas and J. J. Garcia-Luna-Aceves. A receiver-initiated collision-avoidance protocol for multi-channel networks. In *Proc. of IEEE Twentieth Annual Joint Conference of the IEEE Computer and Communications Societies (INFOCOM)*, Anchorage, Alaska, USA, Apr. 2001.

[48] T. van Dam and K. Langendoen. An adaptive energy-efficient MAC protocol for wireless sensor networks. In *Proc. of ACM International Conference on Embedded Networked Sensor Systems (Sensys)*, Los Angeles, CA, USA, Oct. 2003.

[49] R. Vedantham, S. Kakumanu, S. Lakshmanan, and R. Sivakumar. Component based channel assignment in single radio, multi-channel ad hoc networks. In *Proc. of the international conference on Mobile computing and networking (Mobicom)*, Los Angeles, CA, USA, Sept. 2006.

[50] A. J. Viterbi. *CDMA: Principles of Spread Spectrum Communicagtion*. Addison Wesley, 1995.

[51] T. Voigt, F. Osterlind, and A. Dunkels. Improving sensor network robustness with multi-channel convergecast. In *Proceedings of 2nd ERCIM Workshop on e-Mobility*, Tampere, Finland, May 2008.

[52] H. Wang, H. Zhou, and H. Qin. Overview of Multi-Channel MAC Protocols in Wireless Networks. In *Proc. of International Conference on Wireless Communications, Networking and Mobile Computing (WiCOM)*, Dalian, China, Oct. 2008.

[53] J. Wang, Y. Fang, and D. Wu. A Power-Saving Multi-Radio Multi-Channel MAC Protocol for Wireless Local Area Networks. In *Proc. of IEEE International Conference on Computer Communications (INFOCOM)*, Barcelona, Spain, Apr. 2006.

[54] J.-P. Wang, M. Abolhasan, F. Safaei, and D. Franklin. A Survey on Control Separation Techniques in Multi-Radio Multi-channel MAC Protocols. In *Proc. of International Symposium on Communications and Information Technologies (ISCIT)*, Sydney, Australia, Oct. 2007.

[55] C. Won, J.-H. Youn, H. Ali, H. Sharif, and J. Deogun. Adaptive radio channel allocation for supporting coexistence of 802.15.4 and 802.11b. In *Proc. of IEEE Vehicular Technology Conference (VTC)*, Dallas, TX, USA, Sept. 2005.

[56] S.-L. Wu, C.-Y. Lin, Y.-C. Tseng, and J.-P. Sheu. A New Multi-Channel MAC Protocol with On-Demand Channel Assignment for Multi-Hop Mobile Ad Hoc Networks. In *International Symposium on Parallel Architectures, Algorithms, and Networks (ISPAN)*, Dallas, TX, USA, Dec. 2000.

[57] W. Ye and J. Heidemann. Medium Access Control in Wireless Sensor Networks. In C. S. Raghavendra, K. Sivalingam, and T. Znati, editors, *Wireless Sensor Networks*, pages 73–92. Kluwer Academic Publishers, 2004. Chapter 4.

[58] W. Ye, J. Heidemann, and D. Estrin. An energy-efficient mac protocol for wireless sensor networks. In *Proc. of IEEE Conference of Computer and Communications Societies (INFOCOM)*, New York, NY, June 2002.

[59] J. Zhang, G. Zhou, C. Huang, S. H. Son, and J. A. Stankovic. TMMAC: An Energy Efficient Multi-Channel MAC Protocol for Ad Hoc Networks. In *Proc. IEEE International Conference on Communications (ICC)*, Glasgow, Scotland, June 2007.

[60] G. Zhou, C. Huang, T. Yan, T. He, J. A. Stankovic, and T. F. Abdelzaher. MMSN: Multi-Frequency Media Access Control for Wireless Sensor Networks. In *Proc. of IEEE International Conference on Computer Communications (INFOCOM)*, Barcelona, Spain, Apr. 2006.

Chapter 5

Energy Optimization Techniques for the PHY and MAC Layers of IEEE 802.15.4/ZigBee

Al-Khateeb Anwar, Luciano Lavagno

CONTENTS

802.15.4 was developed to meet the needs for simple, low-power and low-cost wireless communication. In the last few years it has been rapidly adopted in industrial, control/monitoring, and medical applications and it has become a popular technology for wireless sensor networks.

Wireless sensor networks are often power and/or energy limited. Thus it is very important to statically and (whenever possible) dynamically tune the communication protocol parameters for given performance and quality of service requirements. This chapter introduces mathematical and simulation models of the PHY and MAC layers for the ZigBee protocol (which are covered by the IEEE 802.15.4 standard) that can be used in a model based design environment to provide opportunities for co-design of the application and the network.

The main goal is energy consumption optimization, within performance constraints, by tuning the parameters of the communication stack depending on various application scenarios. We propose a mathematical formulation and algorithms to optimize parameters such as bit error rate, signal to Noise Ratio (SNR), number of repeater nodes, distance between repeaters at different environmental and interference noise levels. We also show at the MAC level how, by re-using a combination of results from the literature, one can dynamically optimize energy and latency by adaptively changing several parameters (e.g., transmitted power, adaptive backoff delay, minimum backoff exponential value and number of Guaranteed Time Slots (GTS)). We also experimentally show the advantages of cross layer optimization for the ZigBee stack, by adaptively changing transmitted power based on the Link Quality Index (LQI).

5.1 Introduction

ZigBee is an industry standard that is especially suited for low rate sensors and devices used for control applications that do not require a high data rate but must have long battery life. Some of these applications are in the fields of medicine, home/office automation, military, and many others [1,25]. This standard is aimed at providing interoperability between WSN devices from different manufacturers. Energy efficient protocols are one of the most important requirements in designing a WSN that is dependent on battery for supplying power. In particular, the Medium Access Control (MAC) protocol can be designed to reduce the energy consumption in WSNs. Idle listening is often the largest source of energy waste and duty cycling (i.e., periodically putting the radio in a sleep state) is considered one of the best techniques to reduce energy consumption in WSN MAC protocols.

However, excessive sleeping causes an increase of transmission power, due to lost packets and long preambles. Asynchronous Scheduled MAC (AS-MAC) [12,20,21] is a simple but very energy efficient protocol, in which the energy consumption monotonically decreases as the wake-up interval increases. The nodes store the wake-up schedules of their neighbors; therefore they know when their neighbors are ready to receive and do not need to add long preambles at the beginning of each transmission. AS-MAC also asynchronously coordinates the wake-up times of neighboring nodes to reduce overhearing, contention and delays that are typical of synchronous scheduled MAC protocols.

In this chapter we describe a few improvements of AS-MAC, that we collectively call Power Efficient Asynchronous Scheduled MAC protocol (PEAS-MAC). They improve power and performance, by adaptively controlling transmission power, backoff delay, and the number of GTS slots. We also show how to support different throughput performance requirements for individual nodes, which is very difficult to achieve with the current standard, by dynamically adjusting the minimum backoff exponent (minBE) of some nodes to their transmission requirements and channel conditions, in order to shorten the backoff delay of nodes with frequent transmissions.

Finally, for WSNs where the set of destination nodes is separated from the set of sources, tree-based topologies seem to be the most efficient ones. Routing is much simpler, and distributed data aggregation mechanisms are more efficient. If nodes belonging to any level can be assumed to be uniformly distributed over the plane, then the topology is optimized simply by selecting the average number of children per parent and the number of levels in the tree. Thus we discuss how to incorporate past work on the optimum design of tree-based topologies, by selecting judiciously tree height, which is related to energy consumption, and the average number of children per parent.

In this chapter we present an overview of ZigBee / IEEE 802.15.4 Protocol, focusing on PHY and MAC characteristics. Then we discuss energy efficient techniques to implement both PHY and MAC layers. For the PHY layer, we introduce the Layer Energy Consumption Minimization Model, where energy consumption can be optimized at a given SNR range for single-hop and multi-hop transmission, both with a known and an unknown number of hops, both at the same and at different distances between nodes. Then we describe the PEAS-MAC optimizations to the AS-MAC layer, including both the initialization and the periodic phases.

5.2 Overview of ZigBee / IEEE 802.15.4 Protocols

The ZigBee/IEEE 802.15.4 protocol is the standard that covers Wireless Personal Area Networks (WPAN), and that was approved and published in 2006 [1,25]. It defines the characteristics of the physical and MAC layers.

The main functions of the physical layer are spreading, de-spreading, modulation and demodulation of the signal, activation and deactivation of the radio transceiver, energy detection (ED), link quality indication (LQI), channel selection, clear channel assessment (CCA), and transmission as well as reception of packets across the physical medium.

At the physical layer, ZigBee operates in the ISM band within three different frequency ranges. There is a single channel between 868.0 and 868.6MHz, Ch 0, 10 channels between 902 and 928 MHz, Ch 1-10, and 16 channels between 2.4 and 2.4835 GHz, Ch 11-26. ZigBee uses direct sequence spread spectrum (DSSS) as a spreading technique. DSSS is used to increase the frequency of the signal in order to increase its power and reduce the influence of noise from nearby networks. The 2.4 GHz band uses the orthogonal quadrature phase shift keying (OQPSK) technique for chip modulation and DSSS. Each 4-bit symbol is mapped into a 32 chip PN sequence. In the 915 MHz and 868 MHz bands each one-bit symbol is mapped into a 15 chip PN sequence, and uses the binary phase shift keying (BPSK) technique for modulation. The ZigBee standard specifies a receiver sensitivity of -92 dBm in the 868/915 MHz bands and -85 dBm in the 2.4 GHz band. Having several channels in different frequency bands makes it possible to relocate them within the available spectrum [1,2,11].

The MAC sublayer provides two services: the MAC data service and the MAC management service interfacing to the MAC sublayer management entity (MLME) service access point (SAP) (known as MLME-SAP). The MAC data service enables the transmission and reception of the MAC protocol data units (MPDUs) across the PHY data service. The MLME-SAP allows the transport of management commands between the next higher layer and the MLME [1,6].

The features of the MAC sublayer are beacon management, channel access, guaranteed time slots (GTS) management, frame validation, acknowledged frame delivery, association, disassociation, security mechanisms and collision avoidance. The MAC protocol supports two operational modes selected by the coordinator; the non beacon-enabled mode, in which access is controlled by non-slotted CSMA/CA and the beacon-enabled mode, in which beacons are periodically sent by the coordinator to synchronize associated nodes.

Figure 1.1 shows the basic procedure of the CSMA/CA Algorithm. The backoff algorithm is processed inside the CSMA/CA algorithm. The initial value of BE (Backoff Exponent) is given as macMinBE and the system selects a backoff time as a random number between $[0:2^{BE}-1]$. The fixed backoff range given in the IEEE 802.15.4 standard [1,25] limits the default value of the macMinBE (minimum BE) as 3, and macMaxBE (maximum BE) as 5, giving only the range of $[0:2^3-1]$ to $[0:2^5-1]$ for randomly selecting the actual backoff delay value [1,15].

Once the backoff time is set, the system starts the backoff random delay [step (2) in Figure 1.1]. After the backoff time is over, the system performs CCA (Clear Channel Assessment) to check if the channel is busy or not [step (3) in Figure 1.1]. If the channel is sensed busy, the value of BE is increased by one and a new random number is selected from the new range for another backoff period [step (4) in Figure 1.1]. If NB, the number of times the CSMA-CA algorithm was required to backoff while attempting the current transmission, is greater than macMaxCSMABackoffs, the CSMA-CA algorithm terminates with a channel access failure status [1,17]. For the IEEE 802.15.4 standard $macMaxCSMABackoffs = 4$. If the channel is sensed idle for two consecutive CCAs (i.e., when $CW = 0$) the channel is considered idle and the system can start transmission [step (5) in Figure 1.1].

In beacon-enable mode the format of the super-frame is defined by the coordinator and is shown in Figure 1.2. It is bounded by the beacons and is divided into 16 equally sized slots. Optionally, the super-frame can have active and inactive parts. The beacon is transmitted in the first slot of each super-frame and is used to synchronize the attached devices and to describe the super-frame structure. During the contention access period (CAP), a device competes with other devices using a slotted CSMA/CA mechanism. For applications having specific latency and bandwidth requirements, the coordinator may dedicate portions of the active super-frame to each device. These portions are called guaranteed time slots (GTSs) and form the contention-free period (CFP), which always appears at the end of the active part following the CAP. The coordinator may allocate up to seven of these GTSs, and a GTS may occupy more than one slot period, up to seven [1,14].

The structure of the super-frame is described by two parameters: beacon order (BO) and super-frame order (SO). BO describes the interval at which the coordinator shall transmit its beacon frames, i.e., Beacon Interval (BI). SO describes the length of the active portion, i.e., Super-frame Duration (SD). The relationship between BO-BI and SO-SD is given by Equations (1.1) and (1.2) respectively. The parameter aBaseSuperframeDuration depends on thc frcqucncy range [1,7].

In the rest of this chapter we will discuss modifications of the CSMA/CA algorithm based on the so-called AS-MAC [12] aimed at increasing the efficiency of the MAC in reducing power consumption and improving performance.

$$BI = aBaseSuperFrameDuration * 2^{BO} \quad 0 \leq BO \leq 14 \tag{5.1}$$

$$SD = aBaseSuperFrameDuration * 2^{SO} \quad 0 \leq SO \leq BO \leq 14 \tag{5.2}$$

5.3 Physical Layer Energy Consumption Model

5.3.1 Physical Layer Energy Consumption Minimization Model

Using the scheme of Figure 1.3 the energy consumption of the Physical layer for transmitting and receiving data, including retransmission energy, can be written similar to [5,8,10] as:

$$E'_{Data} = \left[SNR_R\gamma(\mu d^2 - 1) + \beta\right]\frac{(L+a)}{R} \tag{5.3}$$

$\gamma = P_N \cdot F_N$, $\mu = \sqrt{\frac{4.\pi}{\lambda.G_T.G_R}}.(1+\eta)$, $\beta = P_{O,TX}.P_{O,RX}$

Where

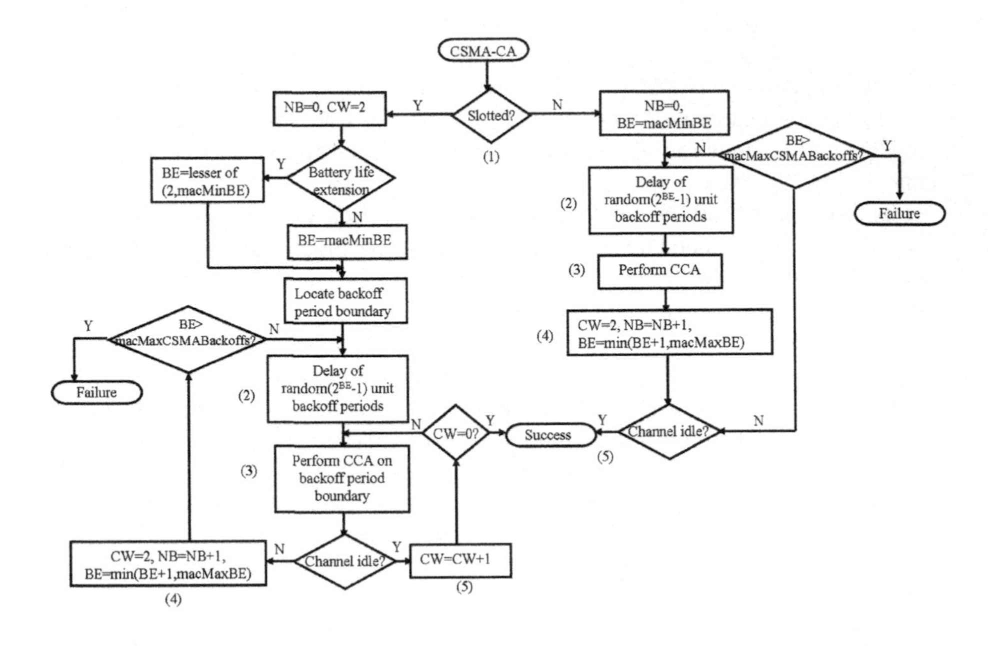

Figure 5.1: The CSMA-CA procedure in the IEEE 802.15.4 standard.

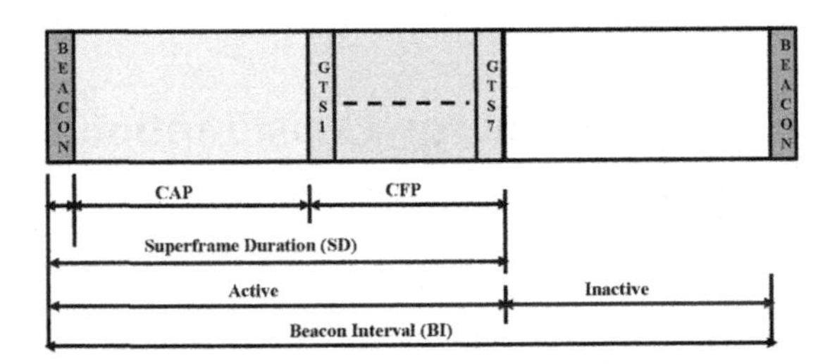

Figure 5.2: The super-frame structure in the IEEE 802.15.4 standard.

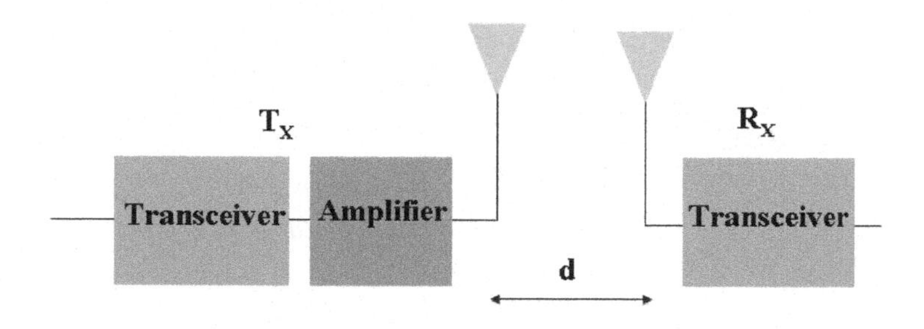

Figure 5.3: Wireless sensor communication model.

$P_{O,TX}$ =transmit overhead power, is the power consumed by all the transmit chain circuitry except for the amplifier
$P_{O,RX}$ = receive overhead power, is the power consumed by all the receive chain circuitry
η = transmit amplifier efficiency
L = size of payload
a = size of header
R = data rate
$G_{T,R}$=transmit/receive antenna gain
d=transmitting distancc
P_N=noise power
SNR_R=signal-noise power ratio at the receiver
F_N=noise figure
In order to consider retransmissions, the average energy must be multiplied by a factor that accounts for the retransmission overhead. This factor can be determined by writing the average consumption energy with retransmissions as [5,9]:

$$E_{Data} = E'_{Data} + E'_{Data}P_{err} + E'_{Data}(P_{err})^2 + \dots \tag{5.4}$$

Where P_{err} is the probability of data transmission failure for a packet. Since Equation (1.4) is a geometric series, it can be simplified to:

$$E_b = \frac{1}{(1-P_b)^{L+a}} \left[SNR_R \gamma(\mu d^2 - 1) + \beta\right] \frac{(L+a)}{RL} \tag{5.5}$$

Where
E_b =average energy consumption per bit
P_b= bit error rate
P_b is related to SNR_R and for the IEEE 802.15.4 physical layer with OQPSK modulation and operating frequency 2.4GHz, can be calculated as follows [3,10,11]:

$$P_b = \frac{M}{2M(M-1)} \sum_{k=2}^{M} (-1)^k C_M^k \exp\left(5SNR_R\left(\frac{1}{k} - 1\right)\right) \tag{5.6}$$

Now we proceed to minimize the energy per bit E_b by taking the log of Equation (1.5) and setting its derivative with respect to SNR_R equal to zero.

$$\begin{aligned}\frac{\partial \log E_b}{\partial SNR_R} &= -(L+a)\frac{1}{(1-P_b)}\frac{\partial P_b}{\partial SNR_R} + \\ &\quad \gamma(\mu d^2 - 1)/\left(SNR_R\gamma(\mu d^2 - 1) + \beta\right)\end{aligned} \tag{5.7}$$

where

$$\frac{\partial P_b}{\partial SNR_R} = \frac{M}{2M(M-1)}\sum_{k=2}^{M}(5(\frac{1}{k}-1)))(-1)^k C_M^k.exp(5SNR_R(\frac{1}{k}-1)) \tag{5.8}$$

Now let,

$$\frac{\partial \log E_b}{\partial SNR_R} = 0 \tag{5.9}$$

Then we get the following relation:

$$\frac{\gamma(\mu d^2 - 1)}{[SNR_R\gamma(\mu d^2 - 1) + \beta]} = (L+a)\frac{1}{(1-P_b)}\frac{\partial P_b}{\partial SNR_R} \tag{5.10}$$

$$SNR_R\gamma(\mu d^2 - 1) + \beta = \frac{\beta}{1 + (L+a)\frac{1}{(1-P_b)}\frac{\partial P_b}{\partial SNR_R}} \tag{5.11}$$

Substituting Equation (1.11) in Equation (1.5) yields

$$E_b = \frac{1}{(1-P_b)^{L+a}}\frac{\beta}{1 + (L+a)\frac{1}{(1-P_b)}\frac{\partial P_b}{\partial SNR_R}}\frac{(L+a)}{RL} \tag{5.12}$$

From Equation (1.12) we find E_b as a function of L,P_b and SNR_R where β ,R and a are constants. β can be computed from the WSN node data sheet, while R and a are taken from IEEE 802.15.4 for 2.4GHz and OQPSK modulation as 250KBps and 16 Bits respectively. The minimum value of E_b can now be computed for a given SNR_R range, which again can be obtained from the expected operating conditions of the WSN node.

5.3.2 Energy Consumption for Multi-hop Transmission with Unknown Number of Hops

Assume that a wireless transmission needs to cross a relatively long distance. Then it is often useful to use repeater nodes to reduce the overall energy, within latency constraints. If we consider a multi-hop link with $N+1$ nodes and we assume a uniform noise level along the link (the next section will relax this assumption), then we can also assume that all nodes use the same transmission (and reception) power, since this will lead to a uniform battery life for all nodes. The total energy consumption is:

$E_{b,tot} = (N+1)E_{b,hop}$. If the hop distance is d then the total distance is $D = Nd$. In order to minimize $E_{b,tot}$ we need to minimize $E_{b,hop}$ which is:

$$E_{b,hop} = \frac{(L+a)}{RL(1-P_{b,hop})^{L+a}}\left[SNR_{R,hop}\gamma(\mu\frac{D^2}{N^2}-1)+\beta\right] \tag{5.13}$$

We calculate the derivative of $\log E_{b,hop}$ with respect to N

$$\frac{\partial \log E_{b,hop}}{\partial N} = -(L+a)\frac{1}{(1-P_{b,hop})}\frac{\partial P_{b,hop}}{\partial N} + \tag{5.14}$$

$$\frac{\frac{\partial \log SNR_{R,hop}}{\partial N}h(N) - \frac{2SNR_{R,hop}\gamma\mu D^2}{N^3}}{SNR_{R,hop}h(N)+\beta} \tag{5.15}$$

where $h(N) = \gamma(\mu D^2 N^{-2} - 1)$. Also we have:

$$\frac{\partial SNR_{b,hop}}{\partial N} = \frac{2SNR_{b,hop}}{N} \tag{5.16}$$

$$\frac{\partial \log E_{b,hop}}{\partial N} = -(L+a)\frac{1}{(1-P_{b,hop})}\frac{\partial P_{b,hop}}{\partial N} + \frac{\frac{-2\gamma SNR_{R,hop}}{N}}{SNR_{R,hop}\gamma(\mu\frac{D^2}{N^2}-1)+\beta} \tag{5.17}$$

Again, if we let $\partial \log E_{b,hop} = 0$ then $E_{b,hop}$ will be minimum and we get:

$$SNR_{R,hop}\gamma(\mu\frac{D^2}{N^2}-1)+\beta = \frac{\beta}{1-2(L+a)\frac{1}{(1-P_{b,hop})}\frac{\partial P_{b,hop}}{\partial N}\frac{h(N)}{\gamma N}} \tag{5.18}$$

Substituting Equation (1.17) in Equation (1.13) yields:

$$E_{b,hop} = \frac{L+a}{RL(1-P_{b,hop})^{L+a}}\left[\frac{\beta}{1-2(L+a)\frac{1}{(1-P_{b,hop})}\frac{\partial P_{b,hop}}{\partial N}\frac{h(N)}{\gamma N}}\right] \tag{5.19}$$

From Equation (1.18) $E_{b,hop}$ is a function of N at a given L, R and β.

5.3.3 Energy Consumption for Multi-hop Transmission with Known Number of Hops and Different Distances

We now relax the assumption on uniform noise levels along the link. In order to make the problem tractable, we will use the same number of nodes as estimated in the previous section, using as uniform noise level the average noise along the link itself. Thus we assume to have $N+1$ nodes all with equal transmission power (again to keep the battery life uniform). At different noise levels, we can obtain a good connection with high capacity by keeping the individual SNRs equal [5,6,10]:

$$SNR_{R,i} = SNR_{R,i-1} \tag{5.20}$$

For $i = 1,...,N$. If the total distance is D and the distance between nodes i and $i-1$ is d_i then $\sum_{i=1}^{N} d_i = D$. The distance d_i will be:

$$d_i = d_{i-1}\sqrt{\frac{P_{noise,i-1}}{P_{noise,i}}} \tag{5.21}$$

When $P_{noise,i} = P_{noise,i-1}$ the distance between nodes will obviously be equal (as in the previous section). The total distance D will be:

$$D = d_1 + d_1\sqrt{\frac{P_{noise,1}}{P_{noise,2}}} + d_1\sqrt{\frac{P_{noise,1}}{P_{noise,3}}} + ... + d_1\sqrt{\frac{P_{noise,1}}{P_{noise,N}}} \tag{5.22}$$

and we can obtain the individual inter-node distances as:

$$d_k = \frac{D}{\sum_{i=1}^{N}\sqrt{\frac{P_{noise,k}}{P_{noise,i}}}} \tag{5.23}$$

For $k = 1,...,N$. Note that since in general $P_{noise,i}$ depends on the location of node i, we need to iterate the computation in order to converge to the optimum distance starting from a uniform distribution of the nodes as in the previous section.

5.3.4 Energy Consumption for Multi-hop Transmission with Unknown Number of Hops and Different Distances

Let us consider the same scenario as in the previous section (i.e., same transmission power and different noise levels at the various nodes), but let us fix the "optimum" transmit power to be the same as in the uniform noise case, and let us consider a scenario in which we may need to add nodes if the local noise levels are very different from the average. If we denote again the total distance as D, we again can write:

$$d_i = d_1\sqrt{\frac{P_{noise,1}}{P_{noise,i}}} \tag{5.24}$$

Let us assume that: $d_i >= 1$ for $i = 1,2,....N$. We also use as initial value the distance computed in the case of uniform noise:

$$d_i = d_{average}\sqrt{\frac{P_{noise,average}}{P_{noise,i}}} \tag{5.25}$$

then use it to compute d_i for $i >= 2$ until the summation of d_i is greater than or equal to D. This is sub-optimal with respect to the last link, since only one node "moves" at each iteration, depending on its local noise level. This is the best approach with many hops and/or local noise levels that are very different from the average.

5.4 Power Efficient MAC Protocol

The basic idea of the Asynchronous Scheduled WSN MAC protocol (AS-MAC) [12,21] is that nodes wake up periodically, but asynchronously from their neighbors, to receive packets. Nodes intending to transmit wake up at the scheduled wakeup time of the intended target node. This simple but effective scheme can be improved in several ways, by exploiting results from a number of previous papers, resulting in what we call here the Power-Efficient Asynchronous Scheduling MAC (PEAS-MAC) protocol. Let us consider the two phases (initialization and periodic) of AS-MAC more in detail.

5.4.1 Initialization Phase

When a new node joins a WSN, it performs the initialization phase, in which it builds the neighbor table that stores scheduling information from its neighbors, and then chooses and announces its own unique offset for periodic wake-up [12]. Existing nodes may be in the initialization phase or the periodic listening and sleep phase. Nodes in the periodic listening and sleep phase perform Low Power Listening (LPL) at every wake-up interval, I_{wakeup}, and send a Hello packet at every Hello interval, I_{hello} (which is an integer multiple of I_{wakeup} and is common to the whole network). Hello packets are used to publish scheduling information: I_{wakeup}, I_{hello} and offset of the periodic wake-up, O_W. Figure 1.4 shows the flowchart of the initial phase.

First, the tree level and number of children for each coordinator node are calculated using the optimum tree design algorithm which is described in the following section, since this is essential to creating the look-up table. The protocol designer has two options: on one hand, to avoid the collision-prone contention phase, only GTSs might be used: as a result, at each level in the tree a maximum number of seven children nodes can be attached to a given parent node, thus leading to a topology based on hard capacity limitations. On the other hand, to maximize radio channel exploitation, the CAP of the super-frame can also be used; in this case, no hard capacity limitations are posed by IEEE 802.15.4, and the packet success rate depends on the number of nodes accessing the channel at each level of the tree and competing for the channel. The latter case is known as the soft capacity case. In the following section, the hard capacity limitation is used, like in the standard.

5.4.2 Optimum Tree Network Design

The scenario considered in this section, which summarizes the results of [13,15], consists of sources (sensor nodes) and sinks (gateways) both assumed to be uniformly distributed over the infinite bi-dimensional plane with densities ρ and ρ_o respectively, with the latter much smaller than the former. Denote as $\Delta = \rho/\rho_o$ their ratio. For the sake of simplicity [15] we assume that the ratio between the node density at a given level and the one at the next higher level is set to a common value η, except for the N-th level that will include the remaining nodes. Formally [15],

$$\rho_i/\rho_{i-1} = \eta \quad i = 1,\dots,N-1 \;\; \textit{and} \;\; \rho_N/\rho_{N-1} \leq \eta \tag{5.26}$$

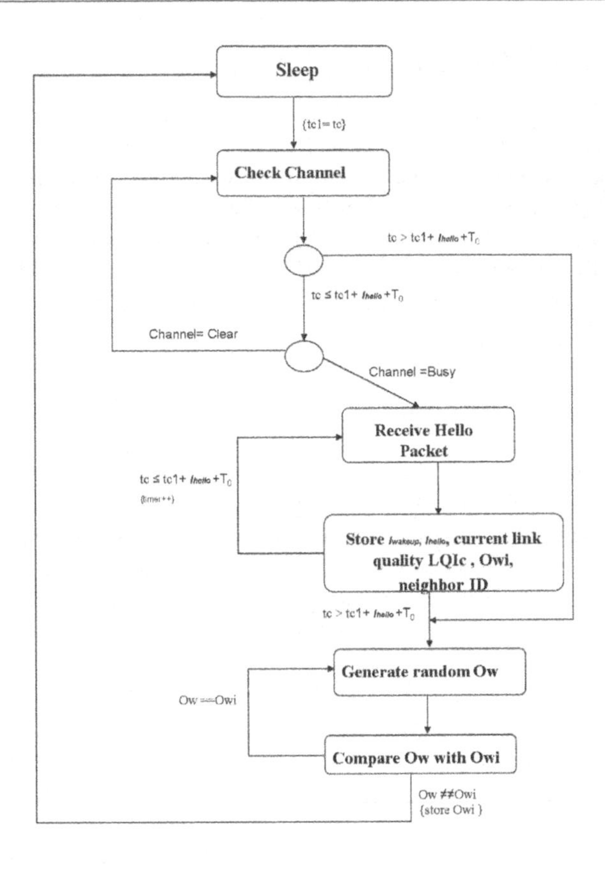

Figure 5.4: Initialization phase of PEAS-MAC protocol.

Thus, the probability of blocking (i.e., the transmission of the samples collected to the higher level is not possible because of capacity limits, or collisions) will be the same at all levels from $N-1$ to 1. As a result, N and η should be fixed according to Equation (1.25). Clearly, when N increases, the minimum value η will decrease [15]:

$$\sum_{i=1}^{N-1} \eta^i + \eta^{N-1}.\rho_N/\rho_{N-1} = \Delta \tag{5.27}$$

Once N is fixed, increasing η increases the connectivity at each layer. On the other hand, if η significantly exceeds the air interface capacity (either in the hard or soft capacity cases), the probability of blocking will increase and the average number of samples received by a given sink will decrease.

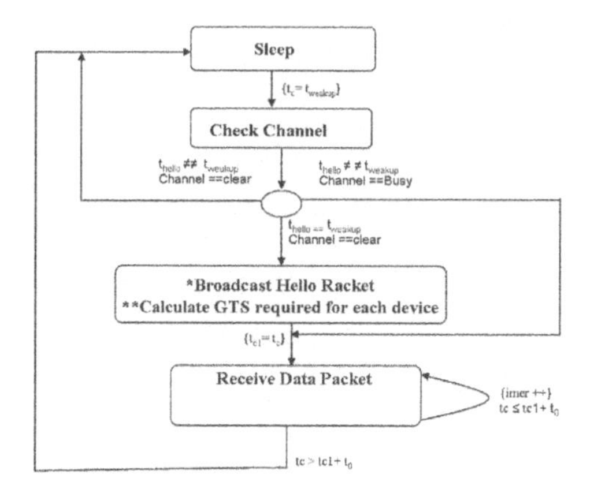

Figure 5.5: The receiver part of PEAS-MAC protocol.

5.4.3 Periodic Listening and Sleep Phase

After the initialization phase, a node performs the periodic listening and sleep phase. The node starts the periodic listening and sleep phase by setting the wake-up interval, I_{wakeup}. A node performs low power listening (LPL) at every I_{wakeup} timeout to receive an incoming packet [12,20]. If the channel is busy, the node receives the incoming packet. If the wake-up time of the node is also Hello time, the node receives the packet after sending a Hello packet. When a node has a packet to send, it waits in the sleep state until the receiver is scheduled to wake up, and it wakes up when the receiver does. If the wake-up time of the receiver is Hello time, it receives the Hello packet and then sends the packet. If not, it directly sends the packet with the preamble compensating clock drift. We first describe the operation of a node acting as a receiver, and then the operation as a sender.

As a receiver, the flowchart of the periodic listening and sleeping phase is shown in Figure 5.5. If the wake-up time is also a Hello time, and the channel is clear, the node broadcasts a Hello packet.

After that, the node waits to receive a packet until a timeout, t_o. If a packet is sent before t_o, the node receives it. The value of t_o should be only slightly longer than the maximum backoff time of the senders.

After receiving a packet, the receiver will calculate the number of Guaranteed Time Slots (GTS) required for the device depending on the data size, as we explain in detail in Section 5.4.4. After that the receiver goes back to sleep. When the wake-up time is not a Hello time, if the channel is silent, the node immediately goes back to sleep. If the channel is busy, the node stays in the listen state and receives the incoming packet. After reception, the node returns to sleep.

As a sender, the flowchart of the periodic listening and sleeping phases is shown in Figure 5.6. The node does not wake up at neighbors' wake-up time if it has nothing to send. If a node has a packet to send, it waits in the sleep state until the receiver's

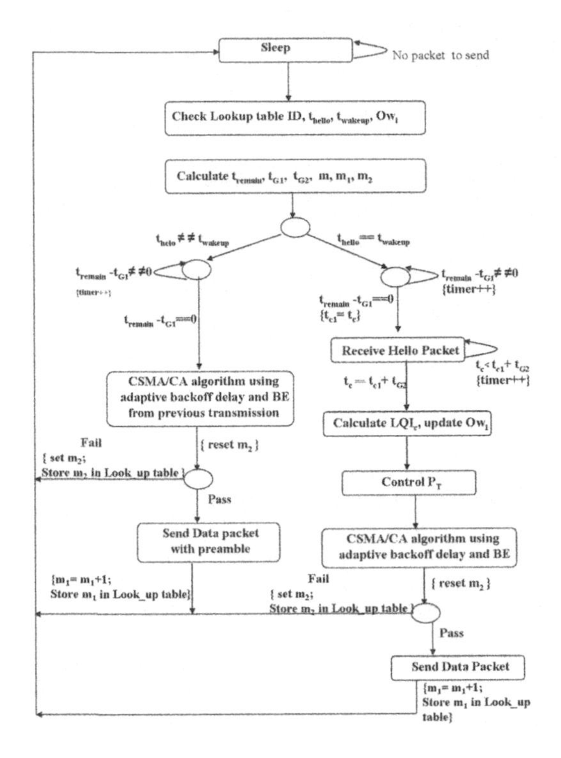

Figure 5.6: Sender part of modified PEAS-MAC protocol.

wake-up time. Every node stores its neighbors' scheduling information in its neighbor table; therefore it can predict the remaining time, t_{remain}, from the current time to the upcoming wake-up time of the receiver [12,21]:

$$t_{remain} = I_{wakeup(i)} - (t_c - O_W(i)) \tag{5.28}$$

where t_c is the current time, and i is the ID of the receiver. When the receiver's wake-up time is also a Hello time, in order to compensate for the potential clock drift between the sender and the receiver, the sender wakeup earlier than t_{remain} by a guard time, t_{G1}:

$$t_{G1} = 2.C_{drift}(i) - (t_c - O_w(i)) \tag{5.29}$$

where C_{drift} is the maximum clock drift amount. The sender waits for the receiver's Hello packet with a timeout t_{G2}:

$$t_{G2} = 4.C_{drift}(i) - (t_c - O_w(i)) + t_{lpl} \tag{5.30}$$

where t_{lpl} is the time for LPL. If the sender does not receive the Hello packet before t_{G2} seconds elapse, it postpones the transmission to the next wake-up time of the receiver. If the sender receives the Hello packet from the intended receiver, it updates the receiver's O_W in its neighbor table to the start time of the reception of the Hello packet to compensate clock drift.

In AS-MAC to avoid collisions with other potential senders, the sender performs collision avoidance backoff and carrier sensing by randomly selecting a slot within the fixed contention window. But for PEAS-MAC, we calculate the link quality indicator from the Hello packet, then compare it with the previous one which was calculated in the initial phase for a new node or from the previous transmission with its parent (assuming a symmetric channel). If the current LQI is greater than the previous one, then the channel is better and we decrease the power. If not, then we increase the power, which helps to succeed in successive transmissions.

After that, AS-MAC uses the CSMA-CA algorithm to send the packet. If it loses the contention, the sender postpones the transmission to the receiver's next wake-up time. If it wins the contention, it sends the packet, and then goes back to sleep. In PEAS-MAC the backoff delay is adaptive depending on a comparison of current LQI calculated from the Hello packet and the previous one as follows [16]:
$Backoffdelay = Rand(0, 2^{BE} - 1)$
If $LQIp \succ LQIc$
then Adaptive Backoff delay= Backoff delay+(PRNG/N_{HOP})
If $LQIp \prec LQIc$
then Adaptive Backoff delay= Backoff delay-(PRNG/N_{HOP})
$1 \prec PRNG \prec Backoffdelay$
$LQIp$ is the previous and $LQIc$ is the current value of Link Quality Indicator (obtained from the radio). N_{HOP} and $PRNG$ are the number of hops to the sink node and a parameter which determines the Backoff delay respectively. $PRNG$ should be large for a network with high traffic variations, while when $PRNG$ is very small, PEAS-MAC operates like the CSMA/CA mechanism. We increase $PRNG$ by the BE value when the channel condition seems to be poor and there are several retransmissions [16,17]:

The Adaptive-Backoff delay is used in step [2] of the CSMA/CA algorithm as shown in Figure 5.1. The backoff waiting time is reduced if the channel is good and reduces the energy consumption in listening. If the channel is bad then waiting will give more chances to transmit successfully and reduces the number of retransmissions.

Note that Adaptive-Backoff delay using LQI can be considered as an example of cross-layer optimization between PHY layer and MAC to improve the performance of the system.

The other modification of AS-MAC on the sender side is to use a minimum value of BE (which in the standard is 3) between 1 and 3, which changes flexibly as the condition of the node changes. Each node has three states: noData, postData, sendData, with each state having the default minBE value of 3, 2 and 1 respectively as shown in the following steps [16,17]:

1. The nodes are initialized to the standard BE ($minBE = 3$). The state is initialized to noData.

2. In state **noData**, when the node detects that a data packet is ready for transmission, the node changes its state to postData and modifies $minBE = 2$. The lower minBE value will give the node a higher chance of selecting a lower backoff value and quickly transmit the data packet.

3. State **postData** includes two cases:

   (a) After two successive transmissions of data packets without any error, the state changes to the sendData state and $minBE$ is modified to 1.

   (b) After two continuous beacon frames with no transmission of data packets, the state changes to the noData state and $minBE$ is modified to 3.

4. In state **sendData**, when no data packets are transmitted for two continuous beacon frames the node changes its state to postData and modifies $minBE$ to 2.

In the flowchart shown in Figure 5.6, $m1$ is the number of successfully transmissions during a hello interval and $m2$ is 1 if the last transmission failed, else it is 0. If $m1 > 2$ and $m2 = 0$ then $minBE = 1$ otherwise $minBE = 2$ when the node has data to send and $minBE = 3$ when it has no data to send.

Now consider the other part of Figure 5.6, when the receiver's wake-up time is not a Hello time. The sender should perform the collision avoidance backoff to avoid collision with other potential senders; therefore the guard time t_{G3} is longer than t_{G1} by the maximum contention window time. In this case the Adaptive Backoff delay does not change depending on the LQI value calculated using the Hello packet (because there is no such packet at this time) but depending on the previous transmission.

In our case m, which is the number of packets transmitted from each child to its parent, during a Hello interval, is saved and $m2$ is 1 if a transmission failed else it is 0. If m and $m1$ are greater than 1 and $m2 = 0$, then the previous transmissions to the parent were successful, hence the channel status is good. Then the node reduces the transmission power, and the Adaptive Backoff delay, and sets $minBE = 2$, which is reduced to 1 when $m1 > 2$ and $m2 = 0$.

If it wins the contention, the sender sends the data packet with a preamble that is longer than t_{G2} by the remaining contention time. If it loses the contention, the sender postpones the transmission to the receiver's next wake-up time.

5.4.4 Selection the Number of Required GTS Slots

AS-MAC is based on IEEE 802.15.4 which fulfills many of the WSN requirements. But still there are some limitations especially with respect to energy and bandwidth critical applications. Some of these limitations are [17]:

- The first and foremost problem with the current GTS allocation is bandwidth under-utilization. Unfortunately, the standard only supports values of BO and SO that are powers of two and the slot length must be $1/16$ of SD. Most of

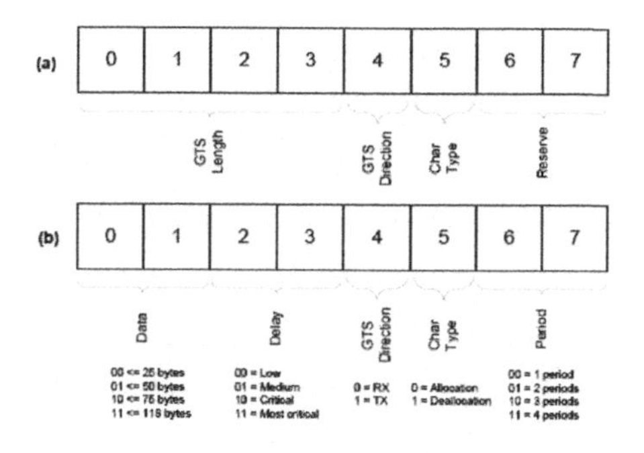

Figure 5.7: (a) Original GTS characteristics field. (b) Modified GTS characteristics field.

the time, a device uses only a small portion of the allocated GTS slots. This creates an empty hole in the CFP, like the memory fragmentation problem in operating systems.

- The protocol only supports explicit GTS allocation and hence a maximum of seven GTS descriptors can be allocated in each super-frame.
- A device uses its GTS Characteristics field, as shown in Figure 5.7(a), to request the number of GTS slots it wants. A device can request for all seven GTS slots, even if they are not really needed. Such unbalanced slot distribution can prevent other devices from taking advantage of the guaranteed service.
- Even if the CFP is not present in the super-frame, beacons transmitted by the coordinator always use unnecessarily one byte for the CFP, resulting in energy waste.
- The current super-frame structure must contain at least a constant size CAP, while strict real time applications, may need a variable size CAP.

The PEAS-MAC uses the advantages of enhanced IEEE 802.15.4 proposed in [17], thus dealing with most IEEE 802.15.4 limitations listed above. The bit structure of the 8*bit* GTS characteristic field changes, because each device, rather than sending a fixed slot length, sends its data and delay specification to the coordinator. The original GTS characteristics field is shown in Figure 5.7(a), whereas, the new one is shown in Figure 5.7(b). The coordinator, thus decides the slot length for each device.

In this way, bandwidth under-utilization and the supporting of at most seven GTS requests are avoided. To avoid the constant GTS expiration, the coordinator uses the period bits of the GTS characteristics field and performs GTS expiration dynamically. Rather than using one whole byte in the super-frame to indicate the presence of the CFP part, it uses one available reserved bit in the super-frame.

Before choosing the number of CFP slots needed by the devices, the coordinator first calculates the maximum number of CFP slots. The MAC super-frame is divided into 16 equal slots $(0-15)$. The beacon is always transmitted in the first slot and CAP should have at least the size of aMinCAPLength, which is equal to 440 symbols (1 slot = 60 symbols). Then the maximum number of available CFP slots can be calculated as [1,17,25]:

$$MaxCFP_{avail} = \frac{15 - aMinCAPLength/60}{aBaseSlotDuration * 2^{SO}} \tag{5.31}$$

The maximum number of available CFP slots from the above equation is approximately seven. The coordinator allocates GTS slots to each device depending on its data requirements, considering that the maximum total length of data packets is 127 bytes. From the standard, 11 bytes are required for an (optional) acknowledgment request. To receive acknowledgment, each device needs a TurnaroundTime (12 symbols) to change its radio from the TX to RX mode (or vice-versa). Additionally, we have to consider the inter-frame spacing (IFS), which separates two successive frames sent by the device [1,17]. Its length is dependent on the size of the frame that has just been transmitted, as shown in Equation (1.31). The total number of CFP slots required for the device is shown in Equation (1.32).

$$IFS = \begin{cases} 12 & MPDU \leq 18Bytes \\ 40 & MPDU \succ 18Bytes \end{cases} \tag{5.32}$$

$$CFP_{req} = \frac{2 * Data + IFS + aTurnaroundTime}{aBaseSlotDuration * 2^{SO}} \tag{5.33}$$

5.5 ZigBee Cross Layer Optimization

Cross-layer design is a good way to maximize the use of limited resources in wireless sensor networks. In traditional networks, each layer obeys a strict compartmentalization codified in protocol stacks [22]. This means that it can only communicate with the layers above and below it in the stack. This simplifies coding, debugging, evolution and interoperability, but makes higher efficiency and lower power consumption much harder to achieve. Cross-layer design means that any layer of the system can provide and utilize any useful information within the stack, in order to implement a global adaptive optimization strategy [23].

This does not change the traditional five layer model, but allows more communication throughout its dynamic structure. Each layer will respond to its own changes and to changes of other layers and of the operating environment. Cross-layer design leads to improved performance and lower resource usage, for example in terms of computing ability, storage, transmission power and battery energy or power [24].

In our case, we chose the link quality indicator (LQI) as a key metric to optimize the transmission power. The link quality indicator (LQI) measurement is a characterization of the strength and/or quality of a received packet. The LQI value, as defined in [22], is limited in the range between 0x00 and 0xff. A larger LQI value indicates better link quality [23].

This section describes an example of cross-layer optimization that minimizes energy under QOS constraints. A key aspect when using the LQI metric to optimize an actual link is to define the appropriate LQI threshold, above which transmission power should be decreased, and below which it should be increased. In order to keep the optimization protocol stable, we actually define a "dead band" of LQI, within which no change in transmission power is required, delimited between Low LQI Threshold (LLT in the following) and High LQI Threshold (HLT in the following).

From the result of Section 5.3, we can find the minimum energy consumption which gives the optimum LQI (also called OL in the following) at optimum energy consumption, maximum SNR and minimum Packet Error Rate (PER). The IEEE 802.15.4 standard defines a 0.01 maximum PER [1,25,26], thereby we determine the LLT as the LQI which corresponds to the maximum PER, since any LQI higher than this will necessarily have a smaller PER. The choice of the HLT is more arbitrary, since it is not tied to the standard, but only needs to be greater than the optimum LQI. In this example, we choose it to be at the same distance from the optimum LQI as the Low LQI Threshold, i.e.,

$$HLT = OL + (OL - LLT) \tag{5.34}$$

As mentioned above, if the measured LQI is higher than HLT, we should decrease the transmission power, while if the LQI is lower than LLT, we should increase the transmission power, in order to keep the link working in good condition. In this way we can save power and energy, within the appropriate link quality constraints.

5.6 MATLAB® / Simulink® Modeling

MATLAB / Simulink provide the designer with an environment to model and simulate the key characteristics of an embedded system. In particular, we want to use them to enable the simultaneous design of a WSN application and the optimization of the underlying protocol parameters and network topology. This means that the model has to be able to provide as output both the functional data (e.g., sensed, elaborated and transmitted data) and the interesting parameters of the underlying network and protocol (e.g., position of the nodes, modulation algorithm, MAC parameters).

The results of a simulation allow the designer to verify that the desired functional behavior is obtained, and at the same time to evaluate, for example, energy consumption under given bit error rate and number of hops constraints. The latter can be used to validate and refine the results of the analytical optimization techniques presented in the previous section.

We created three Simulink models for the three physical layer bands in the standard, including the following major building blocks: a spreader, a de-spreader, a modulator, a demodulator, and a channel. Noise is added to the modulated stream using various models: Gaussian, Rayleigh and Rician, for different multi-path propagation

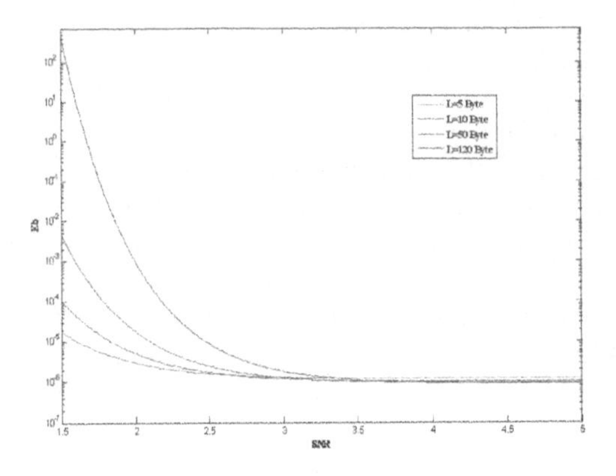

Figure 5.8: Energy consumption per bit versus SNR at different payload sizes and distance=100m.

conditions. In the envisioned design flow, the best model for the channel will be selected based on the representation of the planned WSN deployment location and of its propagation characteristics. Also the effects of communication stack parameters on energy consumption were simulated. We used the energy-optimal number of hops for uniform noise distribution as computed from the previous sections. The distance between the nodes was determined for both a known and unknown number of hops.

5.6.1 Simulation Results

Simulations were performed in [18] to find the optimal value of the 802.15.4 Physical layer energy consumption depending on parameters such as modulation type, BER, number of bits per symbol (M), SNR, channel model, number of nodes and distance between them. In simulation we check and validate the optimal number of nodes that gives the minimum energy consumption and the distance between the nodes for uniform and non-uniform noise distributions.

The energy consumption per bit E_b shown in Equation (1.5) is affected by the value of L. E_b decreases with decreasing L as shown in Figure 5.8 with different L, 5,10, 50 and 120 bytes, and distance $= 100m$ and different SNR. Also at each value of L the E_b will decrease with increasing SNR while P_b decreases. The second simulation compares E_b at different distances from 1 to 150m and five different values of SNR (-10, -5, 0, 5 and 10 dB) at L=10 bytes as shown in Figure 5.9. E_b increases with decreasing SNR and increasing the distance at given L.

Decreasing BER by using BPSK versus OQPSK decreases E_b at different SNR, D=100m and L=8 bytes as shown in Figure 5.10. In the same manner E_b increases with increasing M and decreasing SNR for all three channel models and increases

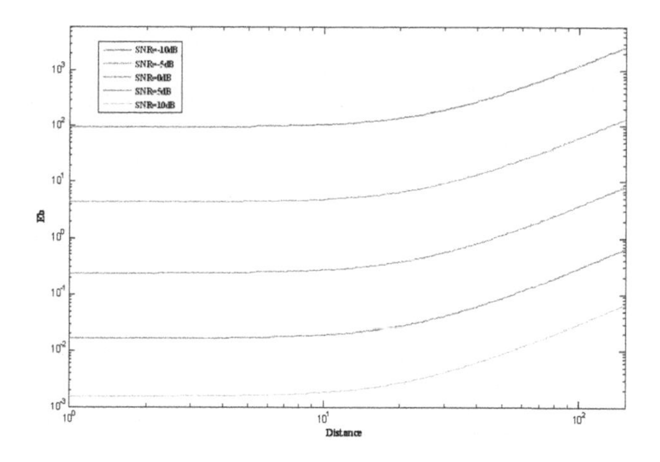

Figure 5.9: Energy consumption per bit versus distance at different SNR and payload=10 bytes.

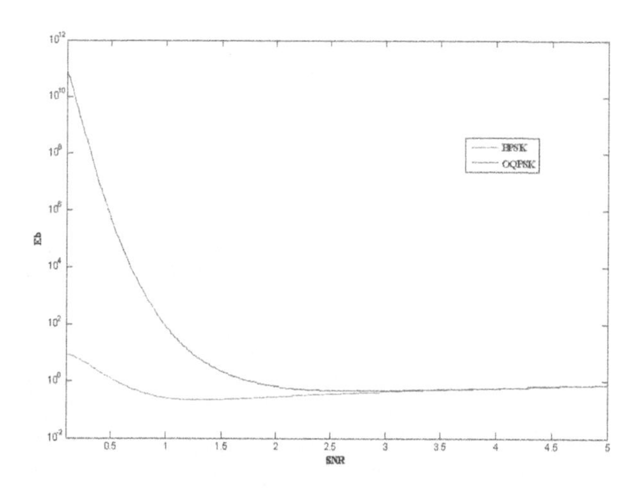

Figure 5.10: Energy consumption per bit versus SNR for different types of modulation, total D=100m and payload size=10 bytes.

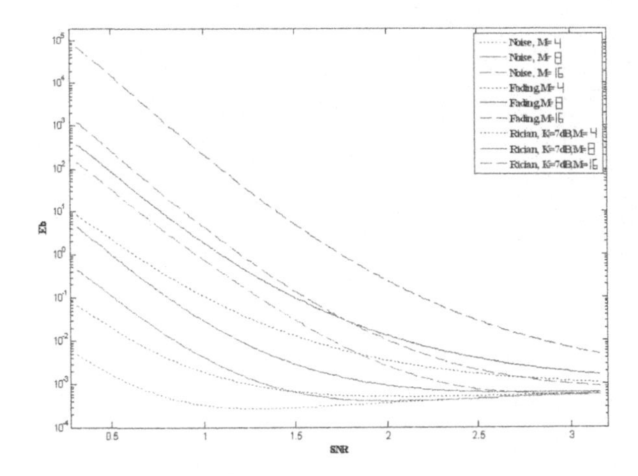

Figure 5.11: Energy consumption per bit versus SNR for different M, channel model and K-factor, D=400m and payload size=10 bytes.

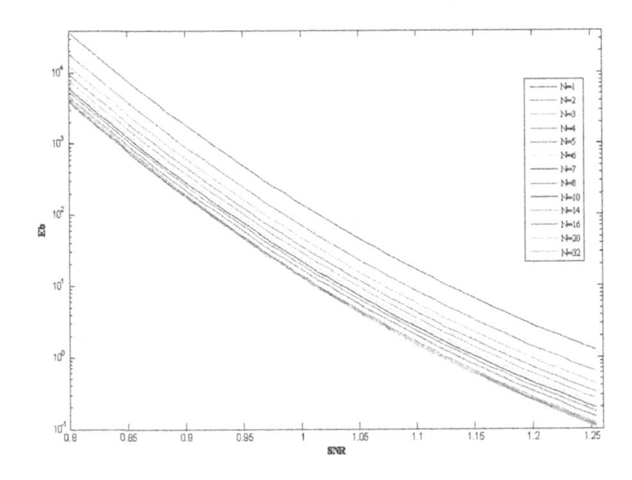

Figure 5.12: Energy consumption per bit versus SNR for different number of hops N, D=200m and payload size=10 bytes.

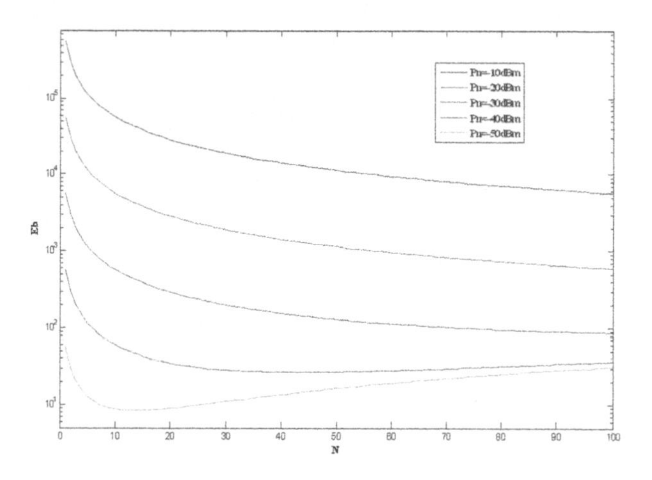

Figure 5.13: Energy consumption per bit versus number of hops N for different uniform noise power, total D=400m and payload size=10 bytes.

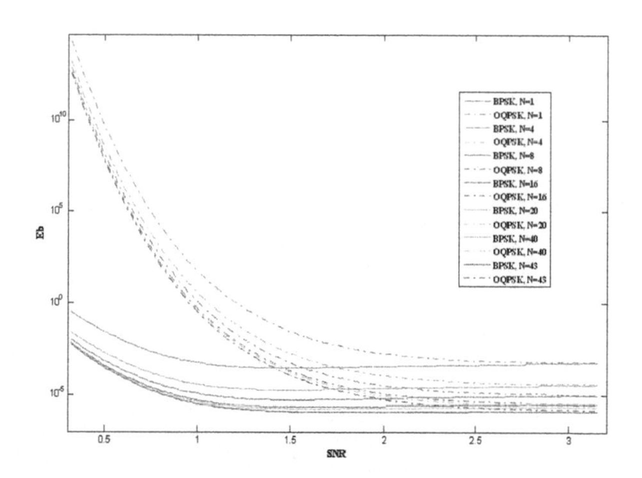

Figure 5.14: Energy consumption per bit versus SNR for different types of modulation and N, total D=400m, payload size=10 bytes.

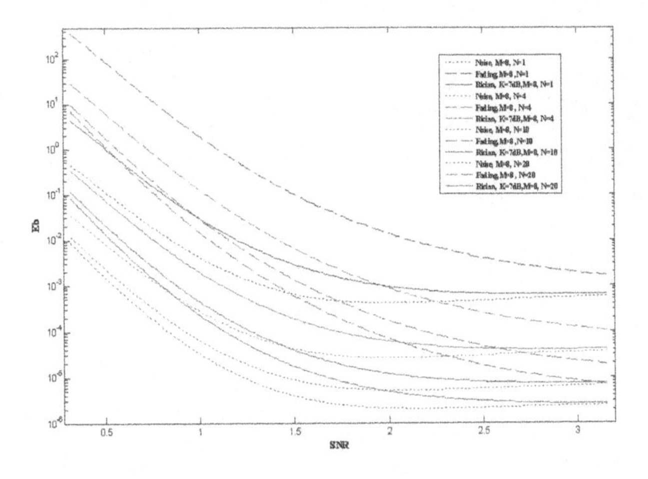

Figure 5.15: Energy consumption per bit versus SNR, for different N, channel model and K-factor, M=8, D=400m and payload size=10 bytes.

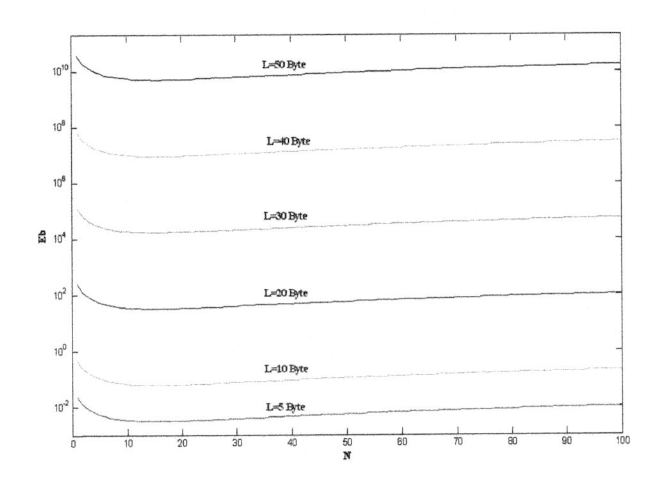

Figure 5.16: Energy consumption per bit versus number of hops N at different payload sizes and total D=400m.

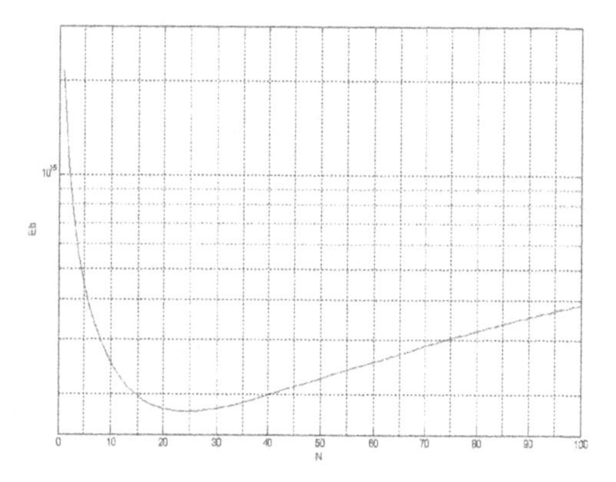

Figure 5.17: Energy consumption per bit versus number of hops N at L=10 bytes and distance=400m.

with decreasing K-factor at given M, as shown in Figure 5.11 with M= 4, 8 and 16, L=10 bytes, K-factor=7dB, D=400m and different SNR.

The next simulation compares E_b with different numbers of hops (1, 2, 3, 4, 5, 6, 7, 8,10,14,16, 20 and 32 nodes) L=10 bytes, D=200m and different SNR as shown in Figure 5.12. Eb decreases with increasing number of nodes (N) at given SNR. Using multi-hop will decrease E_b as shown in Figure 5.12 for different numbers of hops N, D=200m and payload size=10 bytes.

The sixth simulation was run changing the noise power for different numbers of nodes from 1 to 100 and finding E_b, as shown in Figure 5.13, where noise power increases from -50dBm to -10dBm in steps of 10dBm at L=10 bytes and D=400m. E_b increases with increased noise power for all N.

In the seventh simulation two different modulations, BPSK and OQPSK, are used with multi-hop processing at different numbers of hops (4, 8, 16, 20, 40 and 43 nodes) and SNR, at L=10 bytes and D=400m, as shown in Figure 5.14. E_b for all SNR and N is less when using BPSK versus OQPSK and decreases with increasing SNR and N. We now show in Figure 5.15 the effect of multi-hop transmission, considering the effect of three channel models (AWGN, Rician and fading) on E_b for different SNR and N (1, 4, 10 and 20 hop) for L=10 bytes, D=400m, K-factor=7dB and M=8 bits/symbol. E_b increases when decreasing the K-factor, from ∞ when using AWGN, to 0 when using the fading channel, while using the Rician model at intermediate values of K.

The eighth simulation was run changing payload size (5, 10, 20, 30, 40 and 50 bytes), showing E_b for the multi-hop case with N varying from 1 to 100 hops at a fixed D=400m as shown in Figure 5.16. Note how E_b increases with the payload size at all N. In Figure 5.17 E_b is shown at different N for SNR=-5dB, L=10 bytes and D=400m. The minimum E_b was found at N=24 hops, which is the optimum number of nodes computed by the theoretical analysis above in this case, assuming that the noise is uniform and hence that the distance is the same between nodes.

5.7 Conclusion

Wireless Sensor Networks (WSN) are a new generation of embedded computing systems with a wide range of applications. Most nodes in a WSN are battery operated and need to maximize the battery lifetime. It is important to design energy efficient protocols to reduce unnecessary energy consumption in order to increase the battery lifetime. This chapter provides mathematical and simulation models of the PHY and MAC layers for the ZigBee protocol, and illustrates some energy optimization techniques. We propose a mathematical formulation and an algorithm to optimize parameters such as bit error rate, SNR, number of repeater nodes, distance between repeaters at different environmental and interference noise levels.

We also introduce the PEAS-MAC protocol, which gives further advantages in terms of power and performance, by coupling transmitted power control, adaptive backoff delay, changing minimum backoff exponential value and dynamically choosing the number of GTS slots required depending on data size, traffic, delay, and energy constraints.

References

[1] IEEE, IEEE Std 802.15.4-2006, Part 15.4: Wireless Medium Access Control (MAC) and Physical Layer (PHY) Specifications for Low-Rate Wireless Personal Area Networks (WPANs), 2006.

[2] Marina Petrova, Janne Riihijarvi, Petri Mahonen and Saverio Labella,"Performance Study of IEEE 802.15.4 Using Measurements and Simulations", IEEE Wireless Communications and Networking Conference (WCNC06) proceeding, 2006.

[3] Sangjin Han, Sungjin Lee, Sanghoon Lee and Yeonsoo Kim, "Coexistance Performance Evaluation of IEEE 802.15.4 Under IEEE 802.11B Interference in Fading Channels", 18th Annual IEEE International Symposium on Personal, Indoor and Mobile Radio Communications (PIMRC'07), September 2007, Athens, Greece.

[4] Jennifer A. Hartwell, Geoffery G. Messier and Robert J. Davies, "Optimization Physical Layer Energy Consumption for Wireless Sensor Networks", 65th IEEE Vehicular Technology Conference, April 2007, Dublin.

[5] Jonqyin Sun and Irving S. Reed, "Performance of MDPSK, MPSK, and Noncoherent MFSK in Wireless Rician Fading Channels", IEEE TRANSACTIONS ON COMMUNICATIONS, VOL. 47, NO. 6, JUNE 1999.

[6] S.C. Ergen, C. Fischione, D. Marandin and A. Sangiovanni-Vincentelli, "Duty Cycle Optimization in Unslotted 802.15.4 Wireless Sensor Networks", Globecom 2008, December 2008. (Extended version in preparation to be submitted to IEEE Transactions on Mobile Computing.)

[7] N. Golmie, D. Cypher and O. Rebala. "Performance evaluation of low rate WPANs for sensors and medical applications", Military Communications Conference (MILCOM04) Proceeding, November 2004.

[8] B. Bougard, F. Catthoor, C. Daly, A. Chandrakasan, and W. Dehaene, "Energy efficiency of IEEE 802.15.4 Standard in dense wireless microsensor network Modeling and improvement perspectives", Design, Automation and Test in Europe (DATE) Proceeding, 2005.

[9] Sankarasubramaniam, I. F. Akjildiz and S.W. McLaughlin, "Energy Efficiency based Packet Size Optimization in Wireless Sensor Networks", Proceeding of IEEE International Workshop on Sensor Network Protocols and Applications, USA, 2003.

[10] X. Li, "Efficient algorithm for hop optimization in multi-hop ad hoc wireless networks," Proceedings of 2008 ICST Second International Conference on Networks for Grid Applications and Wireless Grids, Beijing, China, Oct. 8-10, 2008.

[11] Khaled Shuaib, Maryam Alnuaimi, Mohamed Boulmalf, Imad Jawhar, Farag Sallabi and Abderrahmane Lakas, "Performance Evaluation of IEEE 802.15.4: Experimental and Simulation Results", JOURNAL OF COMMUNICATIONS, VOL. 2, NO. 4, JUNE 2007.

[12] Beakcheol Jang, Jun Bum Lim, Mihail L. Sichitiu "AS-MAC: An Asynchronous Scheduled MAC Protocol For Wireless Sensor Networks", Journal of Network and Computer Applications, Vol. 31, No. 4, Nov. 2008, pp. 807-820.

[13] Chiara Buratti, Francesca Cuomo, Sara Della Luna, Ugo Monaco, John Orriss, Roberto Verdone, "Optimum Tree-Based Topologies for Multi-Sink Wireless Sensor Networks Using IEEE 802.15.4", IEEE Vehicular Technologhy Conference, USA, October 2007

[14] Pardeep Kumar, Mesut Gunes, Abd Al Basset Al Mamou and Intesab Hussain, "Enhancing IEEE 802.15.4 for Low-latency, Bandwidth, and Energy Critical WSN Applications", International Conference on Emerging Technologies, Pakistan, Oct. 2008

[15] Chiara Buratti, John Orriss, Roberto Verdone, "THE DESIGN OF TREE-BASED TOPOLOGIES FOR MULTI-SINK WIRELESS SENSOR NETWORKS" IEEE Personal Indoor and Mobile radio Communication, Greece, Sept. 2007

[16] Jeong-Gil Ko, Yong-Hyun Cho and Hyogon Kim, "PERFORMANCE EVALUATION OF IEEE 802.15.4 MAC WITH DIFFERENT BACKOFF RANGES IN WIRELESS SENSOR NETWORKS", International Conference on Communication Systems, Singapore, Oct. 2006.

[17] Young-Duck Kim, Won-Seok Kang, Dong-Ha Lee and Jae-Hwang Yu, "Distance Adaptive contention Window Mechanisms for WSN", The 23 International Technical Conference on Circuits/ Systems, Computers and Communications (ITC-CSCC2008), Japan, July.

[18] Al-Khateeb Anwar, Luciano Lavagno, "Simulink Modeling of the 802.15.4 Physical Layer for Model-Based Design of Wireless Sensor Networks", The Third International Conference on Sensor Technologies and Applications, Greece, June 2009.

[19] Ilker Demirkol, Cem Ersoy, and Fatih Alagz, "MAC Protocols for Wireless Sensor Networks a Survey", IEEE Communications Magazine, April 2006.

[20] Rong Zheng, Jennifer C. Hou and Lui Sha, "Asynchronous Wakeup for Ad Hoc Networks", The Fourth ACM International Symposium on Mobile Ad Hoc Networking and Computing, MobiHoc03, USA, June 2003.

[21] Jaehyun Kim, Jeongseok On, Seoggyu Kim and Jaiyong Lee, "Performance Evaluation of Synchronous and Asynchronous MAC Protocols for Wireless Sensor Networks", Second International Conference on Sensor Technologies and Applications SENSORCOMM08, Cap Esterel, France, 2008.

[22] Andreas Lachenmann, Pedro Jose Marron, Daniel Minder, Kurt Rothermel, "An Analysis of Cross-Layer Interactions in Sensor Network Applications", International Conference on Intelligent Sensors, Sensor Networks and Information Processing Conference Proceeding, 2005.

[23] Qi Wang and Mosa Ali Abu-Rgheff, "Cross-Layer Signalling for Next-Generation Wireless Systems", IEEE Communications Magazine, 2005.

[24] Michael I. Brownfield, Almohanad S. Fayez, Theresa M. Nelson, and Nathaniel Davis, "Cross-layer Wireless Sensor Network Radio Power Management", IEEE Wireless Communications and Networking Conference (WCNC06) proceeding, 2006.

[25] "ZigBee specification", http://www.zigbee.org

Chapter 6

Security Issues in ZigBee Networks

Sudip Misra, Sumit Goswami

CONTENTS

6.1 Introduction

ZigBee technology is developed to support automation, wireless remote control, monitoring and personal area network applications. The IEEE 802.15.4 committee was formed a short while later after inception of ZigBee to work on a low data rate standard. Thereafter the ZigBee Alliance and the IEEE joined together for further development of the standard and the technology was commercially named as ZigBee [1]. It is developed keeping those applications in mind which have a relaxed bandwidth requirement, require a self organizing short-range wireless communication network and are unable to handle the battery discharge to power the heavy protocol stack. IEEE 802.15.4 is a global standard that was developed for Low-Rate Wireless Personal Area Network (LR-WPAN) to cater to the requirements of slower data transmission rates, low power requirements, durability, maintenance-free operation and cheaper cost. The target of this standard was to provide networking at an inexpensive cost among the devices which may be fixed or mobile in a home or industrial network. It provides connectivity to equipment that does not require a high data transfer rate like those in Bluetooth but should sustain its battery life for a few years. ZigBee is based on IEEE 802.15.4 standard which is very less complex than the Bluetooth due to reduced Quality of Service (QoS) [2] [3]. 802.15.4 defines only 45 MAC primitives and 14 PHY primitives [4][5]. ZigBee addresses the unique needs of sensor network applications by providing logical network, interoperable data networking, security services, wireless home and building control solutions, interoperable compliance testing and other application layer primitives in the upper protocol layer which has the IEEE 802.15.4 at the lower layers. IEEE 802.15.4 provides specifications for the physical and data link layer and ZigBee alliance provides standards from network layer to application layer. While ZigBee determines the contents of the transmitted message, the 802.15.4 standard provides details about the robust radio hardware (PHY) and medium access control (MAC) to use for transmission [6].

6.1.1 ZigBee and Its Applications

There are two types of devices in ZigBee: the Fully Functional Devices (FFD) and the Reduced Function Devices (RFD). A network should include at least one FFD, which should take the role of PAN coordinator. An FFD can discover [7] and communicate with another FFD as well as an RFD. In order to provide seamless communication in the network based on the characteristic features of these devices, ZigBee supports two basic network topologies which are single-hop star between the RFD and FFD and the multi-hop peer to peer mesh which is between the FFDs. A cluster tree topology [8] is also supported which is a hybrid of star and mesh. In the static star network, ZigBee uses a master-slave configuration where slaves are comprised of many infrequently used devices that talk through small data packets to the co-ordinator, which acts as the master. ZigBee allows up to 254 nodes which generally spends most of its time snoozing. An FFD supports 45 MAC primitives, operate in any of the two network topologies and can take the role of a PAN coordinator, ZigBee coordinator (ZC), ZigBee router (ZR), ZigBee trust center (ZTC), Gateway or a ZigBee End Device (ZED) [9]. However, an RFD should support a minimum of predefined 38 MAC primitives, operate only under a star topology and acts as a network device communicating only with an FFD. The nodes transmit using CSMA so as to avoid any collision and detect when a node can transmit. The nodes can transmit up to a distance of 75 meters [1][4]. A device that acts as an RFD is generally in a sleep mode to save battery power. However, ZigBee has a low latency and after the node is powered down, it can wake up and start transmitting packet in about 15ms which is an important feature for its use in near real-time applications like detecting a catastrophe. Each application of ZigBee demands a specific topology which better suits the requirement. Star topology is better suited for applications like building automation, connecting PC peripherals, docking gaming devices, toys, smart badges [7] etc., which generally has a central PAN coordinator which has an abundant power source. Peer-to-peer topology supports ad hoc, self healing and self organizing networks. This topology is applicable in asset tracking, inventory control, container security, marine wireless, and equipment monitoring and industry control. Some of the common applications of ZigBee are as mentioned:

Smart Grid

- Automation and control of devices in factories, buildings, depots and stock houses

- Entertainment by integration of sensors in interactive toys, PC gaming peripherals.

Smart Home

- Home control system comprising of automated and remotely controlled lighting, temperature control, smoke and poisonous gas detector, curtain rollers and door closers. ZigBee remotes can serve to control more than one device after memorizing' access codes. It also assists in automated meter reading, personal health monitoring, automated telephone dialer for elderly and disabled and integrated wireless home security [6][10][11][12].

e-Healthcare

- Monitoring of public health, environmental data, natural catastrophe and infrastructure safety.

Emergency and Rescue Grid

- Emergency response network in a disaster application like earthquake integrating sensor nodes with mobile devices to support search and rescue operations. It helps in tracking and monitoring rescue personnel, firefighters, victims, and patients [13].
- Deployment of autonomous robot based sensor system in emergency application. It senses the environment and generates knowledge and situational awareness building a map of sensor information which guides rescue personnel to important tasks and avoids dangerous areas [14].

Soldier System

- Personal Area networks in a soldier where various equipment and wearable are made smart by embedded sensors and these sensors communicate using ZigBee with each other or with those of nearby soldier or field sensors in a battlefield formation. An example of the various processes initiated by a smart weapon guided by multiple sensors is depicted in Figure 6.1.

6.1.2 ZigBee Network Security

Wireless network security is brought at par with the high level of security in wired networks by implementing encryption, transmission security and authentication protocols. Protocols have been developed to prevent eavesdropping, spoofing, rogue node, theft of service and denial of service. The security protocols in ZigBee are lightweight solutions requiring less computational power, minimal energy requirement, smaller number of message exchanges, less storage requirement and minimal bandwidth requirement. A sensor node in ZigBee is comprised of a processor, memory and transmitter. Implementing the security protocols should be achievable with slower processing speed and weak signal transmission. There are four major security issues in ZigBee networks [15].

Authentication:

Mutual authentication of the coordinator and the device is important. It is beneficial for both, the cluster head, which is an FFD acting as the coordinator or router, and the sensor node. It protects the coordinator from unauthorized intrusion and prevents unauthorized service access by the node. Mutual authentication also allows a node to authenticate a coordinator so as to prevent any malicious node to act as a coordinator and then forward the data to the coordinator. This requires an authentication protocol to run at each coordinator or from a centralized node providing the authentication service which is contacted by the nodes to authenticate the data. However, the delay in authentication should be minimized and appear as a transparent process to the connecting nodes. This also depends on the link between the mutually authenticating nodes. The authentication process may be denied to the nodes by bringing down the

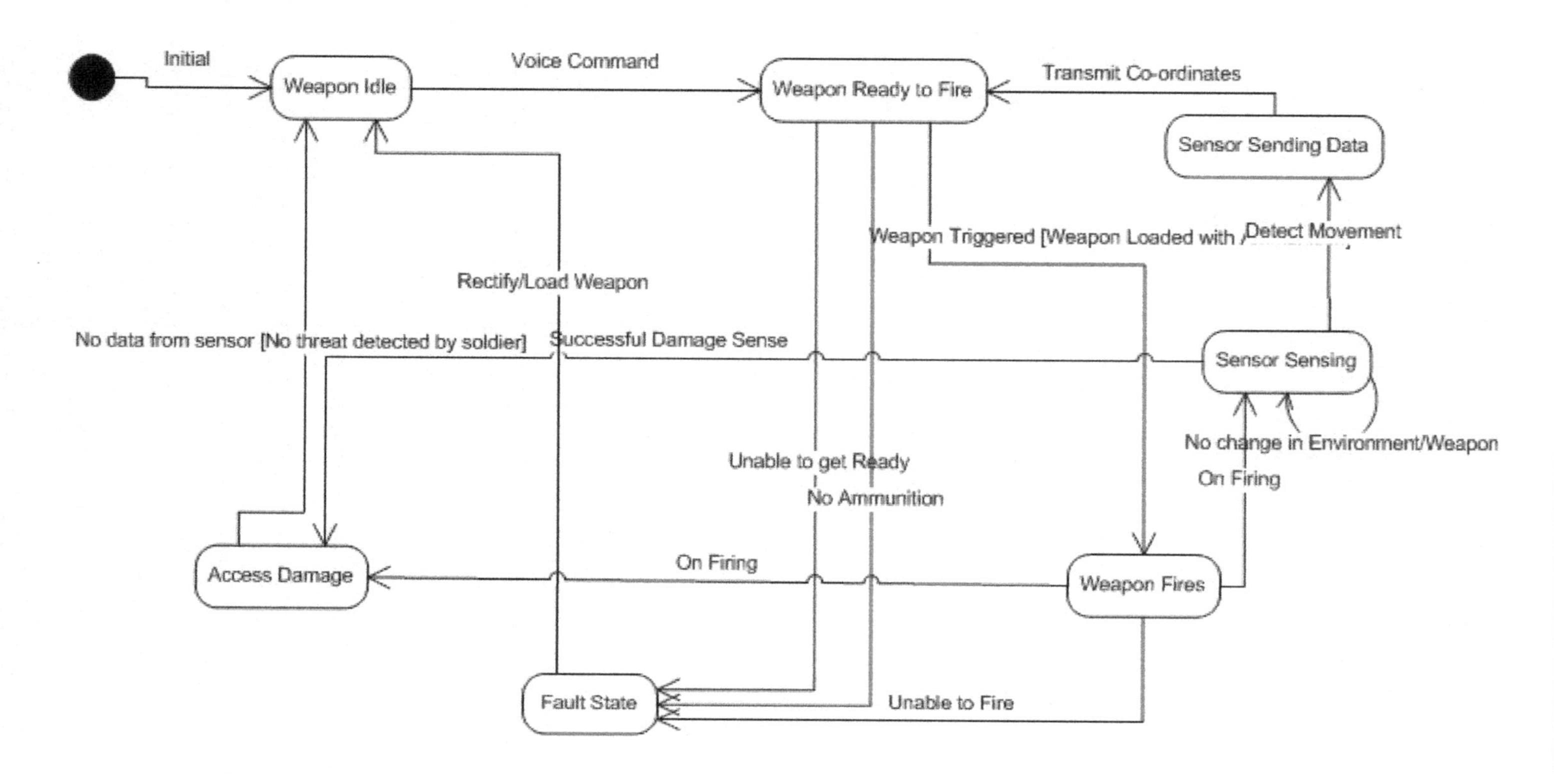

Figure 6.1: A state chart diagram of a sensor based weapon system controlled by sensors.

authentication server by cutting off the communication link between the server and the nodes by flooding the network.

Privacy and Anonymity:

A sensor node involved in a strategic network may require privacy of its identity, activity, location and mobility path. Providing such security features is difficult in a wireless network as the communication channel is vulnerable to eavesdropping and tapping. Anonymity can be ensured by preventing association of users with message received/transmitted and communication session that they participate.

Device Vulnerability:

The size and mobility of a wireless ZigBee sensor node makes it vulnerable to misplacement, theft or loss. Even though no confidential data may be stored in it, its loss can lead to operational ineffectiveness and may cost in terms of delayed or no information. Even though the data in the sensor may be of no importance for any active attack by an intruder, it may lead to some indirect information like awareness regarding deployment of a monitoring system, the type of sensor in usage and the range of deployment. Such type of information can be critical in defense applications. A malicious user may use the sensor to launch an attack on the network by injecting this compromised node back into the network and accessing data or injecting malicious information in the network. One method to overcome such vulnerability is to encrypt the entire data and configurations stored in the ZigBee node at the expense of computational power and energy consumption.

Domain Crossing:

A security cluster consists of a set of nodes under a cluster head acting as coordinator and under one security policy. As the nodes may be mobile, it may disassociate with one coordinator and associate with another while it is mobile. While it enters a new cluster domain, it has to authenticate itself to ascertain the trustworthiness of one another. Similarly an FFD has to authenticate itself with other FFDs acting as routers, coordinators so as to be a part of the ZigBee multi-hop mesh. A level of trust can be assigned to each node which determines the level of services that should be provided to the node, and the type of information that can be shared with it.

6.1.3 ZigBee Security Problems

The security protocols are designed to assist in authentication, non-repudiation, privacy, prevent network attacks, detect malicious nodes and provide seamless communication channel. As mentioned in [4][15], it can be divided in four domains:

1. Device domain: It ensures that only an authorized user can access the device and can program, modify or delete data from a device.
2. Network access domain: This ensures that only authorized devices can connect

to the network. This ensures that the network services are available only to the legitimate nodes. It also guarantees data privacy and integrity over the wireless link.

3. Network domain: This ensures the security of the network infrastructure like the devices that are acting as coordinator, PAN coordinator or the router in the ZigBee network. It also supports the network security across various connecting communication channels like the data transmission from a ZigBee network to a wired network with a WLAN in between and thus caters to the security requirements across multiple carriers.

4. Application domain: This ensures that only safe and trusted applications run on the device and the communication between the applications running on the nodes and the coordinator across a communication channel is safe, secure and maintains integrity.

Providing a secured communication for ZigBee nodes is more challenging as it has relatively low processing capability, bandwidth and battery power availability. Being smaller in size, they are susceptible to thefts which demands secured data storage in the devices. Securing a ZigBee device and communication channel invites the following security challenges [4][15]: Security processing gap: The ZigBee devices being inexpensive and pre-configured have very limited computational capabilities. However, the security protocols are putting up a huge computational requirement on these devices. This leads to a poor performance by the device as well as more utilization of the communication channel making it busy and slowing down the data transmission rates and shortening the battery life. Battery life: Though the battery capacity is increasing with time, the power consumption for the new security protocols are increasing at a much faster rate. Enhancement of security features which plug the security flaws that had been found in the previous protocols would lead to widening of the battery gap. Flexibility: A security protocol may specify a number of cryptographic algorithms at various layers of network communication and a few of them might be device specific. A node might be required to execute different protocol at different layer of the network protocol stack and to communicate with a specific device. The tradeoff between such security protocols and the flexibility of its usage in terms of computational overhead, energy consumption and channel utilization should be considered. Tamperproof implementation: A ZigBee device may fall into the reach of an adversary due to its size, critical applications and mobility. The security risk in such cases should be minimized by encrypting the data and configuration of the device. To maintain interoperability between the devices in a wired network and ZigBee, a few wired protocols might be required to be supported in ZigBee. But such protocols should be stripped off unwanted features and trimmed for usage in the ZigBee network to reduce the processing load. Some of the security protocols may be termed as prerequisites for a secured communication in a particular ZigBee network and this can lead to attaching cryptographic accelerators to the basic core of the device.

6.2 Overview of ZigBee Security

ZigBee technology has a security layer which incorporates the security functionality of varying levels. The security functionality makes use of symmetric keys, asymmetric keys, challenge authentication procedure or the public key infrastructure. The keys may be preconfigured within the device before deployment to enhance security or may be generated during its field operations for establishing a session and communicating with a node. ZigBee networks are vulnerable to security attacks as the transmission takes place in the open medium. Though there is a central PAN coordinator, the network may reorganize itself changing the PAN coordinator and the cluster coordinators leading to an interrupted availability of monitoring station and nodes keep on joining and leaving the network. There are communications overheads in detecting optimum routes with power saving, detection of malicious nodes or captured nodes. ZigBee networks may have to encrypt the backhaul communication, which represent the critical part of the network infrastructure [16]. The wired networks are generally secured using firewalls and encryption devices and monitored using proxy, intrusion detection system and routers. However, such legacy devices are not available within the ZigBee framework. The intrusion detection techniques used in the wired networks are not effective in the ZigBee network, as unknown nodes may join the network at a point of time. The algorithms for asymmetric cryptography have to be modified and customized to make it usable within ZigBee. The Public Key Infrastructure (PKI) is difficult to implement with ZigBee as they generally lack connectivity with Internet for authentication from the Certifying Authority (CA). However, it has to be customized with respect to an ZTC. The joining of new nodes under a cluster too low in the network hierarchy leads to delayed transmission and makes it difficult to distribute its public key to the other nodes in ZigBee by the ZTC using a secured channel. The security algorithms for ZigBee networks are designed to reduce the computational requirement of the nodes to save on the power. This is in addition to the prime requirement of keeping the dependency of the algorithm on a central node to a minimum. A centralized node is generally required in the security protocols for administrative as well as repository applications. The frequent topological changes in the network design, alive but hibernating nodes and compromised node makes it difficult to model the network for simulation and analysis of security protocols for ZigBee. The routing service, which supports topological change, can be attacked due to its property of accepting changes. A compromised node or a malicious node, can camouflage itself to be a self declared coordinator or PAN coordinator or ZTC of the network by generating exceptionally high or low traffic, beacon signals [17] and thus misleading the other nodes creating an inaccurate representation of the network and leading to a denial of service attack. The density of nodes in a ZigBee network is another factor generally ignored in designing security mechanisms. There are a few critical factors such as transmission power level, bandwidth used for each node and the rate of movement of the node which must be taken into account while designing the security protocol [18][19][20][21][22][23][24][25][26][27]. The nature of attack on a ZigBee network for civilian application may be entirely different from the security attack on a tactical battlefield ZigBee network.

6.2.1 Security Issues in Wired versus Sensor Network

In the wired networks, the transmission is through a guided medium, but still a number of cryptographic techniques are used to secure the channel from sniffing, traffic analysis and eavesdropping. The nodes can be physically protected against capture and are generally detectable in case of a malicious attack. Contrary to this, the sensor networks transmit in the open and adversaries can listen to the transmission. The nodes in sensor networks are prone to capture or attacks, as they are mobile. Thus, the data storage and transmission framework in a sensor network should be resistant to failure of multiple nodes. The sensor networks also call for a high degree of fault tolerance as it operates in an unpredictable environment and prone to non malicious faults like fading energy level of nodes, fluctuating signal strength with change in distance between nodes and varying transmission and reception level due to a variety of electromagnetic interference. Wired networks have a continuous connectivity while the sensor networks suffer from topological variations with time and thus leads to disconnection and reconnection of nodes [28]. The security algorithm should also cater the scenarios of sleeping nodes, which is rarely the scenario in a wired network. Thus, a key management framework for sensor network should support regular disruptions in connectivity, high resilience towards compromised nodes, sniffing and mobility associated issues in transmission. Security is relatively simpler to enforce by a fixed central authority that is generally in charge of securing the system under consideration [29]. As the ZigBee network is dynamic and open, nodes leave or join the network at any time and communicate through the publicly accessible electromagnetic spectrum and, thus, open for eavesdropping or injection of fake packets by the adversary. In such scenarios, selfish and malicious nodes can easily creep into the network [30]. The multi-hop communication depends heavily on the co-operation among nodes and the selfish behavior of a few nodes can affect the speed and reliability of the communication. A node may desire to obtain the services from other nodes but may not participate in routing or coordinating and thus break the communication link or increase the path length. Various algorithms, which generate wave like data transmission pattern where a receiver node communicates with other nodes in its vicinity in a multi-hop mesh to forward the information may fail or become less effective with such selfish nodes. It can also lead to the initiation of a polling algorithm to elect another PAN coordinator or cluster coordinator if the FFD turns out to be selfish even for a small amount of time or the selfish node disrupts communication path to the leader. Generally, leader election algorithms cost heavily on CPU usage, network traffic and power consumption. Lack of authentication also leads to provision of free-of-charge services to malicious nodes.

6.2.2 Design Issues

A sensor network is generally established under the following conditions [4]:

- **Ad hoc deployment:** The sensors are generally deployed in varying density in a region. Their distribution is not uniform which leads to varying signal strength, collisions and propagation effects. The connectivity is never seamless and the node associates and disassociates with the network due to mobility of nodes, varying obstructions, non-cooperative behavior of the intermediate nodes and security attacks.

- **Energy constraints:** The sensor nodes have to be durable, maintenance free and energy efficient as they are rarely redeployed with energy recharge. They operate generally on non-replaceable battery power and thus their energy consumption has to be minimized to prolong the life of the sensor.
- **Unattended operation:** A sensor network generally is comprised of a few hundred sensor nodes or more and thus it is generally not possible to manually configure them. The nodes are generally configured during design time after duly incorporating the environmental dynamics.

The nodes in ZigBee network have a varying data transmission pattern with unique characteristics depending on the application it supports. The data transmission pattern can be categorized in three distinct patterns [31]:

1. **Periodic data:** This type of data transmission is the characteristics of a wireless sensor based metering system which wakes up at a set time; checks for the beacon signal for the availability of the coordinator and exchanges data with it and again snoozes. In an automatic meter, reading of electricity, gas, water is transmitted by a sensor fitted in the meter. The transmission is forwarded to a centrally powered sensor which thereafter transmits it to the central processing server using a wired or wireless network. The meter generally has an RFD and the central powered sensor is generally the PAN coordinator which transmits the beacon signals.

2. **Intermittent data:** The sensor is activated by an external stimulus or an application. The sensor connects to the network only when data transmission is required and thereafter gets disconnected to save on the energy. Wireless light switches generate such type of intermittent traffic which connects to the network when the lights are to be switched on or off.

3. **Repetitive data:** This type of data transmission pattern is generally the characteristic of a security or a monitoring system. The monitoring system may be enabled for a critical patient or for a precise operation of a device or a factory like a nuclear power plant. These sensors use the guaranteed time slot (GTS) capability which ensures a reliable transmission where a PAN coordinator grants specific time duration in the Super-frame to the node to transmit without any latency or channel contention which is known as the contention free period (CFP).

6.2.3 Types of Keys

In symmetric cryptography, the sender as well as the receiver uses the same keys. When a number of nodes are communicating with each other a few keying models are available to decide who shares what with whom. The keying model that a

Comparison of master key, network key and link key

Type of Key → Features ↓	Master Key	Network Key	Link Key
Protection Type	Device security	Outsider attack	Insider protection
Generation	Factory Installed	Generated by Network Manager	Derived using SKKE
Update Frequency	One time installation	Updated Periodically	Updated Periodically
Visibility	Application Layer	Network Layer	Application layer
Sharing	Unique to device.	Shared between all devices	Shared by a pair of nodes
Memory Requirement	Low	Low	High
Zigbee-2006	Supported	Supported	Not Supported

Figure 6.2: Comparison of master key, network key and link key.

group of nodes use in a network depends on the amount of threat faced by the communicating nodes. The most common keying models in use are the network shared keying, pair-wise keying [32][33], group keying and hybrid keying [34]. In network shared keying each node in the network shares the same key and can use it to communicate with any other node in any session. As the nodes have a single key, the memory requirement is low but it comes at the cost of vulnerability of the nodes in case of a single compromised node where the key can be exposed and leads to the breakdown of the entire encrypted communication in the network. Pair-wise keying can tolerate such compromise of nodes as it will lead to the decryption of messages only between a pair of nodes because in this each pair of nodes share a separate key for encrypted communication. It has excessive memory overhead and computational delay in selecting the appropriate key for a particular node during communication. In group-keying, a group of nodes share a single key and that key can be used for communication between any two nodes in the group. The groups may be formed on the basis of the role, reliability, location or topology. ZigBee security architecture implements security mechanism at MAC, network and Application layer of the protocol stack by using methods like cryptographic key establishment, key exchange, device management and frame protection. Each layer has security protocol in place to ensure secure transmission of the frame using encryption, authentication and integrity check in the corresponding layer. 128-bit advanced encryption standard (AES) is the cryptographic encryption standard used in ZigBee. It uses three types of keys master key, link key and network key to secure the communication channel. The network key is shared between all communicating nodes and is the minimal requirement for secured data transmission between the nodes by encrypting the frames with the network key. This prevents unauthorized access to network resources by malicious nodes and ensures confidentiality. Link keys are the secret keys shared between two communicating devices and are used to secure a session between those two particular devices. The Master Key in the device is used to generate the Link key [9]. An overview of the various features and purpose of the keys is shown in Figure 6.2.

6.2.3.1 *Master Key*

The master key is preinstalled in the ZigBee device and helps in generating the link keys for data exchange between two nodes by symmetric-key key exchange (SKKE). The master key is visible only to the application layer and is the basis for the long term security needs of the device.

6.2.3.2 *Network Key*

It is a 128 bit key that is shared between all the devices in the network. It is generated by the network manager and is updated periodically by it. Each node has to obtain a network key to join the network. When the ZTC creates a new network key for updating, it is passed on to the devices using the old network key and thereafter initializes the frame counter to zero. The role of network manager is generally played by the ZTC but it can be any other dedicated device. Each device joining the network is validated and authenticated by the network key. The network key protects the network infrastructure and the application data from the outside attack. As a form of key rotation, an alternate network key may also be used for key updating purposes.

6.2.3.3 *Link Key*

These are unique secret session keys for each pair of nodes which provides them with end to end cryptographic encrypted channel. The key is visible only to the application layer. They are used to encrypt the data communication between the two devices. These keys can be updated periodically and are used as a confidentiality measure from an insider attack. The link key is generated from the master key of the device. The link key provides security only against the insider attack and it increases the memory requirement. So, ZigBee-2006 recommendations eliminate the use of link key as a cost-security tradeoff measure.

6.2.4 Key Management and Trust Setup

Key management deals with generation, distribution, updating and revocation of keys [35]. The method of key management in a ZigBee network varies among the devices depending on its vendor. However, the initial generation and distribution of the cryptographic key is done in one of the following three ways [9], an overview of which is given in Figure 6.3.

In-band:

The key is transmitted to the device over the normal wireless transmission medium using the same channel over which data communication takes place. The method is used to deliver a key when a node intends to join a network and requests for a key for communication in the network. As the node is not preconfigured to operate in the secured encrypted communication mode, it is a less secured method for delivery of the key. There is a small period of vulnerability before the receipt of the key by the node.

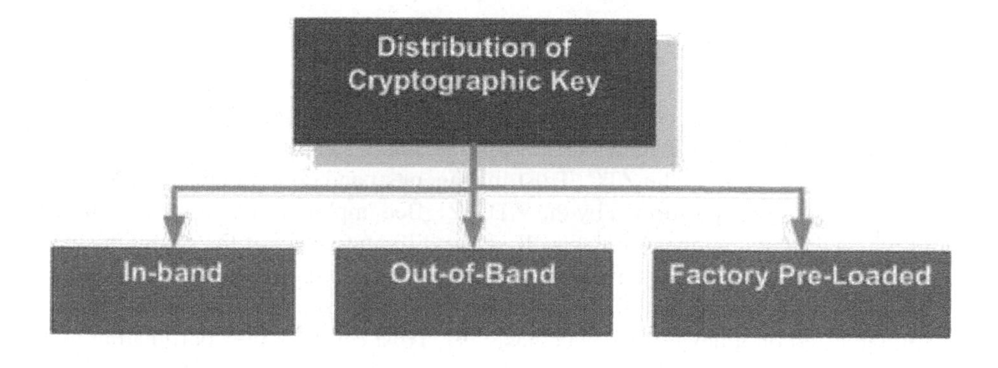

Figure 6.3: Overview of cryptographic key distribution techniques.

Out-of-band:

The key is transmitted to the device by some other communication channel different from the normal wireless communication mode. The key never goes on the same communication channel over which the data packets are transmitted. Some other side channel, which may either be wired or wireless is used for transmitting the key to the device. For, e.g., a key may be transmitted from a laptop to a ZigBee device using Bluetooth or a serial wired connectivity between the laptop and the ZigBee device.

Factory pre-loaded:

The ZigBee device vendor generates the keys and loads them into the device at the manufacturing location and the device is shipped for its usage with the preloaded key. The key value should be conveyed to the user along with the delivery of the device. The approach is secured after deployment as the device may not have the option to alter the key or upload another key. However, the approach is insecure from commercial view as the manufacturer knows the key of all the devices shipped by it and are under deployment. It also has the added task of secured delivery of the key to the user which is prone to confidentiality attack during its transmission to the customer. The most preferred method of key distribution is out-of-band where a user generates the key himself and uploads it to the device himself through an alternate communication channel. The security of an in-band key distribution can be enhanced by delivering the key in a controlled network condition. As the range of the ZigBee is about 75 meters, a secured sanitized key upload area can be setup where the keys are generated by the user and then transmitted to the device in-band in the sanitized area before deployment. The factory loaded key is always known to the vendor and thus can't be relied upon for strategic applications as the key is known to more than one person and may be compromised by the vendor or during its transmission to the user. The key may be stolen from the vendor's key repository or change hands during the sell-out of the manufacturer.

6.2.4.1 Trust Center

The ZigBee trust center is the core component of the security architecture in ZigBee. All the devices in a network authenticate and trust only one ZTC. Though many ZigBee nodes come preloaded with keys for enhanced security, the others are granted keys in the network by the ZTC. Trust management and configuration management are the major roles performed by the ZTC. ZigBee implements an open trust model in which different applications as well as the different layers of the communication stack running on the same device trust each other. The implication of this is that the applications and nodes within a ZigBee network trust each other and the security is implemented on a device-to-device basis [9]. Trust center is the coordinating node for the security implementation in ZigBee network. As the ZTC is trusted by all the nodes in the network and there is only one ZTC in a network, the address of the ZTC should preferably be preloaded in the devices. ZigBee protocol specifies a method by which a node should authenticate itself before joining the network. The node initially associate itself with the network requesting admission from the ZTC. The trust center decides on the permission to be granted to the node to join the network after authenticating it. The authorized nodes possess a network security key which is preloaded in the device. The network key is common to all the devices in the LRWPAN. The network key enables an initial secure association of the node with the ZTC or the ZR instead of an unsecured join in which the node sends an unsecure request for the secured services if it does not possess the network key.

6.3 Security in 802.15.4

Data is transferred between the devices and coordinator or between the devices in the peer-to-peer network topology. Data transfer can be classified into three types [4]:

1. **Direct data transfer:** This is applicable to all modes of data transfer and is done using slotted or un-slotted CSMA/CD. Slotted CSMA/CD is used incase a beacon-enabled mode is used and un-slotted CSMA/CD is used in non-beacon enabled communication mode.

2. **Indirect data transfer:** This type of data transfer takes place from a coordinator to the devices in its cluster. The coordinator has a transaction list and awaits extraction by the devices in its cluster after sending a beacon frame containing the details of pending frame in the coordinator. The devices communicate using slotted or un-slotted CSMA/CD during the process of data extraction.

3. **Guaranteed Time Slot (GTS):** This does not require CSMA/CD and is used for transmission between a coordinator and its devices.

In order to keep the computational overhead and cost low, IEEE 802.15.4 does not specify the key distribution mechanism. This is included in the upper layers of the applications which is standardized by ZigBee. This is required because the 802.15.4

standard does not define the essential security services like cryptographic key management and leaves it to the higher layers.

6.3.1 Security Services

There are four security services defined by the MAC layer in 802.15.4 [9][36] :

1. **Access Control:** This service enables the selection of a device for data communication based on the MAC address of the received frame. To prevent unauthorized access to the secured services available from a device, it provides a list of devices in its access control list (ACL) with which the device can receive data frames.

2. **Data Encryption:** This service provides data privacy by encrypting the MAC frame with a symmetric key. The nodes share a secret key which is used to encrypt and decrypt the message by the communicating devices to ensure data confidentiality.

3. **Frame Integrity:** This service uses a cryptographic message integrity code (MIC) to enable a receiving device to detect the frame integrity. A MIC is generated and added to the MAC frame. MAC frame integrity counters the address spoofing attack by authenticating the MAC layer source address.

4. **Sequential Freshness:** This service adds an ordered sequence number to each MAC frame to prevent replay attack. The sequence number which indicates the freshness of the frame is of the size of five bytes. The receiver checks the frame counter and rejects it if the value is equal or less than the last received frame sequence counter.

6.3.2 Security Modes

IEEE 802.15.4 provides three modes of security [4][36]:

1. **No Security:** This is used in those cases where security is not important or the network is operating in a highly reliable and secured area. Alternately, the upper layers in ZigBee may provide sufficient security and protection and thus the security in 802.15.4 can be ignored. The MAC layer does not provide any security services and the frames are transmitted in clear text without any consideration for data privacy, integrity check or access control.

2. **Access Control List:** The access control list is created in each device and it prevents the unauthorized devices from accessing the network resources and data. This limits the frame reception from only those devices which are in the ACL and thus provide limited security services as the cryptographic protection is not available in this mode. The ACL entry format is comprised of the address, security suite, key, last initial vector (IV) and the replay counter.

However, only the address field in the ACL is used in this mode and the others remain unused.

3. **Symmetric Key Security:** AES with 128 bit encryption is used to encrypt data before transmitting it on the network. It also uses an integrity code (IC) to prevent unauthorized access to data by an impersonating node. In this mode, the device can have any of the four MAC security services enabled depending on the criticality of the application.

6.3.3 Security Suite

A security suite applies the security mechanism defined for the MAC frame. The suite is comprised of the symmetric cryptographic algorithm, the mode of its operation and the key length and these features also decide the name of the security suite which is comprised of these three parts. The security suite generally uses AES as the cryptographic algorithm and the key length may be 32, 64 or 128 bit. The security suite may cater to one or more of the security objectives of access control, message encryption, frame integrity, and sequential freshness, which is represented diagrammatically in Figure 6.4. The Counter (CTR) Mode, Cypher Block chaining with Message Authentication (CBC-MAC) and CTR combined with CBC-MAC (CCM) are the three modes of operation of the security suite [36][37].

6.3.3.1 CTR Mode

The CTR mode provides access control, data encryption and sequential freshness of the frame. In CTR mode, a key stream is generated using a block cipher with a given key and the counter. ZigBee generally uses AES as the block cipher. The output blocks are XORed with plain text and integrity code to produce the cipher text. The counter can be a time stamp or a special marker. The counters must be different in all the encrypted messages with a single key.

6.3.3.2 CBC-MAC Mode

The CBC-MAC does not provide any encryption but is used for integrity check of the data frame. It authenticates the data by checking if the MIC computed by the sender matches with the one generated by the receiver. The integrity code is generated by the block cipher with input as the data frame and the key. The MIC is the leftmost M bits of the output vector where M=8h, where h is an integer and the value of M lies between 32 and 128. AES-CBC-MAC comes with three MIC bits option 32, 64 and 128.

6.3.4 CCM Mode

CCM provides authentication as well as secrecy by using encryption. There are three inputs to the CCM block cipher: the data payload, the associated data like the header

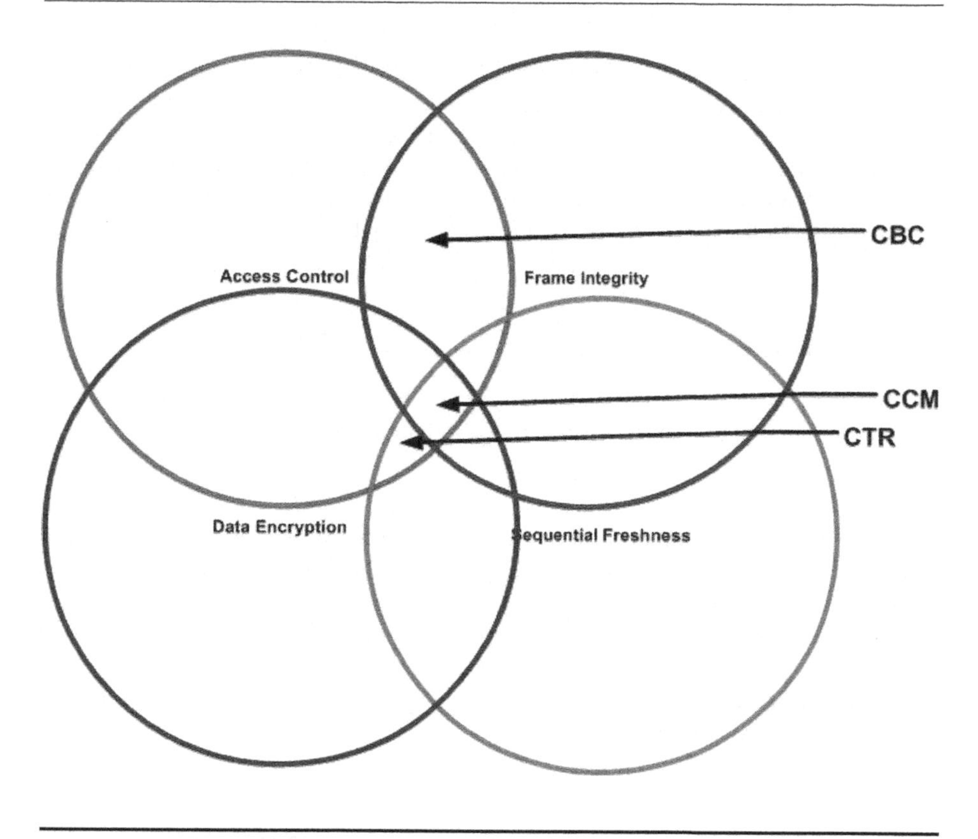

Figure 6.4: A diagrammatic representation of the security services available with security suites.

and the nonce. The payload is encrypted as well as authenticated, the header is only authenticated and the nonce is used for frame freshness. CCM uses CTR for encrypting the data to provide secrecy and it uses CBC-MAC for authentication. The sender first generates the MIC and then encrypts the data as well as the MIC while the receiver first deciphers the cipher text to get the data and the MIC generated by the sender and then recomputes its own MIC and verifies it with the sender's MIC.

6.4 ZigBee Vulnerabilities

Research advancements on ZigBee have progressed notably, gradually perfecting itself towards the goal of providing a cost effective and limited energy, computation, memory and communication capacity services. In a ZigBee, the nodes are distributed, mobile and interconnected through wireless interface. They are self-organizing in nature, as they lack centralized routing, server and administrative infrastructure. Security issues are discussed with reference to services, attacks and security mechanisms

[38]. The services cater for the secure operation of the network, prevention against attacks and the tools and techniques to support the security services and security mechanism is the process of providing secured services. The basic requirement for a secured networking environment is confidentiality, authentication, integrity, non-repudiation and availability. The security mechanisms used for protecting the transmitted data from attacks are encryption, cryptographic hash and digital signatures.

6.4.1 Vulnerability Attacks

A few security vulnerabilities were exposed in IEEE 802.15.4, some of which have been described in [34][36][39] and the major among them are:

Same-nonce attack:

ACL entry format is comprised of address, security suite, key, last IV and replay counter. In the sender's ACL table, if it contains an entry with the same key and nonce, it exposes the device to a security attack. The same nonce can occur due to situations like power failure, snooze mode, etc. If the ACL produces the same nonce and the same key for two different messages, the intruder may be able to regenerate partial information as the XOR of the plain text will be the same as the XOR of the corresponding cipher text.

DoS attack:

The sequential freshness is obtained by preventing message replay in IEEE 802.25.4. The receiver rejects those frames in which the counter value is equal to less than the last received counter value. As shown in Figure 6.5, this replay protection mechanism can lead to Denial of Service attack as the attacker sends a number of frames containing large but different counter values. The receiver on receipt of these frames sets the counter of the last received frame to these exceptionally large values and thus when it receives legitimate frames with a reasonably sized counter value, the frame is discarded by the receiver for the purpose of replay protection and thus leading to denial of service.

ACK attack:

A sender can optionally request for an ACK frame from the receiver. The ACK frames do not have an integrity check. As illustrated in Figure 6.6, an adversary can send a noise to the receiver at the time a legitimate sender is sending a frame to the receiver. This leads to collisions and rejection of frame. But at the same time the adversary sends a forged ACK frame to the sender. Thus the receiver did not receive the message due to interference and the sender did not retransmit after receiving the forged ACK frame and thus the message is not transmitted to the intended destination.

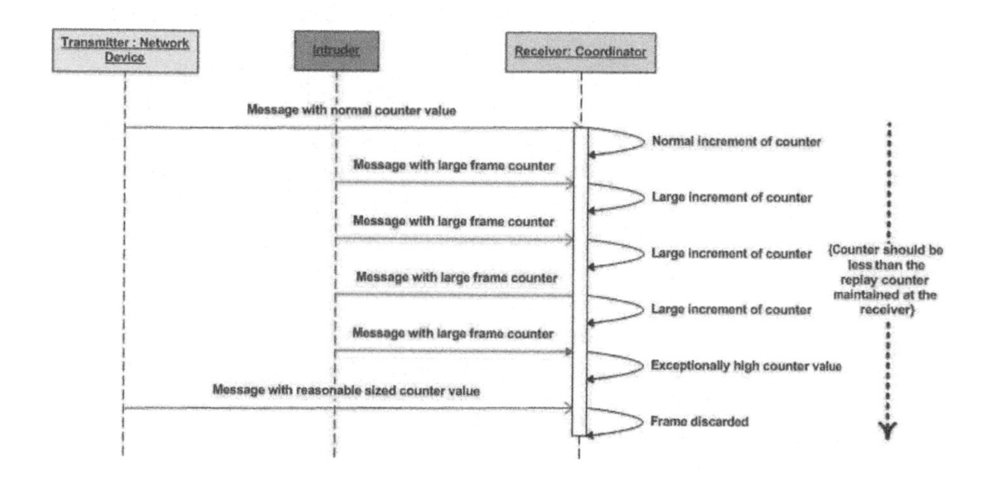

Figure 6.5: A sequential representation of a DoS attack.

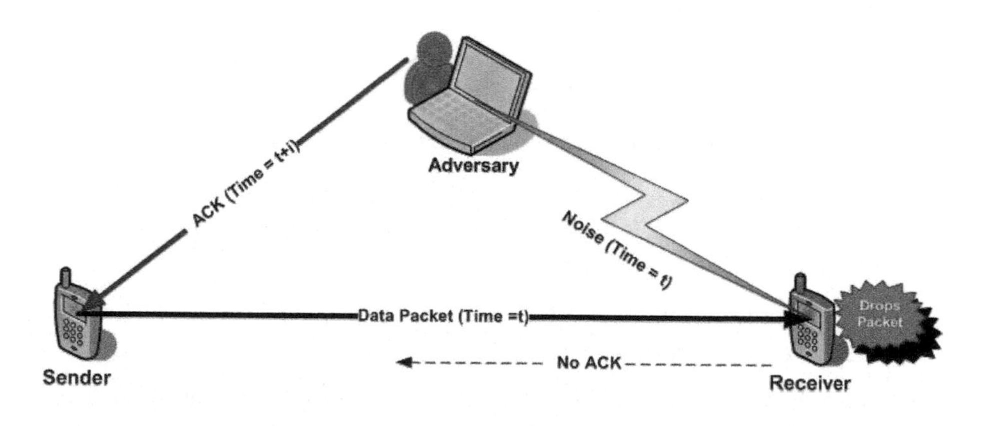

Figure 6.6: Illustration on an ACK attack.

6.4.2 Countermeasures

K. Masica in [9] has elaborated on seven design principles for developing secured ZigBee architecture. The suggested design principles call for applying defense-in depth, analyzing and hardening all components of a system, segmentation of ZigBee network from other networks, restrict traffic in and out of the ZigBee network, enabling 802.15.4 defined security features at the lower layers of the stack, enabling ZigBee defined security features at the higher layers of the stack and development of a security architecture based on maximum protection of the ZTC. These design principles are based on the assumption that the ZigBee network might be integrated with an existing enterprise environment which may have a wired or wireless network connected to the ZigBee network through a gateway. There it suggested segregating the ZigBee network from the wired network through some firewall, bastion host or security gateway. It suggests filtering of the traffic by source and destination address and service port at the gateway level before it enters the ZigBee network. Y. Xiao et al. in reference to [36] have proposed 8 security enhancement techniques in ZigBee to prevent vulnerabilities and attacks. It has suggested the following

- Separating of nonce from frame counter so that both the fields are used. The security is enhanced with this at the cost of introducing a new field.
- As the nonce is separated from the frame counter; the nonce should be generated by a random number generator and not incremented sequentially.
- As most of the attacks are related to frame counter, it would be safer if the time stamp is used for sequential freshness after synchronization of clock with beacon frame.
- Append MIC with ACK frame to ensure authenticity.
- Dividing nonce space in multiple groups to prevent multiple same key entries in the ACL to avoid same nonce attack.
- Eliminate the key sequence counter to save a byte.
- A separate frame counter for each device to prevent replay attack. The disadvantage of this is the increased memory requirement and failure of the sheme on power failure of device restart.
- Supports the usage of CCM* mode as suggested in [39] in which an ACL entry is created in the MAC PIB.

M. Blaser in his paper [40] mentions that security is one of the most important factors that should be analyzed before using ZigBee in industrial and commercial environment. The author advocates usage of Public Key cryptography for ZigBee networks in addition to using the proven cryptographic algorithms like AES and elliptic curve cryptography (ECC). It suggests that the key exchange protocols should be robust enough so as not to expose the keys, the protocols should be designed so as to keep the network always up and networking and increase reliability by using a heartbeat function and management polling. The enrollment and trust establishment of the new devices should be faster after two way identification. The network should be segmented based on the role and reliability of the devices and the security protocols should be able to measure the health of the network and the devices. Tan et al. [41] has presented a lightweight identity based encryption and the associated protocols for it for securing a body sensor network. As the protocol has low computational overhead and data transmission requirement and works in a personal area network it fits well in ZigBee. It allows user based access to data stored in the devices. After the initial setup phase, a public key is generated from an arbitrary string and the secret key is derived from the trusted center. It caters to the security requirements like protect the data and ensure its privacy in the storage area, tolerate compromised nodes, prevent unauthorized access to information and flexibility in granting permission by generating keys on the fly without the intervention of the CA. A few key management problems have been explained by Sastry and Wagner in [34]. The ACL table does not support many keying models properly and thus ZigBee has no support for group keying. The authors also explain that the single network-wide shared key does not protect from replay protection and the pair wise keying supported by ZigBee can be further strengthened by specifying the minimum number of entries that should be available in an ACL.

6.5 Future Directions for Research

There has been limited research on security schemes for ZigBee with consideration for asymmetric keys, PKI, channel capacity and utilization. Various schemes are generally compared based on the CPU utilization, bandwidth requirement and power consumption. However, in the case of ZigBee the communication capacity as well as the energy consumption is turning out to be the bottleneck, rather than the computational power. The number of messages can be equally hampering the performance of the key management algorithm as the size of the message. The future key management schemes should optimally combine the features of bandwidth efficiency, robustness against link failures and power consumption [42]. Most of the recommended key management solutions described in the research papers are theoretical in nature and their efficiency has been proved mathematically or by simulations. Emphasis must be given in implementing these solutions, as there might be various limitations in the real world in implementing these solutions. The storage space in the node of a particular ZED may not be even enough to store a key [43]. Research in the field of enhancing battery life, compact batteries with more power, processors with high computation speed and low energy consumption is also demanding. In addition to detection of malicious nodes, it would be highly beneficial to track the malicious node, trace its communication pattern and ability to detect it on just joining the network. There may be situations when research is initiated to detect malicious node based just on its presence in a ZigBee network of communication even if the malicious node is dormant and not communicating at all. Key management algorithms should evolve in such a way so that a node leaving a network should not be able to carry a key with itself outside the network, get captured or intentionally pass the key to an adversary which in turn joins back the ZigBee network impersonating the friendly node. Process should be evolved to generate a key on the fly based on certain environmental, identity and network parameters at a particular time when a node joins a network and its destruction in case a node leaves the network. Incorporating biometric techniques to encrypt data on a device which is already constrained of memory and computational power is a challenging task. Future research should especially seek techniques for gaining deployment knowledge from transmission pattern as it will support predicting the topological changes and model the behavior of association and dissociation of the devices. Research in the field of hardware as well as algorithms is also required to make the devices tamperproof with the least overhead. As the network architecture is distributed and dynamic, leading to routing and security issues, further decrease in the bootstrapping time required for the network will be a significant contribution to ZigBee in its field of applications. A scheme should also evolve which provides assured node identification.

6.5.1 Conclusion

There is elaborate security architecture and a comprehensive trust management model described by ZigBee standard. The protocol stack features cryptographic frame encryption, authentication and integrity at each layer. The network also has a ZigBee Trust Center which performs the role of key management, node configuration and network management. These security features provide the platform for the application developers to create applications which embed security policies in the ZigBee products and services for a safe deployment. The ZigBee security standard also assists in evaluating the security of a ZigBee based solution for installation in an industrial, home, hospital or disaster management environment. It helps to plan and choose the optimal security based solution for the enterprise network. Users should decide the kind of network usage, specifically the distributed monitoring and control applications that are suited for ZigBee. Security, reliability, privacy and performance are the major challenges for a ZigBee network which employs wireless transmission using RF. Factors like interference, frequency jamming, multipath fading and denial of service attacks by flooding with crafted packets [44][45] are some of the major security issues which can be addressed by selecting suitable topology, using guaranteed time slots and implementation of security measure in the upper layers of the protocol stack.

References

[1] S.C. Ergen, ZigBee/IEEE 802.15.4 Summary, 2004, http://www.sinemergen.com/zigbee.pdf, last accessed on September 23, 2009.

[2] IEEE 802.15.4 WPAN-LR Task Group, www.ieee802.org/15/pub/TG4.html,

last accessed on September 23, 2009.

[3] ZigBee Alliance, www.ZigBee.org, last accessed on September 23, 2009

[4] M. Othman, Principles of mobile computing and communication, Auerbach Publications, New York, 2007, pp.83-106.

[5] LAN-MAN Standards Committee of the IEEE Computer Society, Wireless Medium Access Control (MAC) and Physical Layer (PHY) Specifications for Low-Rate Wireless Personal Area Networks (LR-WPANs), IEEE, 2003

[6] S. Safaric and K. Malaric, ZigBee wireless standard, Proceedings of International Symposium ELMAR-2006, Zadar, Croatia, June 2006, pp. 259-262.

[7] P. Kinney, ZigBee technology: wireless technology that simple works, Communications Design Conference, 2003, http://www.zigbee.org/imwp/idms/popups/pop_download.asp? contentID=5162, last accessed on September 23, 2009

[8] IEEE P802.15 Working Group for WPANs, Cluster Tree Network, April 2001

[9] K.Masica, Recommended practices guide for securing ZigBee wireless networks in process control system environments, California, EUA: 2007, http://csrp.inl.gov/Documents/SecuringZigBeeWirelessNetworksinProcess, last accessed on September 23, 2009

[10] A. Fang, X. Xu, W. Yang and L. Zhang, The realization of intelligent home by ZigBee wireless network technology, Pacific-Asia Conference on Circuits, Communications and Systems (PACCS), Chengdu, China, 2009, pp.81-84.

[11] M.C. Huang, J.C. Huang, J.C. You and G.J. Jong, The wireless sensor network for home-care system using ZigBee, Third International Conference on International Information Hiding and Multimedia Signal Processing (IIH-MSP 2007), 2007, vol. 1, pp.643-646.

[12] E. Callaway, P. Gorday and L. Hester, Home networking with IEEE 802.15.4: a developing standard for low-rate wireless personal area networks, IEEE Communications Magazine, August 2002, vol. 40, no.8, pp.70-77.

[13] K. Lorincz, D.J. Malan, T.R.F. Fulford Jones, A. Nawoj, A. Clavel, V. Shnayder, G. Mainland, M. Welsh, and S. Moulton, Sensor networks for emergency response: challenges and opportunities, IEEE Pervasive Computing, vol. 3, no. 4, 2004, pp. 16-23,

[14] V. Kumar, D. Rus and S. Singh, Robot and sensor networks for first responders, IEEE Pervasive Computing, vol. 3, no.4, 2004, pp.24-33

[15] M.G. Rahman and H. Imai, Security in wireless communications, Wireless Personal Communications, vol. 22, no. 2, 2002, pp.213-228

[16] D.H. Axner, The up side and down side of wireless mesh networks, Business Communications Review, 2006, vol. 36, no. 1, pp.5

[17] M. Jelena, S. Shafi, B.M. Vojislav, Performance of a beacon enabled IEEE 802.15.4 cluster with downlink and uplink traffic, IEEE Transactions on Parallel and Distributed Systems, vol. 17, no. 4, pp. 361-376, Apr. 2006

[18] P. Krishnamurthy, D. Tipper, and Y. Qian, The interaction of security and survivability in hybrid wireless networks, Proceedings of IEEE IPCCC 2004, Phoenix, AZ, April 2004.

[19] S. Misra, K. I. Abraham, M. S. Obaidat and P. V. Krishna, LAID: a learning automata-based scheme for intrusion detection in wireless sensor networks, Security and Communication Networks (Wiley), Vol. 2, No. 2, 2009, pp. 105-115.

[20] S. Misra, S. K. Dhurandher, A. Rayankula and D. Agrawal, Using honeynodes for defense against jamming attacks in wireless infrastructure-based networks, Computers and Electrical Engineering (Elsevier). (In Press).

[21] S. Sarkar, B. Kisku, S. Misra and M. S. Obaidat, chinese remainder theorem-based RSA-threshold cryptography in mobile ad hoc networks using verifiable secret sharing, Proceedings of the 5th IEEE International Conference on Wireless and Mobile Computing, Networking and Communications (WiMob'09), Marrakech, Morocco, October 12-14, 2009.

[22] S. K. Dhurandher, S. Misra, M. S. Obaidat and N. Gupta, An ant colony optimization approach for reputation and quality-of-service-based security in wireless sensor networks, Security and Communication Networks (Wiley), Vol. 2, No. 2, 2009, pp. 215-224.

[23] S. K. Dhurandher, S. Misra, S. Ahlawat, N. Gupta and N. Gupta, E2-SCAN: an extended credit strategy-based energy-efficient security scheme in wireless ad hoc networks, IET Communications, U.K., May 2009, pp. 808-819.

[24] R. Chandrasekar, M. S. Obaidat, S. Misra and F. Pea-Mora, A secure data-centric scheme for group-based routing in heterogeneous ad-hoc sensor networks and its simulation analysis, SIMULATION: Transactions of the Society for Modeling and Simulation International, Vol. 84, No. 2/3, 2008, pp. 131-146.

[25] P. Narula, S. K. Dhurandher, S. Misra and I. Woungang, Security in mobile ad-hoc networks using soft encryption and trust-based multi-path routing, Computer Communications (Elsevier), Vol. 31, No. 4, 2008, pp. 760-769.

[26] S. Misra, S. Roy, M. S. Obaidat and D. Mohanta, A fuzzy logic-based energy efficient packet loss preventive routing protocol, Proceedings of the 2009 International Symposium on Performance Evaluation of Computer and Telecommunication Systems (SPECTS 2009), Istanbul, Turkey, July 13-16, 2009, pp. 185-192.

[27] S. Misra, A. Bagchi, R. Bhatt, S. Ghosh and S. Obaidat, Attack graph generation with infused fuzzy clustering, Proceedings of the International Conference on Security and Cryptology, Part of the International Joint Conference on e-Business and Telecommunications (ICETE 2009), Milan, Italy, July 7-10, 2009, pp. 92-98.

[28] S. Yi and R. Kravets, Composite key management in ad hoc networks, Proceedings of First Annual International Conference on Mobile and Ubiquitous Systems: Networking and Services (MobiQuitous 04), Urbana, USA, August 2004, pp. 52-61.

[29] S. Capkun, J.-P. Hubaux and L. Buttyan, Mobility helps security in ad hoc networks, Proceedings of the 4th ACM international symposium on mobile ad hoc networking and computing, Annapolis, Maryland, USA, 2003, pp. 46-56.

[30] G.F. Marias, K. Papapanagiotou, V. Tsetsos, O. Sekkas and P. Georgiadis, Integrating a trust framework with a distributed certificate validation scheme for MANETs, EURASIP Journal on Wireless Communications and Networking, 2006, Vol 2, pp. 1-18.

[31] W.C. Craig, ZigBee: wireless control that simply works, http://www.zigbee.org/imwp/idms/popups/pop_download.asp?contentID=5438, last accessed on September 23, 2009

[32] D.Liu, P.Ning, and R.Li, Establishing pairwise keys in distributed sensor networks, *ACM Trans. Inf. Syst. Secur.*, February 2005, vol.8, no.1, pp. 41-77,

[33] W. Du, J. Deng, Y.S. Han, P.K. Varshney, A pairwise key pre-distribution scheme for wireless sensor networks, Proceedings of the 10th ACM conference on Computer and Communications Security, October 27-30, 2003, Washington D.C., USA

[34] N. Sastry and D. Wagner, Security considerations for IEEE 802.15.4 networks, Proceedings of the 3rd ACM workshop on Wireless security, October 01-01, 2004, Philadelphia, PA, USA, pp.32-42.

[35] L.Eschenauer and V.D. Gligor, A key-management scheme for distributed sensor networks, Proceedings of the 9th ACM conference on Computer and communications security. New York, NY, USA: ACM, 2002, pp. 41-47

[36] Y. Xiao, H. Chen, B. Sun, R.Wang, and S. Sethi, MAC security and security overhead analysis in the IEEE 802.15.4 wireless sensor networks. EURASIP Journal of Wireless Communication And Networking 2006, vol. 2, pp. 81-92.

[37] M. Khan and J. Misc, Security in IEEE 802.15.4 cluster based networks, Security in Wireless Mesh Networks edited by Y. Zhang, J. Zheng and H. Hu, CRC Press

[38] J. Dong, K. Ackermann, and C. Nita-Rotaru, Secure Group Communication in Wireless Mesh Networks, Proceedings of Ninth IEEE International Symposium on a World of Wireless, Mobile and Multimedia Networks (WOWMOM), Newport Beach, CA, 2008, pp. 1-8.

[39] R. Struik, Security Resolutions 802.15.4, http://www.ieee802.org/15/pub/04/15-04-0540-07-004b-security-motions.ppt, last accessed on September 23, 2009.

[40] M. Blaser, Industrial-strength security for ZigBee: The case for public-key cryptography, Embedded Computing Design, May 2005, http://www.embedded-computing.com/articles/id/?210, last accessed on September 23, 2009

[41] C.C. Tan, H.Wang, S.Zhong, and Q.Li, Body sensor network security: an identity-based cryptography approach, Proceedings of the first ACM conference on Wireless network security, New York, NY, USA: ACM, 2008, pp. 148-153.

[42] M. Eltoweissy, M. Moharrum and R. Mukkamala, Dynamic key management in sensor networks, IEEE Communications Magazine, Vol. 44 (4), April 2006, pp. 122-130.

[43] A. Perrig, R. Szewczyk, V. Wen, D. Culler and J. D. Tygar, SPINS: security protocols for sensor networks, Wireless Networks, Vol. 8 (5), 2002, pp. 521-534.

[44] A.J. Gutierrez, E. Callaway and R. Barrett, Low-rate wireless personal area networks: enabling wireless sensors with IEEE 802.15.4, IEEE Press, 2004. ISN 0-7381-3557-7.

[45] R. Poor, Information on the IEEE 802.15.4-2006 revised standard, http://grouper.ieee.org/groups/802/15/pub/TG4b.html, last accessed on September 23, 2009

[46] S. Ravi, A. Raghunathan and N. Potlapally, Securing wireless data: system architecture challenges, Proceedings of the 15th International Symposium on System Synthesis (ISSS), Kyoto, Japan, 2002, pp. 195-200.

Chapter 7

Coexistence Issues between ZigBee and Other Networks

Dan Keun Sung, Jo Woon Chong, Su Min Kim

CONTENTS

As a ubiquitous communication era is emerging, various types of communication protocols are expected to support diverse services with different requirements. In particular, wireless local area networks (WLANs) like WiFi and wireless personal area networks (WPANs) such as Bluetooth and ZigBee networks may coexist in the same frequency band. However, since there is no central interference coordinator, they may interfere with each other. Especially, since ZigBee devices have relatively lower power emission, compared with other network protocol devices, they may suffer from severe interference, resulting in significant performance degradation. Therefore, we need to solve underlying coexistence problems between ZigBee and other networks. This chapter consists of two main parts. In the first part, the performance analysis of ZigBee networks is described in the presence of interference from heterogeneous communication systems. The performance of ZigBee networks is investigated based on both measurements in a real testbed environment and mathematical analysis using a Markov chain concept. In the second part, two types of interference avoidance/mitigation algorithms are introduced in order to enhance the performance of ZigBee networks in overlaid networks environments. A cognitive radio concept as interference mitigation technology is also discussed.

7.1 Introduction

WLAN, Bluetooth, and ZigBee are widely adopted wireless technologies in current 2.4 GHz industrial, scientific and medical (ISM) band. WLAN technology is deployed in many indoor or outdoor places whereas Bluetooth technology is used for personal wireless headsets or hands-free sets, and wireless peripherals. Since WLAN, Bluetooth, and ZigBee networks operate in the 2.4 GHz ISM band, if they are co-located, then they may interfere with one another. For example, a laptop for Web browsing via WLAN and sensor temperature information gathering through ZigBee devices may operate in the same indoor place. In this environment, the interference among the heterogeneous networks may be severe and the effect of this interference on the system performance is of interest. In addition, some coexistence algorithms among WLAN, Bluetooth, and ZigBee are needed to solve the coexistence problem.

IEEE 802.15.2 working group (WG) proposed coexistence algorithms for WLAN and Bluetooth systems [1]. The coexistence problems between WLAN and ZigBee networks in the 2.4GHz ISM band have been studied. Mutual interference among WLAN, Bluetooth, ZigBee, and microwave oven have been measured in [2, 3, 4, 5]. The interference among ZigBee, Bluetooth and WLAN devices have been analyzed in [6, 7, 8, 9, 10].

7.2 Performance Analysis of ZigBee Networks in the Presence of Interference

As the number of wireless standards increases and they are widely deployed, the unlicensed 2.4 GHz ISM band will be overcrowded more and more. Therefore, the performance of ZigBee networks in the presence of interferers such as WLAN and Bluetooth should be evaluated. In this section, the performance of ZigBee networks in the presence of interference is analyzed in the following two points of view: (1) empirical measurements and (2) mathematical modeling and simulation.

7.2.1 *Empirical Measurements*

A number of empirical measurements have been taken in order to analyze the effect of interference on the performance of ZigBee networks [2, 3, 5, 11]. In these experiments, various wireless devices based on WLAN, Blueooth, and ZigBee, and microwave oven are considered as potential interferers. In this section, we consider a ZigBee network as a target system and WLAN devices and a microwave oven as interferers. Since Bluetooth hops the 2.4 GHz ISM bands for every transmission slot and uses a frequency band among 79 bands in a time slot, it is difficult to capture its effect [12]. Even though Bluetooth is not considered as a potential interferer in this section, its effect will be analyzed through mathematical modeling and simulation in the next section.

From now on, the performance of a ZigBee network is empirically measured in terms of frame error rate (FER) in the following three scenarios:

- ZigBee in the presence of a WLAN interferer
- ZigBee in the presence of a microwave oven interferer
- ZigBee in the presence of both WLAN and microwave oven interferers

Table 7.1: ZigBee Channel Frequencies in 2.4 GHz ISM Band (MHz)

Channel Number	Lower Frequency	Center Frequency	Upper Frequency
11	2404	2405	2406
12	2409	2410	2411
13	2414	2415	2416
14	2419	2420	2421
15	2424	2425	2426
16	2429	2430	2431
17	2434	2435	2436
18	2439	2440	2441
19	2444	2445	2446
20	2449	2450	2451
21	2454	2455	2456
22	2459	2460	2461
23	2464	2465	2466
24	2469	2470	2471
25	2474	2475	2476
26	2479	2480	2481

Table 7.2: WLAN Channel Frequencies in 2.4 GHz ISM Band (MHz)

Channel Number	Lower Frequency	Center Frequency	Upper Frequency
1	2401	2412	2423
2	2404	2417	2428
3	2411	2422	2433
4	2416	2427	2438
5	2421	2432	2443
6	2426	2437	2448
7	2431	2442	2453
8	2436	2447	2458
9	2441	2452	2463
10	2446	2457	2468
11	2451	2462	2473
12	2456	2467	2478
13	2461	2472	2483

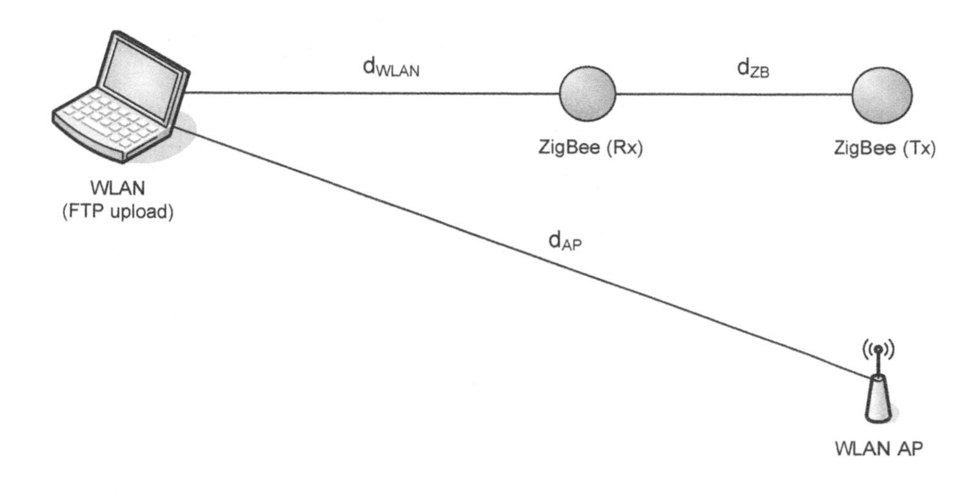

Figure 7.1: Experimental environment of ZigBee devices in the presence of a WLAN interferer.

Tables 7.1 and 7.2 illustrate the channel frequencies of ZigBee and WLAN in the 2.4 GHz ISM band, respectively. For ZigBee based on IEEE 802.15.4 [13], a total of 16 channels (numbered 11 to 26) are available in the 2.4 GHz ISM band. Each channel has a bandwidth of 2 MHz and a channel separation of 5 MHz. On the other hand, the IEEE 802.11b/g WLAN [14, 15] operates in a total of 13 channels (numbered 1 to 13) which is available in the 2.4 GHz ISM band. Each channel has a bandwidth of 22 MHz and a channel separation of 5 MHz. Therefore, only three WLAN channels can be exclusively used in this frequency band without overlapped spectra.

7.2.1.1 ZigBee in the Presence of a WLAN Interferer

Fig. 7.1 shows an experimental environment for ZigBee in the presence of a WLAN interferer. This experiment is performed in an anechoic chamber which shields the unnecessary electromagnetic interference generated from other devices.

In this experiment, d_{ZB} is set to 100 (cm) and d_{WLAN} is a variable. While FTP traffic is uploaded in the WLAN Ch. 9 with center frequency 2.452 GHz, a ZigBee transmitter sends data with constant bit rate (CBR). The operating carrier frequencies of the ZigBee network are set to Ch. 21, Ch. 22, and Ch. 23. Since the center frequency of Ch. 21 for ZigBee is 2.455 GHz, its bandwidth is completely overlapped with the WLAN operation bandwidth. On the other hand, Ch. 22 is located on the boundary of the WLAN bandwidth and Ch. 23 is located outside the WLAN bandwidth.

Fig. 7.2 shows the FER performance of the ZigBee operating at Ch. 21, Ch. 22, and Ch. 23 in the presence of a fixed WLAN operation channel (i.e., Ch. 9). Both ZigBee Ch. 21 and Ch. 22 suffer from FER values of approximately 0.28. On the

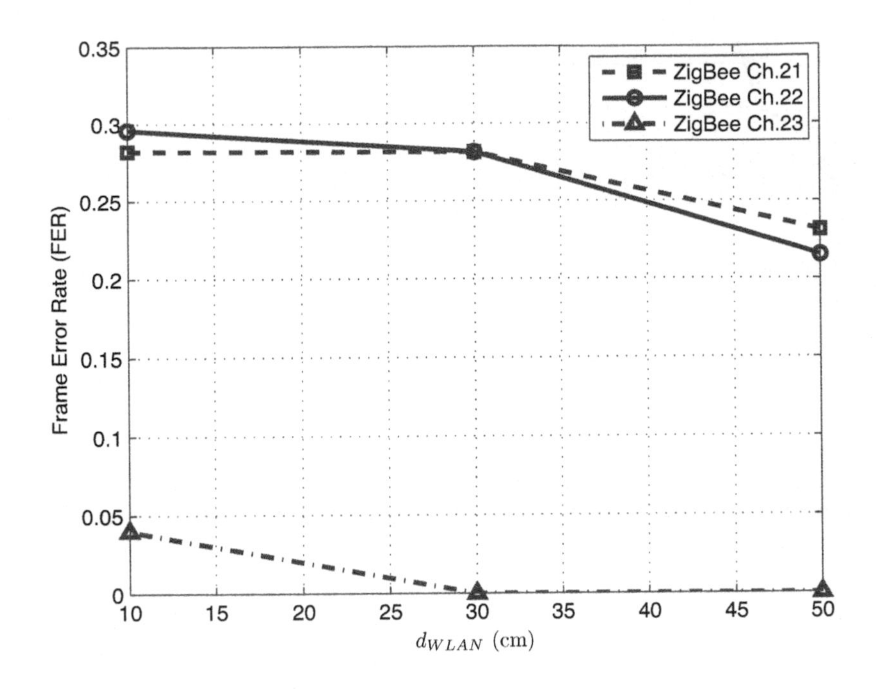

Figure 7.2: FER of ZigBee for varying the value of d_{WLAN} in the presence of a WLAN interferer.

other hand, the FER value is approximately zero regardless of varying distances of d_{WLAN} at the non-overlapped channel (i.e., Ch. 23). This result implies that the key factor of the mutual interference is due to the overlapped region of their frequency bandwidths.

7.2.1.2 ZigBee in the Presence of a Microwave Oven Interferer

A microwave oven is one of the dominant interferers in the 2.4 GHz ISM band. It uses a magnetron tube to generate microwave energy, which in turn heats food. Ideally, the magnetron generates continuous waves centered at 2.45GHz, exactly in the middle of the 2.4 GHz ISM band. In practice, the power spectrum of the microwave oven varies in frequency and exhibits side-bands (or multiple interfering tones) at other frequencies in the 2.4 GHz ISM band. Hence, even though the microwave oven operates mainly in the 2.45 GHz, the operation frequency band varies with time in the 2.4 GHz band.

The experimental environment of ZigBee in the presence of a microwave oven is shown in Fig. 7.3. In this experiment, the effect of interference from the microwave oven to ZigBee is measured in terms of FER for varying the ZigBee channels from Ch. 16 to Ch. 24. The center frequency of ZigBee Ch. 20 is located at 2.45 GHz

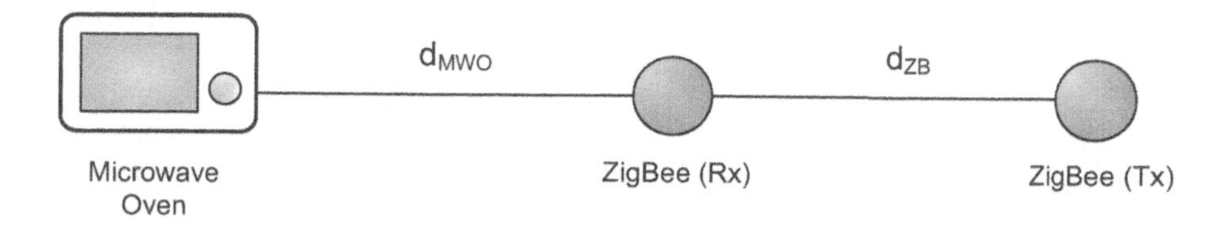

Figure 7.3: Experimental environment of ZigBee devices in the presence of a microwave oven interferer.

which is equivalent to the center frequency of the microwave oven. Both d_{ZB} and d_{MWO} are set to 100 (cm).

Fig. 7.4 shows the measured FER of ZigBee for 9 different channels in the presence of a microwave oven interferer. The result shows double peaks at Ch. 19 and Ch. 22 near 2.45 GHz. This is because the center frequency of the microwave oven is not fixed at 2.45 GHz and varies with time. We can also observe that the FER of ZigBee is not negligible even at the distant channels from 2.45 GHz, e.g., Ch. 16 and

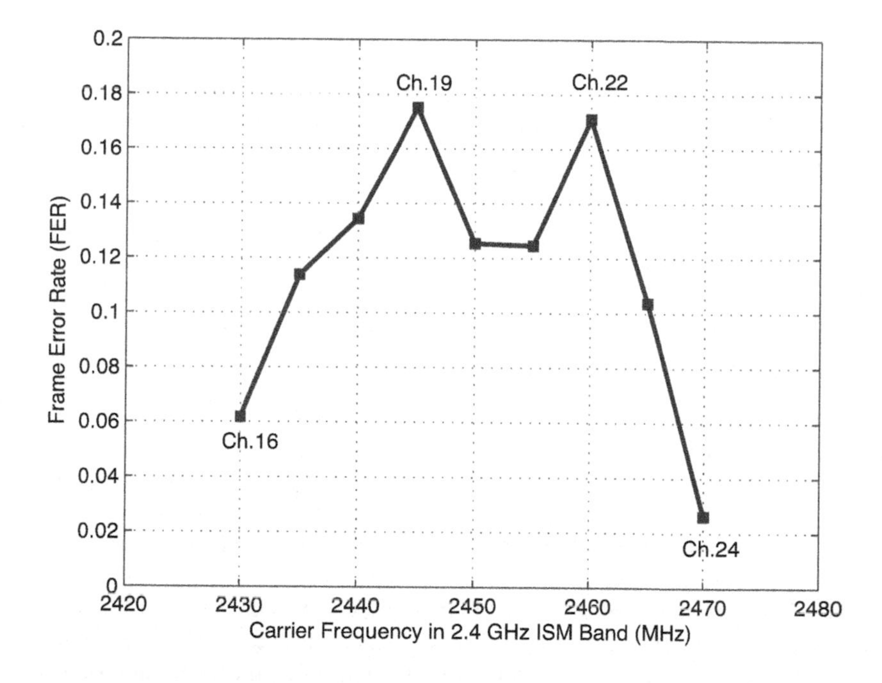

Figure 7.4: FER of ZigBee for varying the ZigBee channels in the presence of a microwave oven interferer.

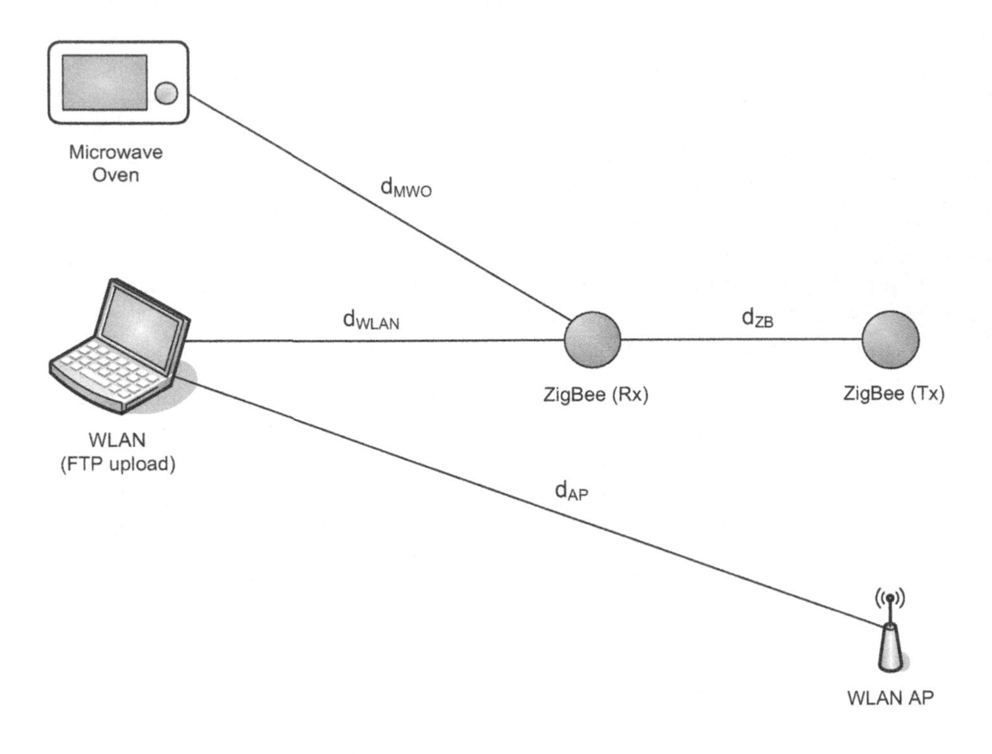

Figure 7.5: Experimental environment of ZigBee in the presence of both WLAN and microwave oven interferers.

Ch. 24. This result implies that the microwave oven may interfere with the ZigBee in a wide range of ZigBee channels near 2.45 GHz.

7.2.1.3 ZigBee in the Presence of both WLAN and Microwave Oven Interferers

From the above experiments, it can be predicted that the mutual interference in the presence of both WLAN and microwave oven interferers is much severer. Fig. 7.5 shows an experimental environment where ZigBee and WLAN operate simultaneously with a microwave oven in the 2.4 GHz ISM band. In the measurement, the center frequencies of WLAN and ZigBee operating channels are set to 2.452 GHz and Ch. 21-23, respectively.

Fig. 7.6 shows the FER of ZigBee in the presence of both WLAN and microwave oven interferers. For Ch. 21 and Ch. 22, the FER values are much higher than those of the cases of the previous two subsections. Especially, the FER values are much higher than 0.1 for Ch. 23, while those of the cases in the previous subsections are smaller than 0.1. It implies that the effect of interference is significant when ZigBee devices are surrounded by strong interferers such as WLAN and microwave oven, compared with the cases of mutual interference between WLAN and ZigBee

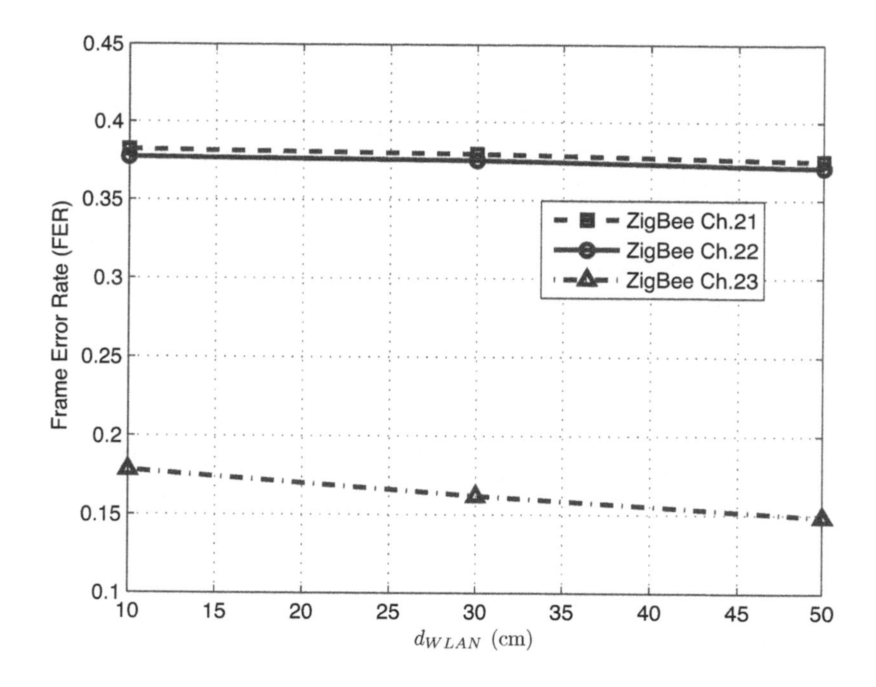

Figure 7.6: FER of ZigBee for varying the value of d_{WLAN} in the presence of both WLAN and microwave oven interferers.

and between microwave oven and ZigBee. Therefore, heterogeneous communication systems or electro-magnetic devices such as a microwave oven are able to severely interfere with ZigBee if there is no central coordination for their coexistence in the unlicensed ISM band.

7.2.2 Mathematical Modeling and Simulation

Fig. 7.7 shows a ZigBee network in the presence of Bluetooth and WLAN interference. Since the transmission power level of ZigBee devices is set tens or hundreds times lower than that of Bluetooth and WLAN devices, it is assumed that Bluetooth and WLAN devices interfere with ZigBee devices, while ZigBee devices do not. Moreover, active Bluetooth or WLAN communication pairs are assumed to interfere with all the ZigBee devices in a ZigBee piconet, as shown in Fig. 7.7.

Fig. 7.8 shows a Markov chain model for the carrier sensing multiple access with collision avoidance (CSMA/CA) operation of a ZigBee device in the presence of Bluetooth or WLAN interference. If a ZigBee device has a backoff stage value of i and a backoff counter value of j, its state is represented as (i, j). If the backoff counter value j of state (i, j) decreases to 0, the ZigBee device performs the first

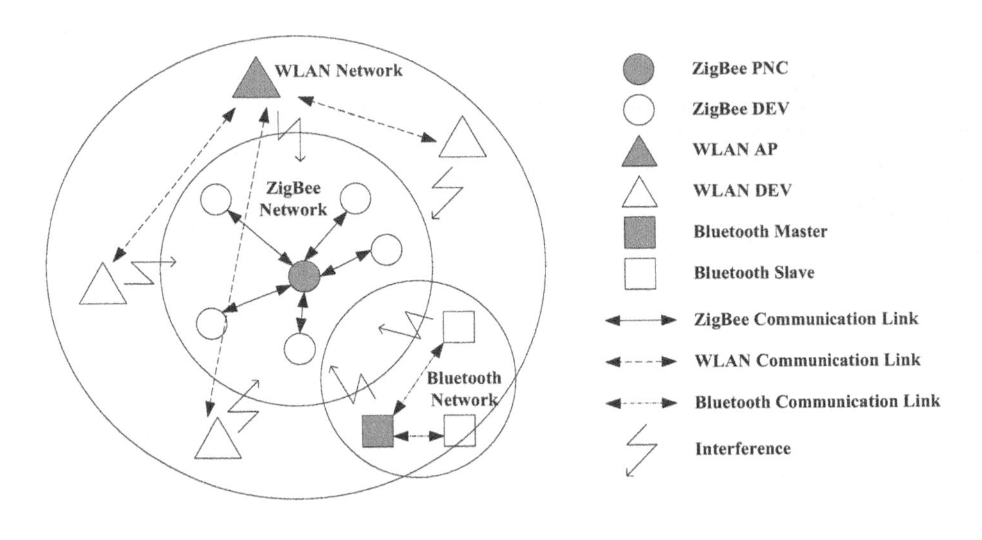

Figure 7.7: A ZigBee network in the presence of bluetooth and WLAN interference.

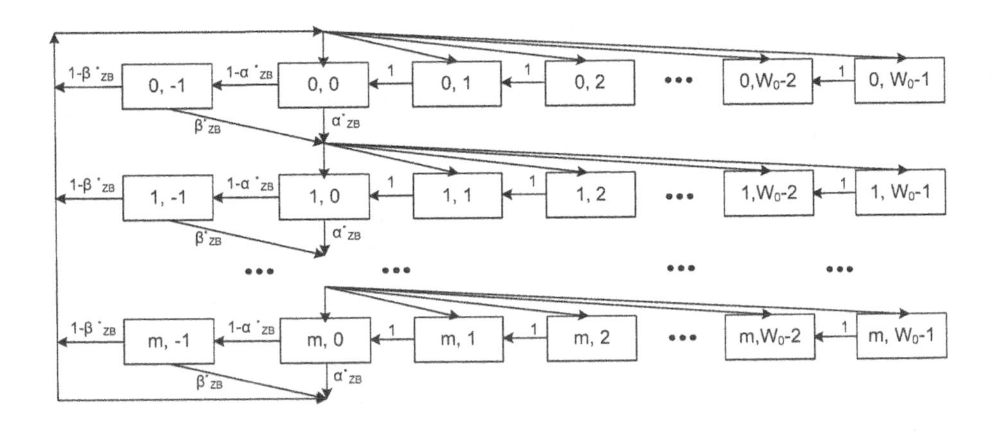

Figure 7.8: A Markov chain model for the operation of ZigBee devices in the presence of bluetooth (or WLAN) interference.

clear channel assessment (CCA). If the ZigBee channel is sensed busy at the first CCA because of transmissions from Bluetooth, WLAN, or other ZigBee devices, then the ZigBee device increases the backoff stage from i to $i+1$ and randomly chooses a new backoff counter value in $[0, W_{i+1}-1]$ where W_{i+1} denotes the backoff window size in the backoff stage $i+1$. On the contrary, if the ZigBee senses the channel idle at the first CCA then it performs the second CCA. In this second CCA, if it senses the channel busy, it performs the same procedure of the first CCA busy case. Otherwise, the ZigBee device transmits a packet.

Assuming that the first and second CCA busy probabilities which are denoted by α_{ZB}^* and β_{ZB}^*, respectively, do not depend on backoff stage values, the transition probabilities in this Markov chain model are expressed as follows [16]:

$$\begin{array}{ll}
P\{b_{i,j}|b_{i,j+1}\} = 1, & i \in (0,m) \quad \textit{and} \quad j \in (0, W_i-2). \\
P\{b_{i,-1}|b_{i,0}\} = 1-\alpha_{ZB}^*, & i \in (0,m). \\
P\{b_{i,j}|b_{i-1,0}\} = \alpha_{ZB}^*/W_i, & i \in (1,m) \quad \textit{and} \quad j \in (0, W_i-1). \\
P\{b_{i,j}|b_{i-1,-1}\} = \beta_{ZB}^*/W_i, & i \in (1,m) \quad \textit{and} \quad j \in (0, W_i-1). \\
P\{b_{0,j}|b_{i,-1}\} = (1-\beta_{ZB}^*)/W_0, & i \in (1,m-1) \ \textit{and} \ j \in (0, W_0-1). \\
P\{b_{0,j}|b_{m,0}\} = \alpha_{ZB}^*/W_0, & j \in (0, W_0-1). \\
P\{b_{0,j}|b_{m,-1}\} = 1/W_0, & j \in (0, W_0-1),
\end{array}$$

where $b_{i,j}$, W_i, and m denote the stationary state probability of state (i,j), the backoff window size in the backoff stage i, and the maximum backoff stage, respectively.

Letting τ_{ZB}^* denote the transition probability from backoff states to the first CCA states, τ_{ZB}^* is derived from the Markov chain model as follows:

$$\begin{aligned}
\tau_{ZB}^* &= \frac{\sum_{i=0}^{m} b_{i,0}}{\sum_{i=0}^{m}\sum_{j=0}^{W_i-1} b_{i,j}} \\
&= \frac{2\{1-(\alpha_{ZB}^*+\beta_{ZB}^*-\alpha_{ZB}^*\beta_{ZB}^*)^{m+1}\}}{\sum_{i=0}^{m}(W_0 2^{min(i,BE_{max}-BE_{min})}+1)(\alpha_{ZB}^*+\beta_{ZB}^*-\alpha_{ZB}^*\beta_{ZB}^*)^i} \\
&\quad \cdot \frac{1}{\{1-(\alpha_{ZB}^*+\beta_{ZB}^*-\alpha_{ZB}^*\beta_{ZB}^*)\}},
\end{aligned} \tag{7.1}$$

where BE_{min} and BE_{max} denote the minimum and maximum values of the backoff exponent, respectively.

In order to model the operation of ZigBee devices in the presence of Bluetooth or WLAN interference, the first and second CCA failure probabilities, α_{ZB}^* and β_{ZB}^* need to be derived considering the effect of Bluetooth or WLAN operation on ZigBee devices.

When ZigBee devices are co-located with Bluetooth (or WLAN) devices, their operation is shown in Fig. 7.9. ZigBee transmission can be affected by interference from simultaneous Bluetooth (or WLAN) transmission (period 4). However, Bluetooth (or WLAN) transmission is not affected by simultaneous ZigBee transmission trial[1].

[1] In this case, ZigBee transmission is deferred after sensing Bluetooth (or WLAN) transmission.

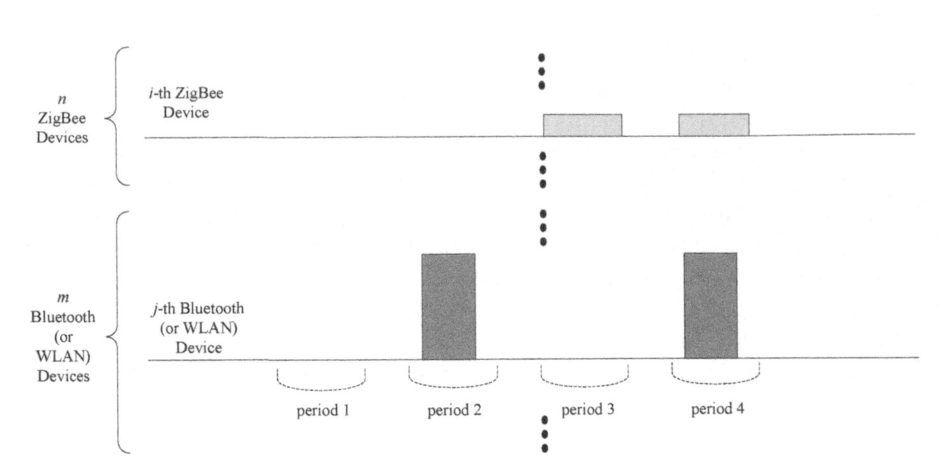

Figure 7.9: Operation of co-located ZigBee and bluetooth (or WLAN) devices.

A given ZigBee device senses the channel busy in the first CCA due to interfering heterogeneous devices (period 2) or other ZigBee (period 3) or simultaneous transmissions of ZigBee and heterogeneous devices (period 4), as shown in Fig. 7.9. Hence, α^*_{ZB} is expressed as a combination of the first CCA busy probabilities of a ZigBee device due to interfering heterogeneous devices and other ZigBee devices as follows:

$$\alpha^*_{ZB} = 1 - (1 - \alpha^{INT}_{ZB})(1 - \alpha^{ZB}_{ZB}), \tag{7.2}$$

where α^{INT}_{ZB} and α^{ZB}_{ZB} denote the first CCA busy probabilities of a given ZigBee device due to only heterogeneous interfering devices[2] and only other ZigBee devices, respectively.

α^{ZB}_{ZB} can be expressed using P^*_{to} and P^*_{so} where P^*_{to} and P^*_{so} denote the transmission and successful transmission probabilities of other ZigBee devices, respectively. The transmission probability P^*_{to}, in terms of a given ZigBee device, is derived as

$$P^*_{to} \approx (1 - \tau^*_{ZB})\{1 - (1 - \tau^*_{ZB})^{n-1}\}(1 - \alpha^{INT}_{ZB})^{\lceil 2\lceil L_{CCA}\rceil \frac{T^{ZB}_u}{T^{INT}_u}\rceil}, \tag{7.3}$$

where $(1 - \tau^*_{ZB})\{1 - (1 - \tau^*_{ZB})^{n-1}\}$ and $(1 - \alpha^{INT}_{ZB})^{\lceil 2\lceil L_{CCA}\rceil \frac{T^{ZB}_u}{T^{INT}_u}\rceil}$ represents the probability that the backoff counter value of at least one ZigBee device (except the given ZigBee device) reaches 0 and two CCAs are successfully performed without heterogeneous network interference with L_{CCA}, T^{ZB}_u, and T^{INT}_u, respectively, representing the length of CCA duration in ZigBee unit backoff slot, length of a ZigBee unit backoff duration in second, and length of a heterogeneous interfering system unit duration

[2]As mentioned, Bluetooth or WLAN devices are considered as heterogeneous network interference devices.

in second. The successful transmission probability P_{so}^* is also derived as

$$P_{so}^* \approx (n-1)\tau_{ZB}^*(1-\tau_{ZB}^*)^{n-1}(1-\alpha_{ZB}^{INT})^{\lceil (2\lceil L_{CCA} \rceil + L_{pkt}) \frac{T_u^{ZB}}{T_{hop}^{INT}} \rceil} / P_{to}^*, \tag{7.4}$$

where $\lceil (2\lceil L_{CCA}^{ZB} \rceil + L_{pkt}^{ZB}) \frac{T_u^{ZB}}{T_u^{INT}} \rceil$ represents the probability that two CCAs are successfully performed and a ZigBee packet with a length of L_{pkt}^{ZB} is successfully transmitted without heterogeneous network interference.

The first CCA busy probability of a given ZigBee device due to only other ZigBee devices can be derived as

$$\alpha_{ZB}^{ZB} = \frac{P_{to}^*\{P_{so}^* L_{bs} + (1-P_{so}^*)L_{bc}\}}{P_{to}^*\{P_{so}^* L_s + (1-P_{so}^*)L_c\} + \tau_{ZB}^* + (1-\tau_{ZB}^* - P_{to}^*)}, \tag{7.5}$$

where L_s and L_c denote the number of backoff slots for successful transmission and collision, respectively. L_{bs} and L_{bc} denote the number of busy slots out of L_s and L_c, respectively. Here, the parameters are set that $L_s = 2\lceil L_{CCA} \rceil + L_{pkt} + \delta + L_{ack}$, $L_c = 2\lceil L_{CCA} \rceil + L_{pkt}$, $L_{bs} = L_{pkt} + L_{ack}$ and $L_{bc} = L_{pkt}$, where L_{ack} and δ represent the acknowledgment (ACK) duration and the ACK wait duration, respectively. Substituting Eqn. (7.5) into Eqn. (7.2), α_{ZB}^* can be derived.

The second CCA busy probability β_{ZB}^* of ZigBee devices in the presence of heterogeneous network devices is expressed as a combination of the second CCA busy probabilities of a ZigBee device due to heterogeneous network devices and other ZigBee devices. Hence, β_{ZB}^* is written as

$$\beta_{ZB}^* = 1 - (1-\beta_{ZB}^{INT})(1-\beta_{ZB}^{ZB}), \tag{7.6}$$

where β_{ZB}^{INT} and β_{ZB}^{ZB} represent the second CCA busy probabilities of a given ZigBee device due to only heterogeneous network devices and only other ZigBee devices, respectively.

Then, β_{ZB}^{ZB} is expressed as

$$\beta_{ZB}^{ZB} = \frac{P_{to}^*\{P_{so}^* L_{is} + (1-P_{so}^*)L_{ic}\}}{P_{to}^*\{P_{so}^* L_s + (1-P_{so}^*)L_c\} + \tau_{ZB}^* + (1-\tau_{ZB}^* - P_{to}^*)} \cdot \frac{1}{1-\alpha_{ZB}^{ZB}}, \tag{7.7}$$

where $L_{is} = 2\lceil L_{CCA} \rceil + \delta$ and $L_{ic} = 2\lceil L_{CCA} \rceil$.

Consequently, the normalized throughput S of a ZigBee network in the presence of heterogeneous network interference is obtained by

$$S = \frac{n\tau_{ZB}^*(1-\alpha_{ZB}^*)(1-\beta_{ZB}^*)\gamma_{ZB}^* l_p}{(1-\tau_{ZB}^*) + \tau_{ZB}^*\alpha_{ZB}^* + 2\tau_{ZB}^*(1-\alpha_{ZB}^*)\beta_{ZB}^* + \tau_{ZB}^*(1-\alpha_{ZB}^*)(1-\beta_{ZB}^*)\{\gamma_{ZB}^* L_s + (1-\gamma_{ZB}^*)(L_c)\}} \cdot \frac{1}{T_u^{ZB} R^{ZB}}. \tag{7.8}$$

where l_p^{ZB} and R^{ZB} in S denote the length of a ZigBee payload in bits and the bit rate of ZigBee, respectively. Heterogeneous interference on ZigBee devices is considered in the calculation of γ_{ZB}^* in S. The term γ_{ZB}^* represents the successful packet transmission probability after two successful CCAs.

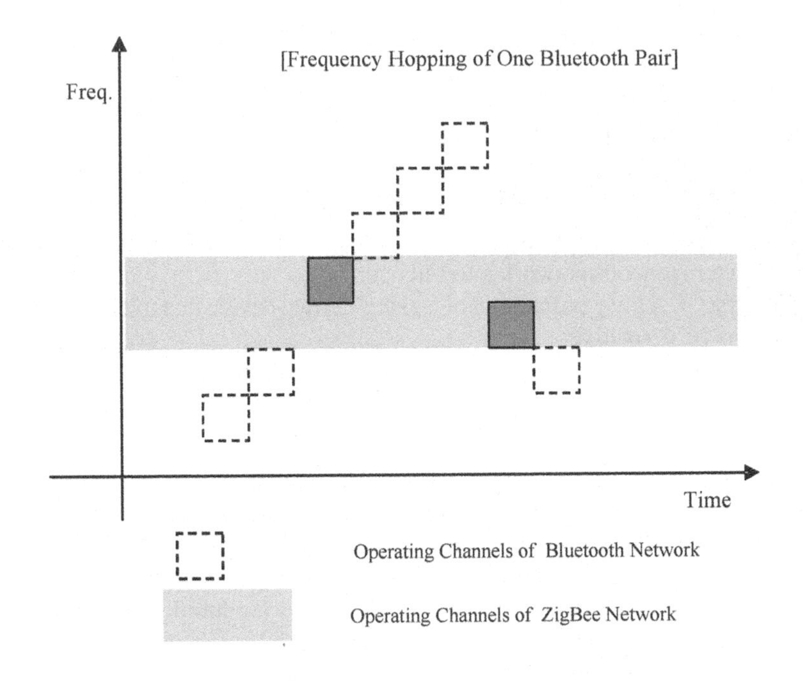

Figure 7.10: Operation of co-located bluetooth and ZigBee devices.

7.2.2.1 ZigBee in the Presence of Bluetooth Interference

Fig. 7.10 shows the operation of co-located ZigBee and Bluetooth devices. Bluetooth devices hop their operating frequencies among 79 channels every 625 μs, while ZigBee devices maintain their operating frequency during transmission. The operation bandwidth of ZigBee is twice as large as Bluetooth. Here, the hopping period of Bluetooth devices is assumed to be twice as large as the length of a ZigBee back-off slot, 320 μs for synchronization between two systems. Hence, for m activated Bluetooth devices, α_{ZB}^{BT} can be derived as

$$\alpha_{ZB}^{BT} = 1 - (\frac{77}{79})^m. \tag{7.9}$$

Substituting Eqn. (7.9) into α_{ZB}^{INT} term of Eqn. (7.2), α_{ZB}^{*} in the presence of Bluetooth interference can be derived.

Since Bluetooth randomly hops among 79 channels, assuming that β_{ZB}^{BT} is equal to α_{ZB}^{BT}, then β_{ZB}^{BT} is expressed as

$$\beta_{ZB}^{BT} = 1 - (\frac{77}{79})^m. \tag{7.10}$$

Substituting Eqn. (7.10) into α_{ZB}^{INT} of Eqn. (7.6), β_{ZB}^{*} in the presence of Bluetooth interference can be derived.

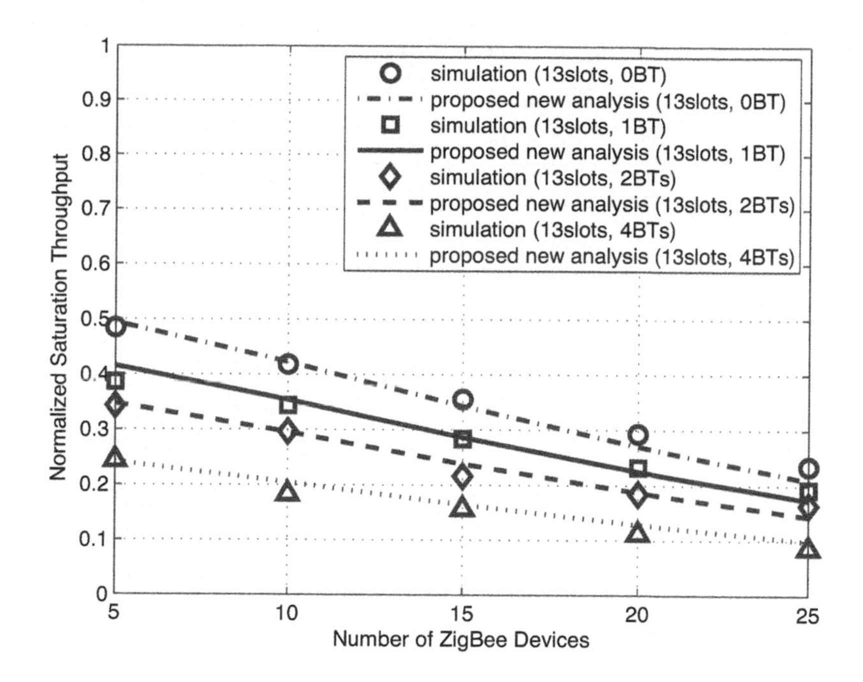

Figure 7.11: Throughput of a ZigBee network in the presence of bluetooth interference.

For a successful transmission after two CCAs, no simultaneous ZigBee transmissions $(1-\tau_{ZB}^{*})^{n-1}$ and no Bluetooth interference during a ZigBee transmission $(1-\alpha_{ZB}^{BT})^{\lceil L_{pkt}^{ZB} \frac{T_u^{ZB}}{T_{hop}^{BT}} \rceil}$ are required. Hence, γ_{ZB}^{*} is written as

$$\gamma_{ZB}^{*} = (1-\tau_{ZB}^{*})^{n-1}(1-\alpha_{ZB}^{BT})^{\lceil L_{pkt}^{ZB} \frac{T_u^{ZB}}{T_{hop}^{BT}} \rceil}, \tag{7.11}$$

where L_{pkt}^{ZB} denotes the length of a ZigBee packet in unit backoff slot.

Fig. 7.11 shows the normalized throughput of a ZigBee network for varying the number of interfering Bluetooth devices. Bluetooth and ZigBee parameter settings are shown in Tables 7.3 and 7.4. The analytical results for a ZigBee network in the presence of Bluetooth interference agree with simulation results. When the number of ZigBee devices is 10, the normalized saturation throughput S decreases from 0.484 to 0.183 if four Bluetooth pairs interfere with the ZigBee piconet.

Table 7.3: Bluetooth Related Parameter Values

Parameter	Value	Description
T_{hop}^{BT}	625 μs	Hopping interval
BW_{BT}	1 MHz	Bandwidth of one operating channel

Table 7.4: ZigBee Related Parameter Values

Parameter	Value	Description
L_{pkt}	3 (or 13 slots)	Length of packet
L_{ack}	1.1 slots	Length of ACK frame
L_{CCA}	0.4 slots	CCA duration
δ	1 slot	ACK wait duration
l_p^{ZB}	120 bits (or 920 bits)	Length of payload (bits)
$l_{p,slot}^{ZB}$	1.5 slots (or 11.5 slots)	Length of payload (slots)
T_u^{ZB}	320 μs	Unit backoff slot (sec)
f_{ZB}	2.415 GHz	Center frequency of operating channel
BW_{ZB}	2 MHz	Bandwidth of one operating channel

7.2.2.2 *ZigBee in the Presence of WLAN Interference*

Assuming that the length of WLAN packets is fixed in an activated WLAN with an aggregated arrival rate of λ_{WL}, α_{ZB}^{WL} is derived as

$$\alpha_{ZB}^{WL} = E(S)\lambda_{WL} = b_{WL}\lambda_{WL}, \tag{7.12}$$

where b_{WL} denotes the fixed length of WLAN packets in seconds. Substituting Eqns. (7.12) into α_{ZB}^{INT} term of Eqn. (7.2), α_{ZB}^{*} in the presence of WLAN interference can be derived.

Moreover, β_{ZB}^{WL} is expressed as

$$\beta_{ZB}^{WL} = \lambda_{WL}\frac{1}{(1-\alpha_{ZB}^{WL})}. \tag{7.13}$$

Substituting Eqn. (7.13) into α_{ZB}^{INT} of Eqn. (7.6), β_{ZB}^{*} in the presence of WLAN interference can be derived.

For a successful transmission after two CCAs, no simultaneous ZigBee transmissions $((1-\tau_{ZB}^{*})^{n-1})$ and no WLAN interference during a ZigBee packet transmission $(exp(1-\lambda\lceil L_{pkt}^{ZB}\frac{T_u^{ZB}}{T_u^{WL}})\rceil)$ is required. Hence, γ_{ZB}^{*} is expressed as

$$\gamma_{ZB}^{*} = (1-\tau_{ZB}^{*})^{n-1}exp(1-\lambda_{WL}\lceil L_S^{ZB}\frac{T_u^{ZB}}{T_u^{WL}})\rceil, \tag{7.14}$$

Table 7.5: WLAN Related Parameter Values

Parameter	Value	Description
T_u^{WL}	20 μs	Unit backoff slot
b_{WL}	1000 μs	Length of WLAN packet
BW_{WL}	22 MHz	Bandwidth of one operating channel

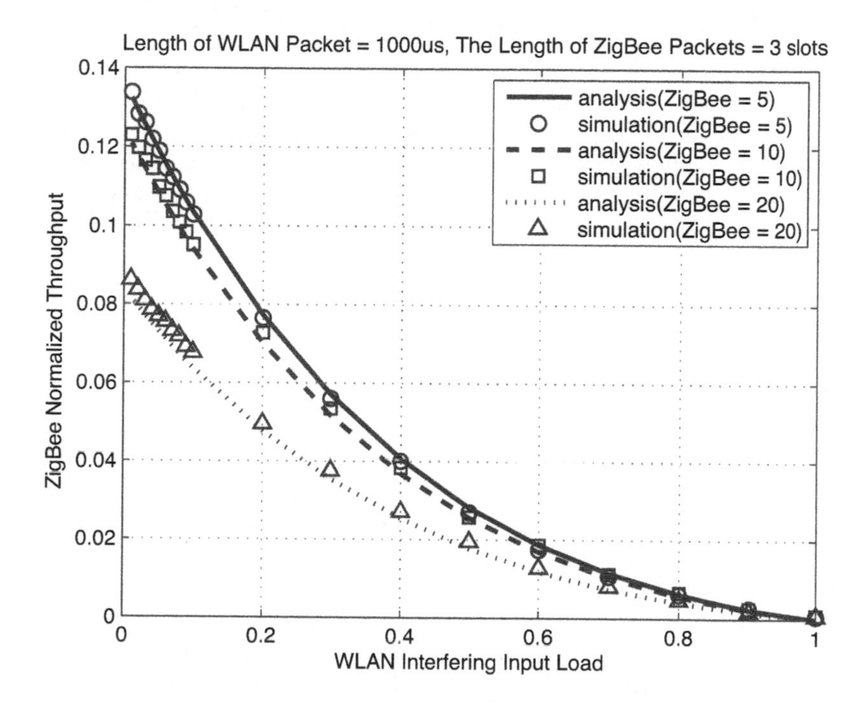

Figure 7.12: Throughput of a ZigBee network in the presence of WLAN interference.

where L_S^{ZB} denotes the length of successful transmission in unit backoff slots.

WLAN devices are assumed to be in an unsaturated mode upon arrivals of WLAN packets in a Poisson manner with input rate λ. The length of every WLAN packet is set to a fixed value[3]. ZigBee and WLAN parameter settings are shown in Tables 7.4 and 7.5 which are adopted in the analysis and simulation. Here, *minBE*, *maxBE*, *maxCSMABackoff* in ZigBee are set to 3, 5, and 4, respectively [13].

Fig. 7.12 shows the normalized saturation throughput of ZigBee for varying input interference load λ_{WL} of WLAN devices. The length of a WLAN packet b_{WL} and a

[3]This model is an *M/D/1* model by Kendall's notation.

ZigBee packet L_s are set to 1000 μs and 3 ZigBee unit backoff slots, respectively. The analytical results for a ZigBee network in the presence of WLAN interference agree well with simulation results. If the number of ZigBee devices is 5, the normalized saturation throughput S drops from 0.1323 to 0.0765 when WLAN interferes with the ZigBee piconet with $\lambda = 0.2$. Moreover, as λ reaches 1.0, S approaches 0.001 and practical ZigBee communication is almost impossible.

7.3 Interference Avoidance/Mitigation Algorithms

ZigBee devices may suffer from significant performance degradation due to interference from other systems operating in the same frequency band since its transmit power is the lowest among communications systems in the 2.4 GHz ISM band. Hence, some resolution schemes for ZigBee devices are needed to communicate data reliably with each other in the presence of interference. In this section, two types of interference avoidance and mitigation algorithms for ZigBee are introduced: (1) self-interference avoidance algorithms and (2) network-aided interference mitigation algorithms.

7.3.1 *Self-Interference Avoidance Algorithms*

In the self-interference avoidance algorithms, ZigBee devices need to overcome mutual interference without any help of other systems or networks. Three different types of self-interference avoidance algorithms for ZigBee have been proposed. First, a *hopping-based self-interference avoidance algorithm* has been proposed to avoid interference from other systems such as WLAN and Bluetooth from the viewpoint of point-to-point communications [2]. Next, the *hopping-based self-interference avoidance algorithm* has been extended to an *adaptive interference-aware multi-channel clustering algorithm* in the ZigBee network with a cluster-tree topology [17]. Finally, an *adaptive and dynamic interference-aware clustering algorithm* has been presented as an advanced interference-aware clustering algorithm [18]. ZigBee devices and networks based on these algorithms can adaptively avoid the existing interference for themselves.

7.3.1.1 Hopping-Based Self-Interference Avoidance Algorithm

According to the specification of IEEE 802.15.4 [13], ZigBee devices cannot dynamically change their allocated channels if pre-determined channels are set to them. Therefore, they may suffer from severe interference from other systems if they operate in the same frequency band.

A *hopping-based self-interference avoidance algorithm* is to adaptively configure the channel of ZigBee devices by considering the current wireless channel status. Hence, the performance degradation due to inter-system interference can be minimized by hopping the current channel to a new unused channel or a low interference channel. However, the ZigBee devices cannot communicate reliably with each other

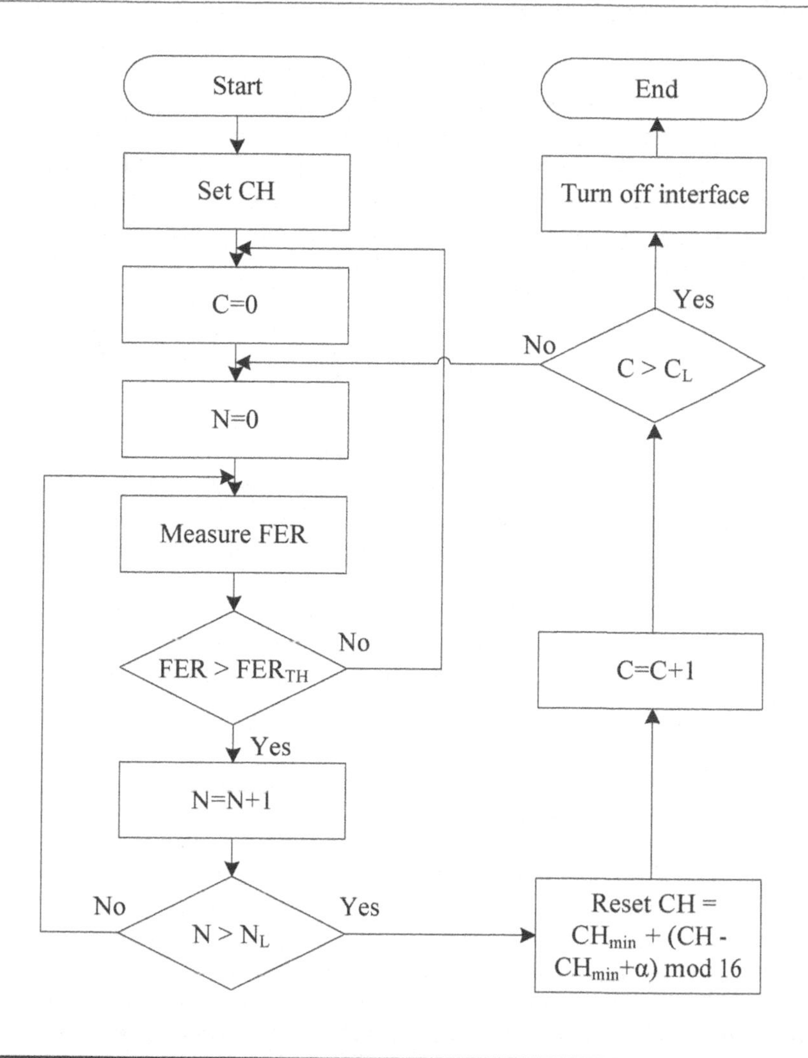

Figure 7.13: Flowchart of a hopping-based self-interference avoidance algorithm.

if the frame error rate (FER) continues to exceed a pre-determined threshold value in spite of adaptively allocating channels during some observing period. In this situation, since all the channels are fully utilized by other systems, the algorithm is able to reduce the unnecessary power consumption, which is caused by failed transmissions, of ZigBee devices by turning off the air interface.

Fig. 7.13 shows a flowchart of the *hopping-based self-interference avoidance algorithm*, where *CH*, *C*, and *N* denote the current channel, the number of channel changes, and the number of counts for exceeding an FER threshold value, respectively. FER is continuously measured during a pre-determined observing period. If the measured FER value exceeds FER_{TH}, then the value of *N* is increased by one. Otherwise, it goes back to the initial state. If the value of *N* is larger than N_L, which is the limit value of N, then the channel CH is changed into $[CH_{min} + (CH - CH_{min} + \alpha) \ mod \ N_{CH}]$, where CH_{min}, α, and N_{CH} denote the minimum channel number, the channel changing interval, the total number of channels, respectively. Then, the value of C is increased by one. Otherwise, FER measurements are continued. If the value of C is smaller than that of C_L, which is the limit value of C, it goes back to the initial state. Otherwise, it determines that ZigBee devices cannot reliably communicate with each other, and the algorithm turns off the air interface to save the energy of the ZigBee device.

In order to evaluate the performance of the algorithm, experiments are carried out in an environment where ZigBee devices, WLAN devices, and microwave ovens coexist. The experimental environment is the same as shown in Fig. 7.5 of the previous Section 7.2.1. In this experiment, both d_{ZB} and d_{MWO} are set to 100 cm and d_{WLAN} is set to 30 cm. FTP upload traffic is considered for WLAN traffic and constant bit rate (CBR) traffic like periodic sensing data is considered for ZigBee traffic.

Input parameters for the performance evaluation are given as follows: the FER measurement interval is set to 30 (sec). FER_{TH} is set to 0.1. N_L and C_L are set to 3 and 5, respectively. CH_{min} and N_{CH} are set to 11 and 16 for ZigBee devices in 2.4 GHz ISM band, respectively [13]. Since the channel spacing of ZigBee is 5 MHz and the channel bandwidth of WLAN is approximately 22 MHz, α is set to 5 [13, 15]. The initial center frequency of WLAN devices is set to 2.453 GHz and the initial ZigBee operating channel is set to Ch. 22 among 16 channels (i.e., Ch. 11 - Ch. 26). In this initial setting, the used frequency band of both devices completely overlaps if they transmit data simultaneously. Therefore, the ZigBee devices can be significantly affected by the WLAN interferer.

Fig. 7.14 shows the FER of ZigBee devices with and without the *hopping-based self-interference avoidance algorithm*. Since the ZigBee channel is overlapped by the channel bandwidth of a WLAN during a time duration of up to 100 (sec), the FER values of both cases exceed FER_{TH}. After 100 (sec), the FER value of the device with the algorithm rapidly decreases. It is because the algorithm changes the operating channel to avoid the interference from the WLAN. On the other hand, the FER value of the conventional operation continues to keep a similar level. When the microwave oven is turned on at 180 (sec), the FER values of both cases become higher than before. This shows that turning on a microwave oven also interferes with the operating channel changed by the algorithm due to the WLAN interferer. After 250 (sec), the FER value with the algorithm rapidly decreases as well since the algorithm changes the operating channel again to avoid the interference from the microwave oven. On the other hand, the FER value of the conventional operation continues to exceed the FER_{TH} constraint. When the microwave oven is turned off, the FER values of both cases are back to the level as shown before turning on it. From these results, the algorithm adaptively changes the ZigBee operating channel to avoid the interference from other systems. It enables ZigBee devices to reliably communicate with each other in the presence of inter-system interference.

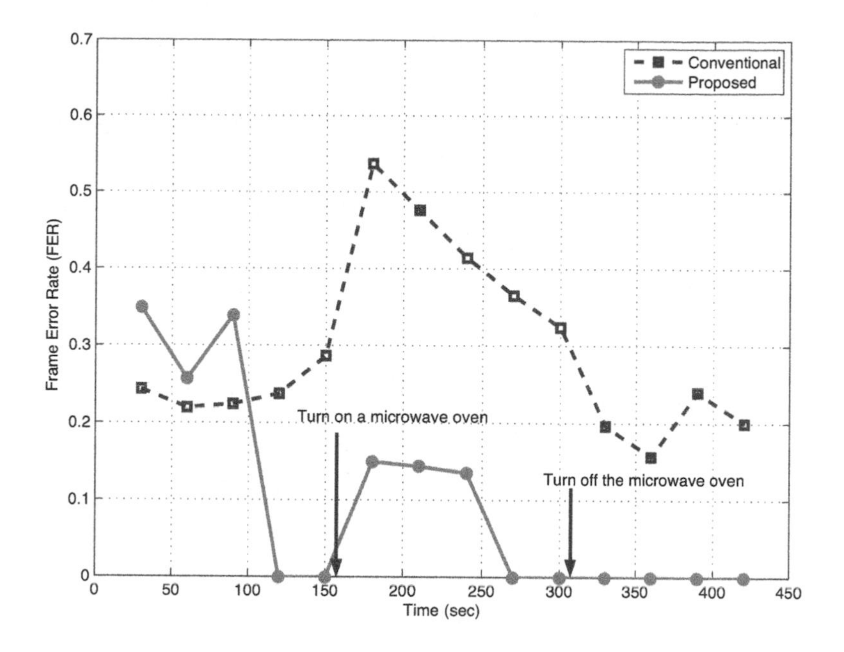

Figure 7.14: FER of ZigBee devices with and without the hopping-based self-interference avoidance algorithm.

7.3.1.2 Adaptive Interference-Aware Multi-Channel Clustering Algorithm

The *hopping-based interference avoidance algorithm* has been developed under the assumption that WLAN and WPAN systems consist of a single pair of communication links. However, a WPAN system generally consists of multiple terminals, which form a tree network or a mesh network, and the interference avoidance algorithms developed in a single link communication environment cannot be adopted when a WPAN forms a network. Thus, a new interference avoidance algorithm for WPAN networks is needed in this case.

An *adaptive interference-aware multi-channel clustering algorithm* has been proposed for IEEE 802.15.4 cluster-tree networks [17]. The cluster-tree network is based on a tree network and it divides its nodes into multiple groups, called *clusters*. A cluster is defined as a basic group of devices that uses the same frequency channel. Devices do not need to exchange information about which channel they need to move to after detecting interference; each device moves to the next channel which is given from a pseudo-random sequence generator whose keys are shared by all devices which use the same channel. This distributed and pre-determined channel

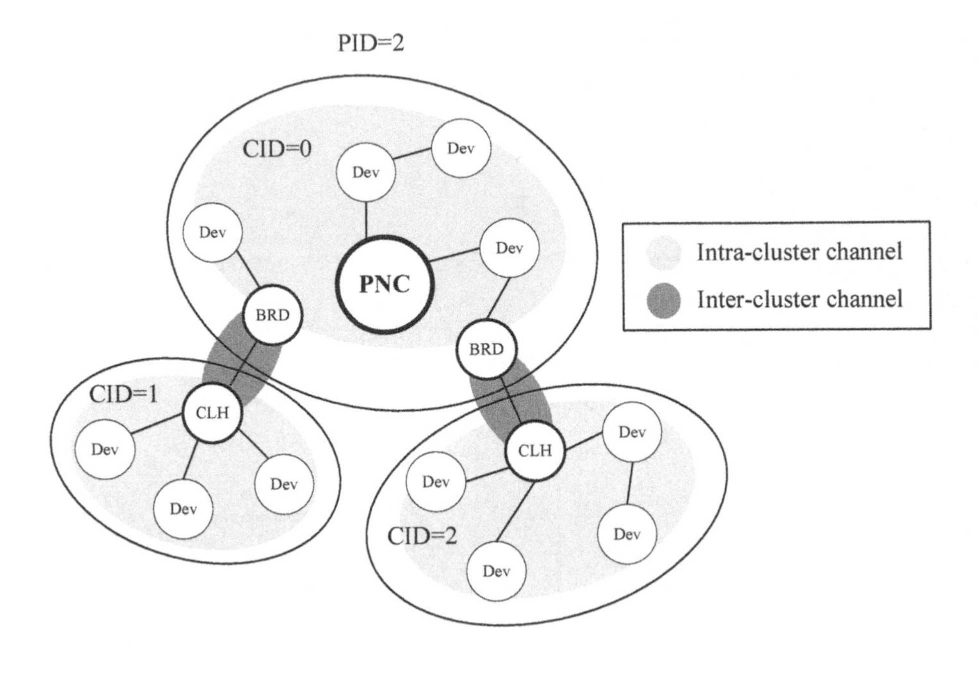

Figure 7.15: An example of a ZigBee cluster-tree network.

change scheme enables IEEE 802.15.4 networks to avoid interference and reconfigure a new cluster-tree network.

In a PAN, there is one PAN coordinator (PNC) which acts as a primary controller of the PAN. Each independent PAN has a unique identifier, called the PAN id (PID). Fig. 7.15 shows an example of general network topology with a cluster-tree network. In the cluster-tree topology, devices are grouped into a cluster, and a cluster head (CLH), which is a local coordinator in the cluster, is responsible for managing the cluster and the cluster identifier (CID) is the shared ID number for all devices in the cluster [13]. The cluster-tree network is widely used to increase the coverage area of the ZigBee network using a multi-cluster structure. A *bridge device* (BRD) is defined as a node that is directly connected to a cluster head of a neighboring cluster. As shown in Fig. 7.15, two types of ZigBee channels can be defined as follows: *intra-cluster channel* and *inter-cluster channel*. An intra-cluster channel is a channel established by devices in a single cluster, while an inter-cluster channel is a channel established by a CLH of one cluster and a BRD of another cluster.

The *adaptive interference-aware multi-channel clustering algorithm* consists of two schemes: an interference detection scheme and an interference avoidance scheme. Fig. 7.16 shows a simplified flowchart for the algorithm. In the *interference detection scheme*, each node intelligently detects the interference from WLAN devices. In this scheme, each node can use an ACK/NACK based interference detection scheme or a beacon-based interference detection scheme. Each node determines

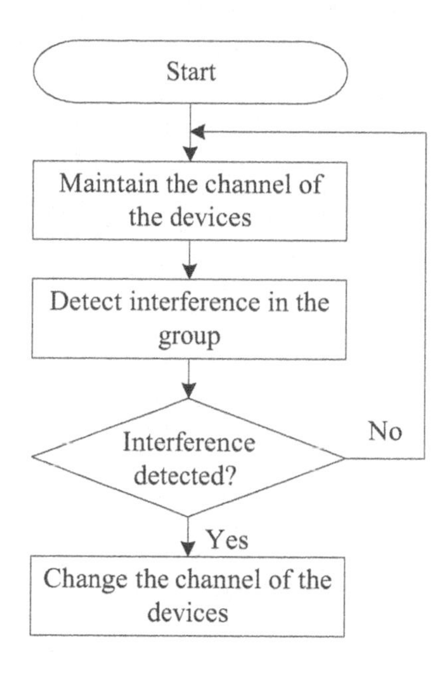

Figure 7.16: Simplified flowchart of the adaptive interference-aware multi-channel clustering algorithm.

whether it is affected by interference from other systems by counting up the number of successive NACKs, or successive beacon loss. The interference detection and avoidance is done by all nodes in a cluster. Thus, all nodes in the cluster should detect the interference at the same time with a short time difference. Broadcasting a channel change broadcast message (CCBM), all the nodes in the cluster can detect the interference at once. In the *interference avoidance scheme*, each node changes its operating channel to the next channel. If all the nodes in the cluster need to communicate with each other after changing their channels, they should share which channel they should change to. However, in many interference dominant environments, it is hardly possible to exchange the information to which channel they should move. Therefore, a pseudo-random based interference avoidance scheme is used [17]. In the pseudo-random based interference avoidance scheme, each node needs to obtain the next channel information from a pseudo-random sequence generator (PRSG), which takes a PAN ID, a cluster ID, a current channel index, and a counter index and returns the next channel information. Using the PRSG, all nodes in the cluster can change their channel to the same channel without exchanging any next channel information to each other.

To evaluate the performance of the algorithm, an experiment was taken on a testbed equipped with 30 ZigBee nodes and 6 WLAN APs. In this experiment, WLAN APs are turned on one by one for every 1000 seconds. Fig. 7.17 shows the FER performance of the *adaptive interference-aware multi-channel clustering*

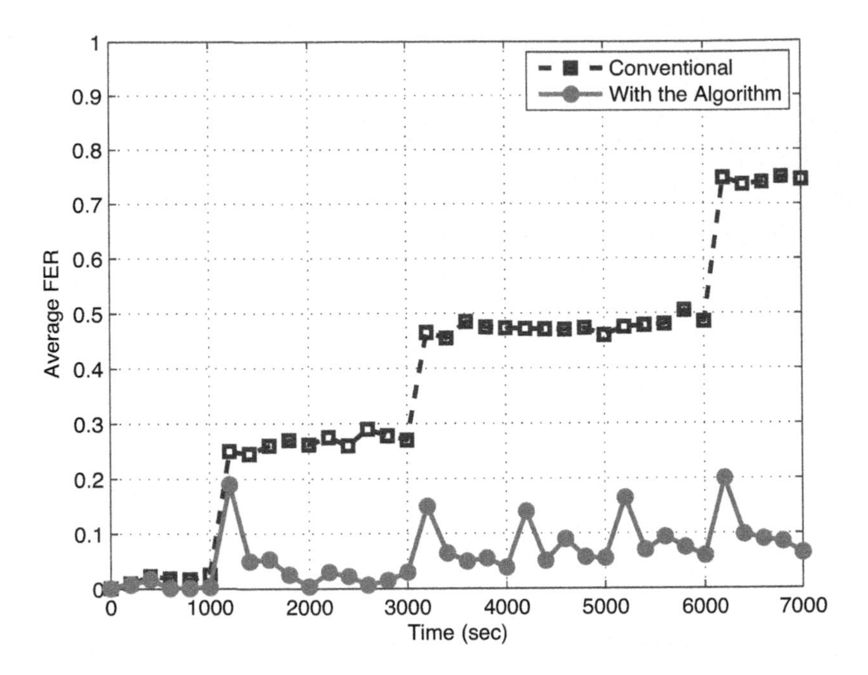

Figure 7.17: FER performance of the adaptive interference-aware multi-channel clustering algorithm.

algorithm. The FER of the ZigBee network remains lower than 10% on average, while the FER of the system with the conventional ZigBee cluster-tree network is extremely high so that it is almost impossible to communicate between ZigBee nodes. From this experimental result, the algorithm enables the ZigBee nodes to reliably communicate with each other in their cluster-tree structured network although several WLAN interferers interfere with a ZigBee network.

7.3.1.3 Adaptive and Dynamic Interference Avoidance Algorithm

An advanced interference-aware clustering algorithm based on an *adaptive interference-aware multi-channel clustering algorithm* was introduced in the previous section. The previously proposed algorithm considered intra-cluster and inter-cluster interference caused by WLAN. Here, *broken clusters* are additionally considered, in which part of or entire cluster members have a difficulty in communicating with their previous cluster head. Three dynamic clustering schemes for the broken clusters of ZigBee networks are proposed and compared. In this scheme, the main objective is to reduce the number of orphan devices for maintaining ZigBee clusters as before.

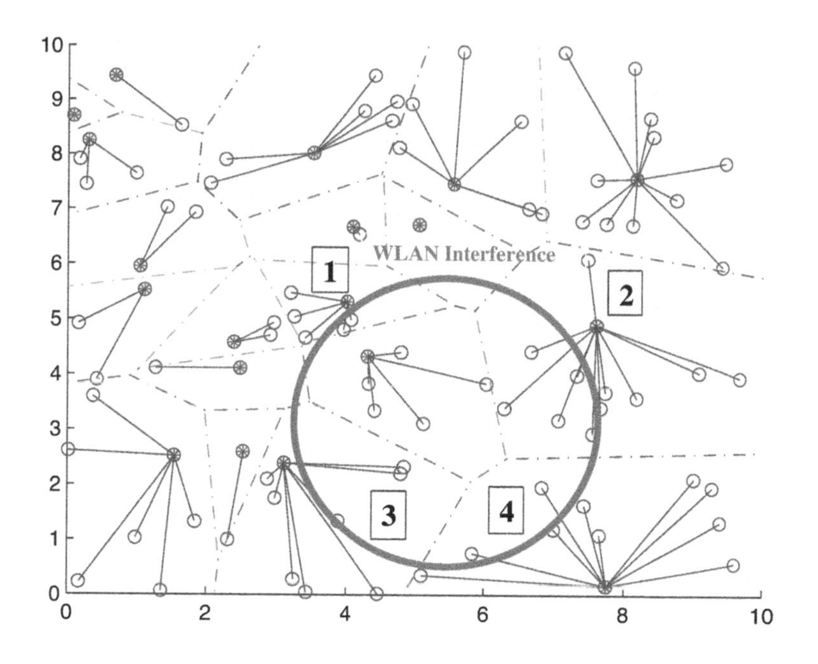

Figure 7.18: ZigBee cluster networks and WLAN interference.

In Fig. 7.18, a ZigBee network region in the presence of WLAN interference is indicated by a big circle and clusters in the circle region may be broken or non-broken. Clusters 1, 2, 3 and 4 are broken clusters. Three dynamic clustering algorithms: a *merged clustering scheme*, a *new clustering scheme*, a *hybrid clustering scheme* are proposed. In the *merged clustering scheme*, each cluster member (CM) breaks its connection in the current cluster and associates with another nearest cluster. Fig. 7.19 shows a flowchart of the *merged clustering scheme*. In the *new clustering scheme*, CMs elect a new cluster head (CH) within CMs and form a new cluster using the previous or new interference-free channel. Fig. 7.20 illustrates the detailed operation of the *new clustering scheme*. The *hybrid clustering scheme* is a combination of the above two schemes. Fig. 7.21 shows a flowchart of the *hybrid clustering scheme*.

Fig. 7.22 shows the number of remaining orphan devices after dynamic clustering. Related system parameters for performance evaluation are listed in Table 7.6. The dynamic clustering schemes are proven to reduce the number of orphan devices, compared to a no-clustering scheme. Among the dynamic clustering schemes, the number of orphan devices in the new clustering and the hybrid clustering schemes is much smaller than that in the merged clustering scheme.

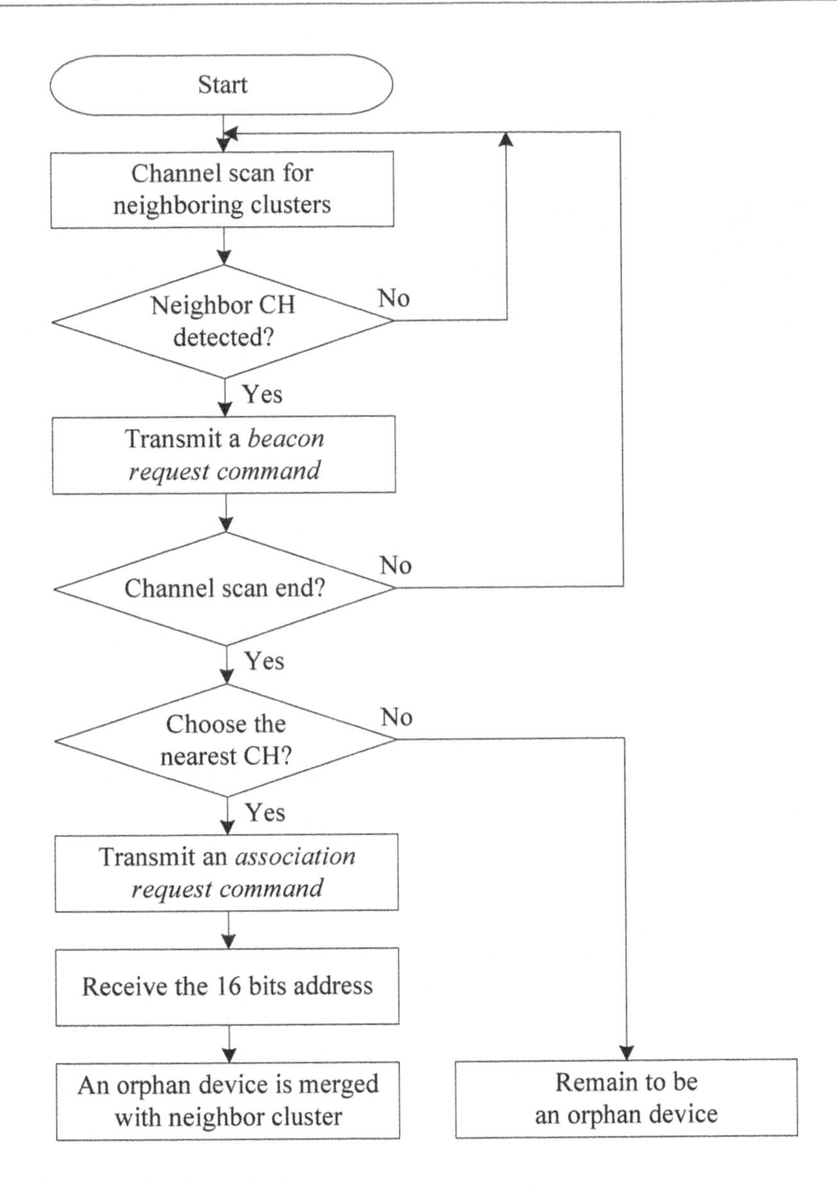

Figure 7.19: Flowchart of the merged clustering scheme.

7.3.2 Network-aided Interference Mitigation Algorithms

Compared with the previous self-interference avoidance algorithms, network-aided interference mitigation algorithms are more aggressive interference mitigation algorithms. Therefore, ZigBee devices or networks can have more reliable communications through the network-aided interference mitigation algorithms even if they

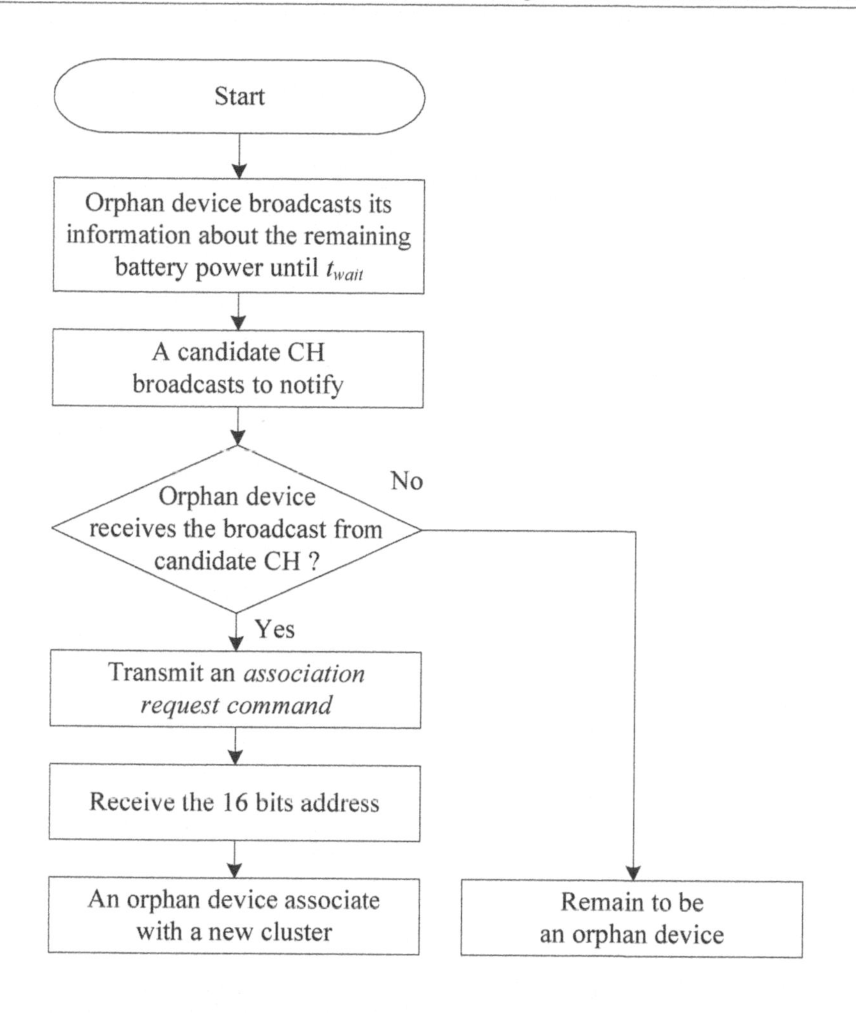

Figure 7.20: Flowchart of the new clustering scheme.

suffer from an inevitable interference situation without any interference-free channel for ZigBee. In this section, two different types of network-aided interference mitigation algorithms are introduced. First, a *portable device aided coexistence algorithm* is presented in an overlaid network of ZigBee and WLAN [19]. This algorithm gives resources to ZigBee nodes by managing a portable device with both types of air interfaces, even if the other systems have already occupied all the resources in the same frequency band. Next, a *network-centric inter-system interference mediation algorithm* using an interference mediator is described. This algorithm can mitigate inter-system interference by adjusting the channel usage through an interference mediator, which is located in the heterogeneous networks, from the perspective of fairness [20].

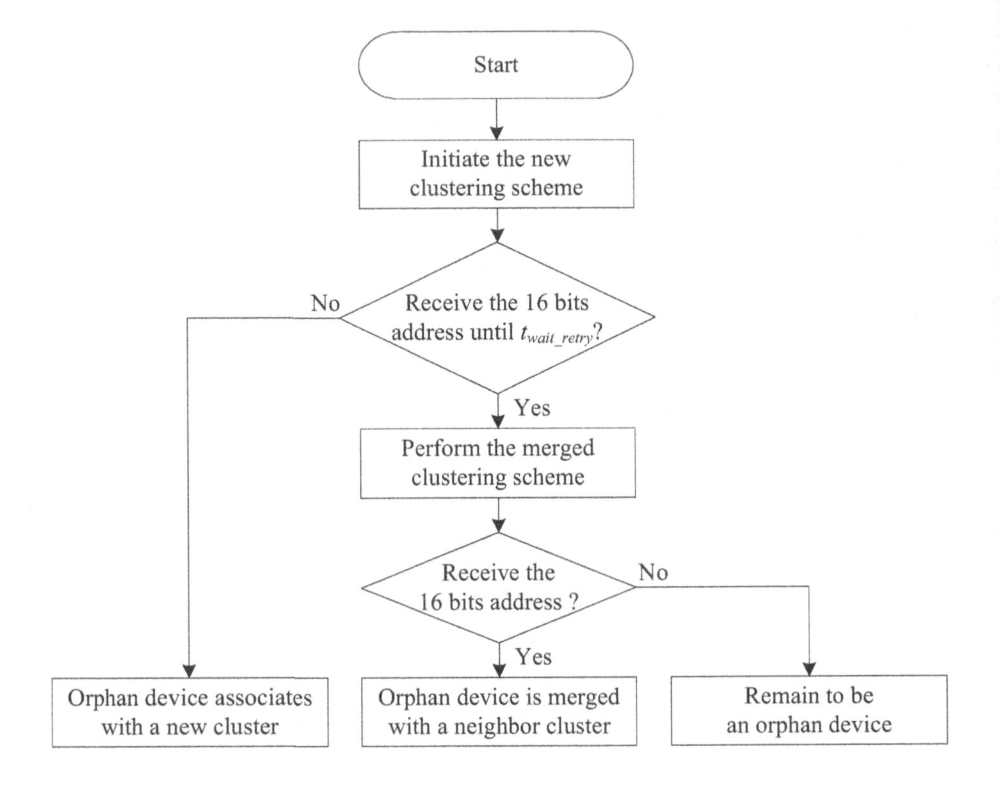

Figure 7.21: Flowchart of the hybrid clustering scheme.

7.3.2.1 Portable Device Aided Coexistence Algorithm

Sometimes ZigBee nodes are hard to start to communicate with each other when they are collocated with WLAN APs or devices because both devices operate in the same frequency band and the transmission power of WLAN is much stronger than that of ZigBee. Fig. 7.23 shows an example of one ZigBee and six WLAN networks in a collocated environment. In this environment, there are no WLAN interference-free channels for ZigBee devices. Hence, there is no way for the ZigBee nodes to communicate with each other due to severe interference from the WLAN APs even though the ZigBee nodes scan all the channels and select one with the least interference.

On the other hand, many recent portable devices have multiple wireless interfaces such as WiFi, Bluetooth, ZigBee, and WiMAX to support various wireless services. Hence, a portable device with both WLAN and ZigBee interfaces can be a good solution to solve the above coexistence problem. In this perspective, a *portable device aided coexistence algorithm* has been proposed [19]. In the algorithm, a portable device helps reliable communications of ZigBee nodes in the presence of WLAN interference using its internal WLAN and ZigBee interfaces.

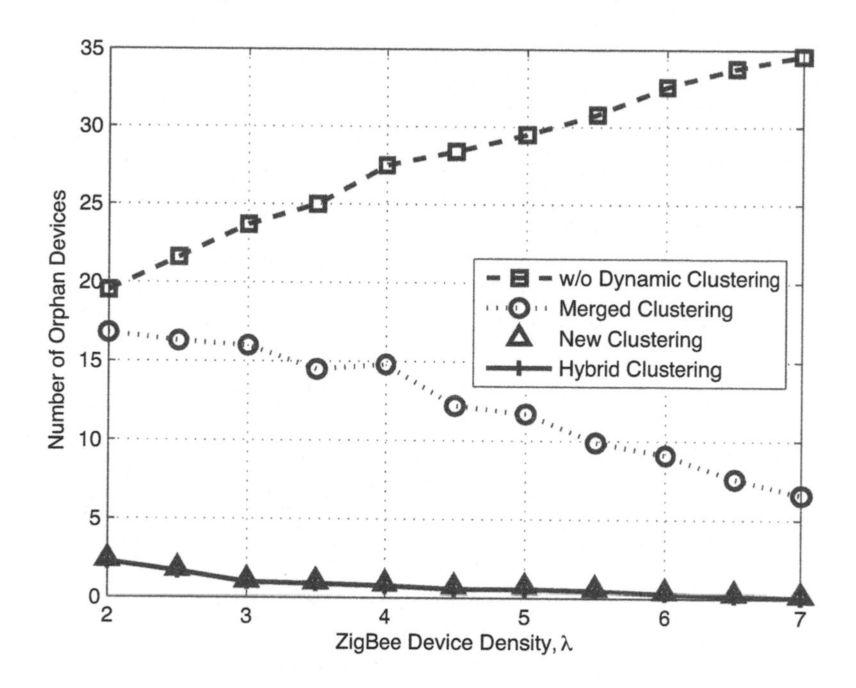

Figure 7.22: Number of remaining orphan devices after dynamic clustering.

Fig. 7.24 shows a block diagram of the *portable device aided coexistence algorithm* which consists of two parts: a *channel status observation part* and an *interference mitigation part*. The *channel status observation part* determines the best ZigBee channel and checks whether the *interference mitigation part* needs to be performed additionally or not. In the same environment as shown in Fig. 7.23, the algorithm finds the least interfered ZigBee channel and lets the WLAN not interfere with the selected ZigBee channel during a specified duration, called the WLAN *interference-free time interval*, for ZigBee data transmission. Therefore, the ZigBee interface in the portable device can transmit or receive data without any interference during the temporary WLAN interference-free time interval.

In the *channel status observation part*, FER is measured by counting the number of ACK/NACK frames from broadcasted messages. If it determines that there are overlapped WLAN channels, the *interference mitigation part* is additionally performed.

In the *interference mitigation part*, a portable device accesses the WLAN channel, which severely interferes with the ZigBee network, using its WLAN interface. After acquiring the channel access, the WLAN interface in the portable device reserves a specified time duration as a WLAN interference-free time interval by sending a request-to-send (RTS) message to the corresponding WLAN AP. Then the

Table 7.6: Related System Parameter Values

Parameter	Description	Value
λ	ZigBee device density	[2 3 4 5 6 7]
$2a$	Width of ZigBee networks	10
p	Probability of becoming CH form a device	0.15
r_{radio}	ZigBee radio range	1.2
R_{radio}	WLAN radio range	3
BE_{min}	Minimum backoff exponent	3
BE_{max}	Maximum backoff exponent	5
CW_{min}	Minimum contention window size	8
CW_{max}	Maximum contention window size	32
$Backoff_{max}$	Maximum number of backoffs the CSMA/CA	4
$slot_{time}$	Time duration of one slot	320 μs
D_{rate}	Data rate	250 kbps
t_{wait}	Time duration of waiting to receive broadcast message in Making New Cluster scheme	14.4 ms
command message	Beacon request command Association request command Association response command	8 bytes 7 bytes 9 bytes
$t_{broadcast}$ t_{ack}	Time that one device transmit command message	simulation results

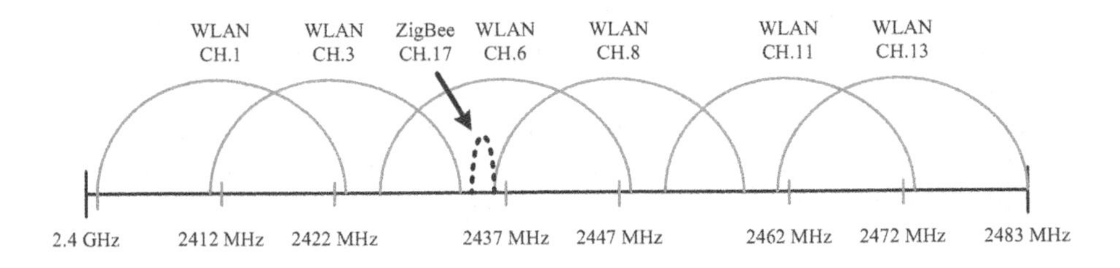

Figure 7.23: An example of one ZigBee and six WLAN networks.

WLAN AP returns a CTS message and the WLAN reserves a time duration for communication between ZigBee devices. Through this operation, the ZigBee nodes are able to communicate with each other without dominant interference from WLAN during the specified duration. Fig. 7.25 shows how to control the channel occupancy time in the *portable device aided coexistence algorithm.*

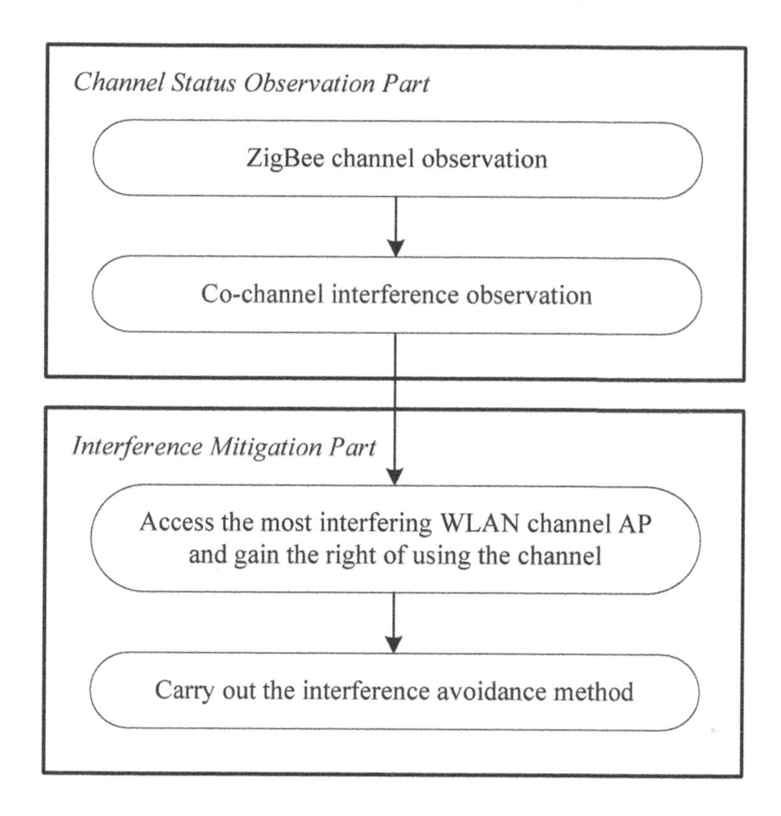

Figure 7.24: Block diagram of the portable device aided coexistence algorithm.

The performance of the ZigBee network is evaluated in terms of FER and goodput, expressed as

$$FER = \frac{\text{number of lost or unsuccessful frames}}{\text{total number of transmitted frames}}, \tag{7.15}$$

$$Goodput = \frac{\text{successfully transmitted data}}{\text{transmission period}}. \tag{7.16}$$

Fig. 7.26 shows an experimental testbed environment for performance evaluations. The distance values of d_{ZB}, d_{AP-STA}, and d_{AP} are set to 1m, 2m, and 5m, respectively. WLAN devices communicate with each other on Ch. 1, Ch. 6, and Ch. 11 among 13 channels with a maximum power of 1000 mW. In this experiment, a portable device plays a role as a piconet coordinator (PNC) for the ZigBee network. However, in general, nodes in the testbed can be replaced by other portable devices with both WLAN and ZigBee interfaces.

Fig. 7.27 shows the FER values of ZigBee data frames for varying the used ZigBee channels. For this result, the used WLAN and ZigBee channels are configured as shown in Fig. 7.28. As the operating center frequency of ZigBee nodes is closer

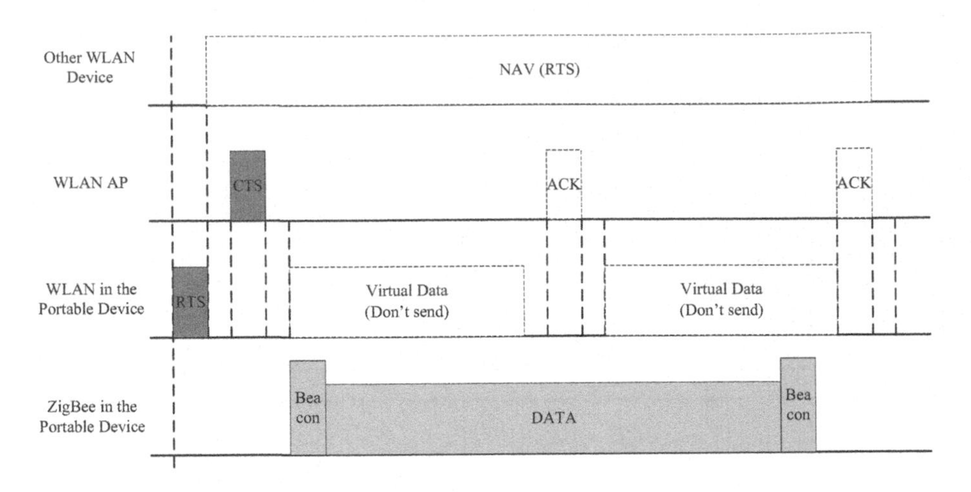

Figure 7.25: Control of the channel occupancy time in the portable device aided coexistence algorithm.

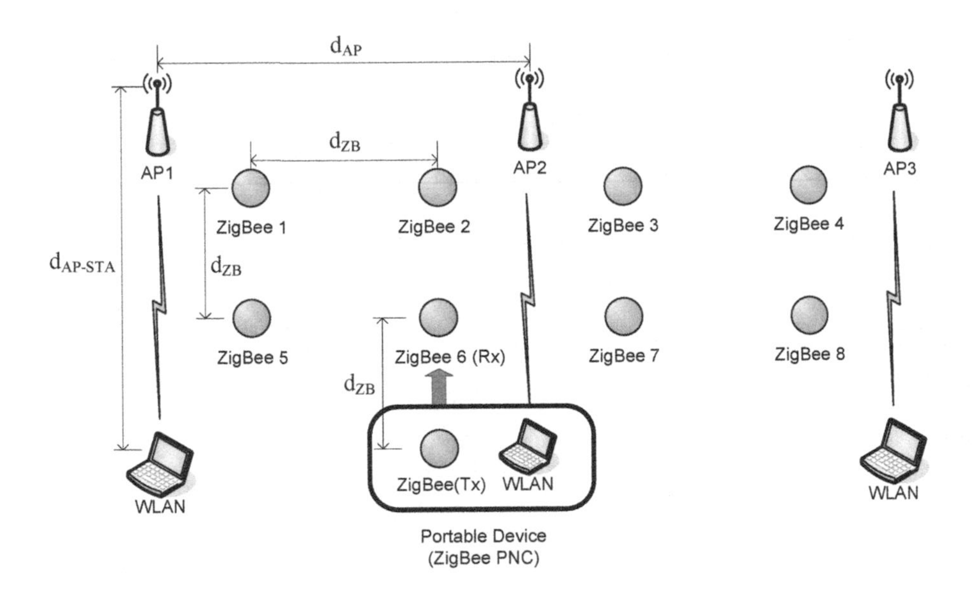

Figure 7.26: Experimental network testbed environment.

to that of the WLAN, the FER performance of the conventional ZigBee nodes becomes worse, while the ZigBee nodes with the algorithm keep the FER to a significant low-level. This result implies that the algorithm enables ZigBee nodes to

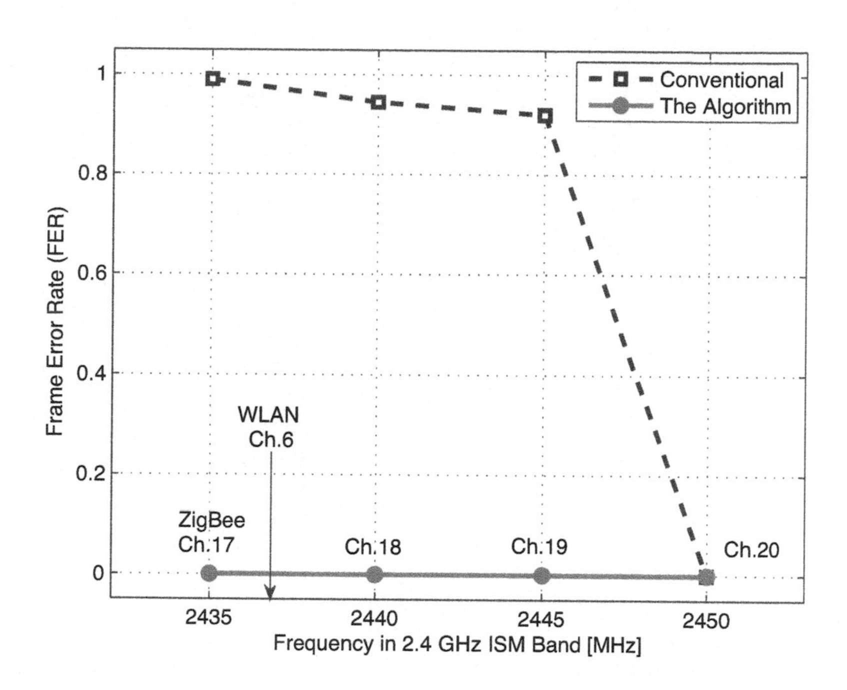

Figure 7.27: FER values of ZigBee for varying ZigBee channels.

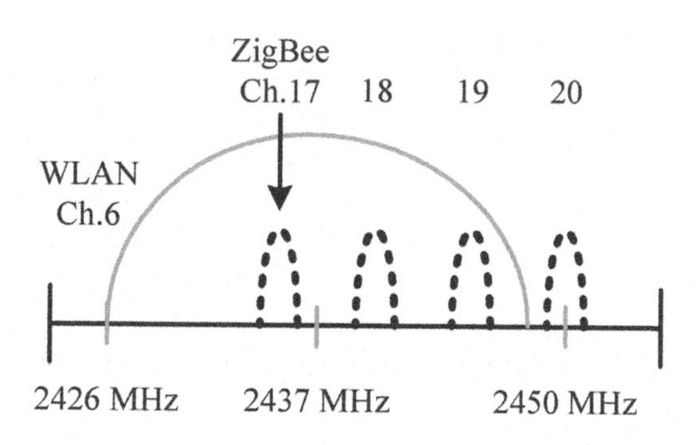

Figure 7.28: Varying ZigBee channels in the presence of WLAN in 2.4GHz ISM band.

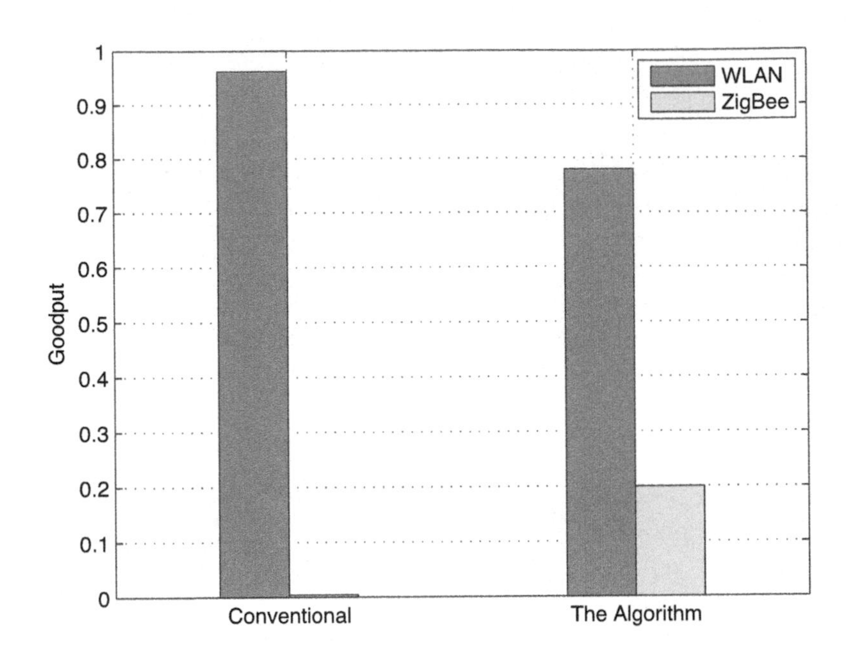

Figure 7.29: Relative goodput for both WLAN at Ch. 6 and ZigBee at Ch. 17.

reliably communicate with each other even if there is no WLAN interference-free channel.

Fig. 7.29 shows the relative goodput for both WLAN at Ch. 6 and ZigBee at Ch. 17. The algorithm yields better goodput for ZigBee nodes and its goodput increases from $(0.004 \times R^{ZB}_{max})$ to $(0.1995 \times R^{ZB}_{max})$ by the relative goodput of WLAN from $(0.962 \times R^{WL}_{max})$ to $(0.7791 \times R^{WL}_{max})$, where R^{ZB}_{max} and R^{WL}_{max} denote the maximum ZigBee data rate and the maximum WLAN data rate, respectively. This result implies that the portable device aided coexistence algorithm guarantees ZigBee transmissions even in a harsh environment due to the severe interference from the overlapped WLAN networks.

7.3.2.2 Network-Centric Inter-System Interference Mediation Algorithm

In the entire network point of view, a coexistence problem between WLAN and ZigBee nodes is similar to a multi-rate problem in WLAN protocol. The multi-rate problem is to fairly support multi-users with different transmission rates. Thus far, in order to solve the multi-rate problem in WLAN protocol, several fairness-based TDMA schemes have been studied [21, 22, 23]. However, physical layer

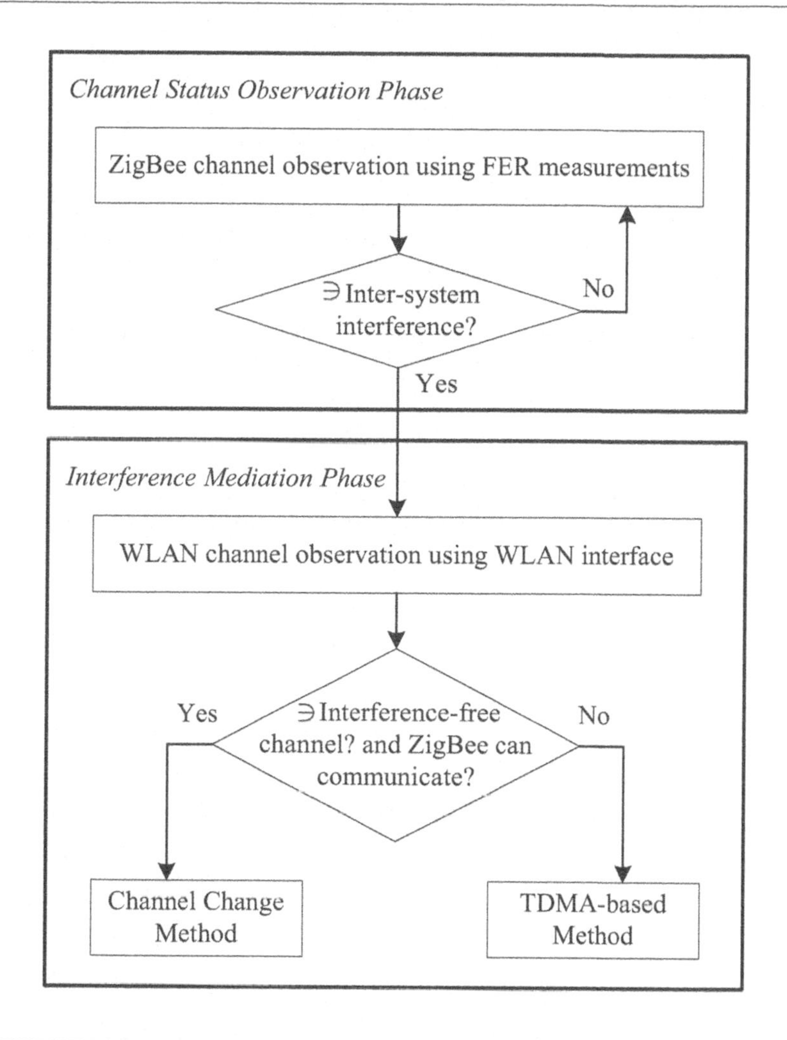

Figure 7.30: Block diagram of a network-centric inter-system interference mediation algorithm.

technologies and MAC structures of the systems in the coexistence problem are different from that in the multi-rate problem.

In order to consider a fairness problem between ZigBee and WLAN networks, a *network-centric inter-system interference mediation algorithm* has been proposed in an overlaid network environment [20]. In this algorithm, an interference mediator (IM) is introduced to mediate inter-system interference between ZigBee and WLAN nodes. The IM is a device which has two types of interfaces and it can coordinate mutual interference by adjusting the shared resource of WLAN and ZigBee in the overlaid networks. Any devices equipped with both air interfaces (e.g., network appliances, desktop PCs, hand-held devices, etc.) can play a role as an IM.

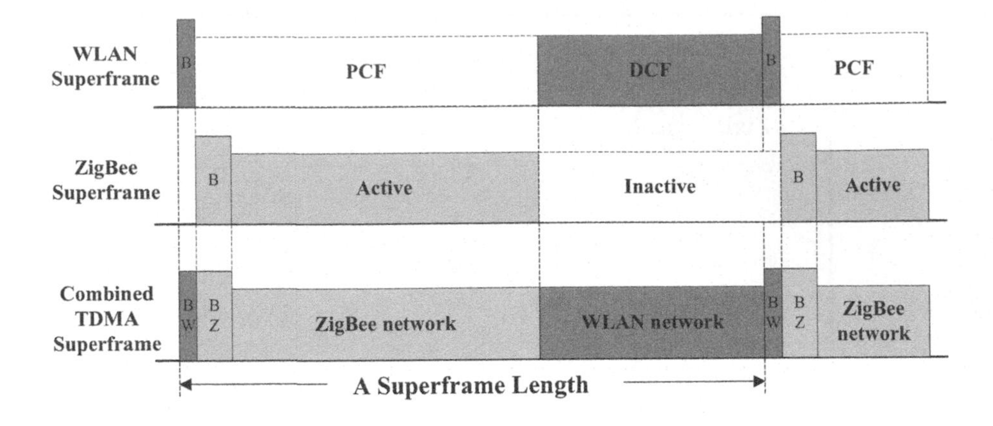

Figure 7.31: Superframe structure of the network-centric inter-system interference mediation algorithm.

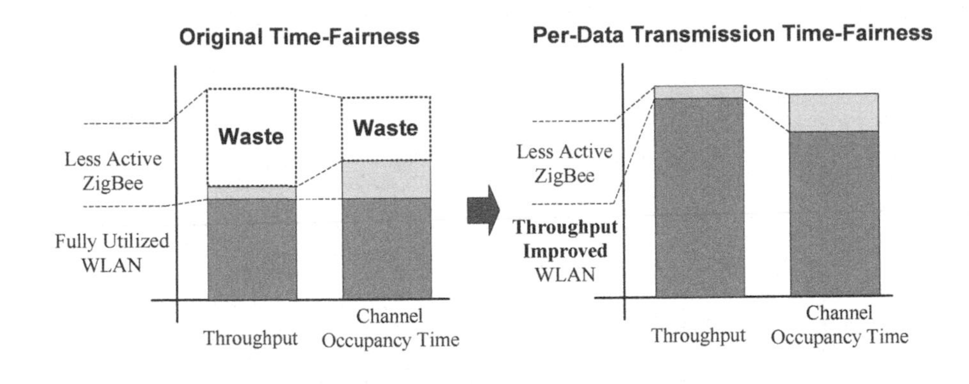

Figure 7.32: An example of the per-data transmission time-based fairness scheme.

Fig. 7.30 shows a block diagram of a *network-centric inter-system interference mediation algorithm* which consists of two phase operations: a channel status observation phase and an interference mediation phase. In the *channel status observation phase*, the IM determines the best ZigBee channel and checks whether the *interference mediation phase* needs to be performed or not, by using the same procedure as shown in the *portable device aided coexistence algorithm*. In the *interference mediation phase*, the IM scans WLAN channels using a WLAN interface and searches an interference-free channel among all the ZigBee channels. If there exists an interference-free channel for ZigBee nodes, then a ZigBee piconet is reconfigured at the new ZigBee channel to avoid the inter-system interference. However, if there is no interference-free channel due to mutual interference from WLAN, the *interference mediation phase* performs a *fairness-based inter-system TDMA scheme* to guarantee ZigBee data transmissions.

Fig. 7.31 shows an integrated superframe structure for a *fairness-based inter-system TDMA scheme*. In this scheme, the IM reserves the WLAN Point Coordination Function (PCF) duration for ZigBee active duration while the remaining ZigBee inactive duration is used for WLAN Distributed Coordination Function (DCF) duration. Actual ZigBee data transmissions are performed in the PCF duration for WLAN by reserving the PCF duration of WLAN MAC protocol and adjusting Contention Access Period (CAP), Contention Free Period (CFP), and inactive interval of ZigBee MAC protocol. In this case, a ZigBee interface in the IM is used to act as a PNC of the ZigBee network, while a WLAN interface does not have to play a role as an AP in the WLAN network. However, the WLAN AP has to support the PCF transmission mode.

In order to control the channel occupancy time ratio between the WLAN and ZigBee networks, the IM can apply several fairness schemes, such as a time-based fairness scheme and a throughput-based fairness scheme. In the time-based fairness scheme, the same channel occupancy time is allocated for both the WLAN and ZigBee networks, while the throughput-based fairness scheme is to achieve the same throughput for both networks.

The channel occupancy time ratio (COR) of each system can be expressed as

$$COR_i = \frac{ChOccupancyDuration_i}{SuperframeLength - Beacon_{WL} - Beacon_{ZB}}, \quad i \in \{WLAN, ZigBee\} \tag{7.17}$$

In order to match the same amount of throughput, the throughput-based fairness scheme provides a much higher COR for ZigBee, compared with that of WLAN. It results in an extremely low total aggregated throughput, compared with the time-based fairness scheme since the transmission rate of ZigBee is much smaller than that of WLAN. Hence, the time-based fairness scheme is a more appropriate solution than the throughput-based fairness scheme.

However, in general, heterogeneous systems have different channel activities according to their applications. For example, WLAN exhibits relatively high-activity for multimedia data transmissions, while ZigBee has low-activity for sensing data transmissions. In this unbalanced condition, the time-based fairness scheme may waste a certain amount of time resource for less active systems such as ZigBee. Therefore, the throughput of the more active system such as WLAN is limited by this resource waste. To overcome this shortcoming, a *per-data transmission time-based fairness scheme* has been proposed to eliminate the waste of time resource caused by the less active system. Fig. 7.32 shows an example of the *per-data transmission time-based fairness scheme*. As shown in the figure, the WLAN network fully utilizes the given time resource while the overlapped ZigBee network does not due to its low-activity. The *per-data transmission time-based fairness scheme* reallocates such waste resource of the ZigBee network to the WLAN network.

Fig. 7.33 shows the throughput of a conventional ZigBee network in the presence of WLAN interference. This result shows that the throughput of ZigBee is sig-

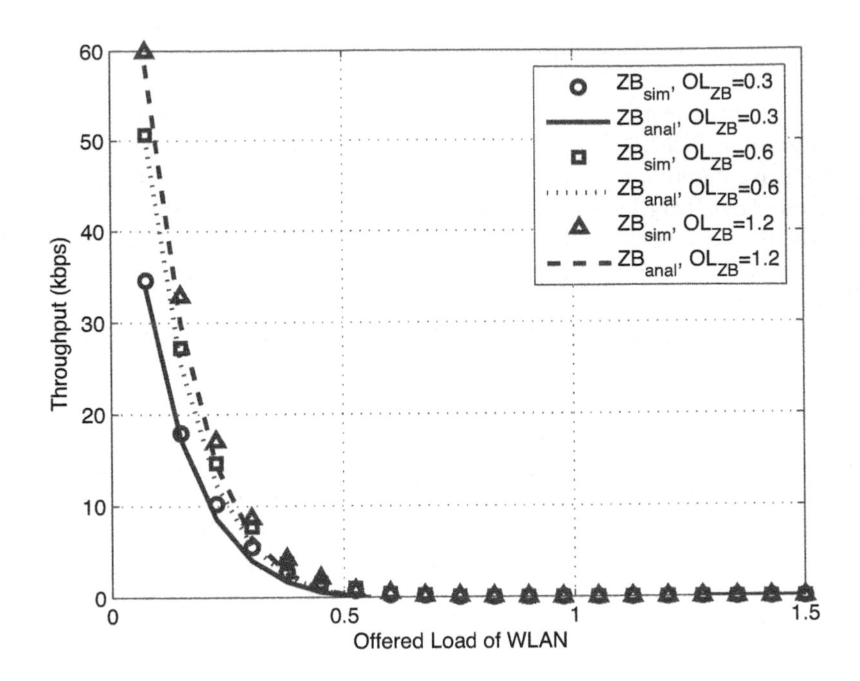

Figure 7.33: Throughput of the conventional ZigBee network in the presence of WLAN interference.

nificantly reduced as the offered load of WLAN, which severely interferes with the ZigBee network, increases.

The offered load (OL) of the i-th system can be defined as

$$OL_i = \frac{\lambda_i \times L_{frame,i} \times N_i}{R_i}, \quad i \in \{WLAN, ZigBee\}, \tag{7.18}$$

where λ_i, $L_{frame,i}$, N_i, and R_i represent the average arrival rate of an active node, frame length, the number of active nodes, and the data rate of the i-th system, respectively.

The performance of the *network-centric inter-system interference mediation algorithm* is investigated in terms of throughput of both WLAN and ZigBee networks through mathematical analysis and simulation. The *per-data transmission time-based fairness scheme* is applied as a time-based fairness scheme for the *fairness-based inter-system TDMA scheme* in the *network-centric inter-system interference mediation algorithm*. The parameters for performance evaluation are configured as listed in Table 7.7.

Fig. 7.39 shows the throughput of both WLAN and ZigBee networks for varying offered loads of ZigBee in the *network-centric inter-system interference mediation*

Table 7.7: WLAN and ZigBee Network Parameters

Parameter	Definition	Value
L_{WL}	Frame length of WLAN	500 bytes
R_{WL}	Data transmission rate of WLAN	11 Mbps
$CW_{min,WL}$	Minimum CW size of WLAN	31
δ_{WL}	Unit slot time of WLAN	20 μs
$Header_{WL}$	Header tx time of WLAN	210 μs
$SIFS_{WL}$	SIFS time of WLAN	10 μs
δ_{WL}	Propagation Time of WLAN	2 μs
ACK_{WL}	ACK tx time of WLAN	304 μs
$DIFS_{WL}$	DIFS time of WLAN	50 μs
$t_{\alpha,WL}$	Average backoff time of WLAN	1500 μs
N_{WL}	Number of active WLAN nodes	10
L_{ZB}	Frame length of ZigBee	105 bytes
R_{ZB}	Data transmission rate of ZigBee	250 kbps
$CW_{min,ZB}$	Minimum CW size of ZigBee	7
δ_{ZB}	Unit slot time of ZigBee	320 μs
CCA_{ZB}	CCA time of ZigBee	320 μs
$Header_{ZB}$	Header tx time of ZigBee	192 μs
ACK_{ZB}	ACK tx time of ZigBee	352 μs
$t_{\alpha,ZB}$	Average backoff time of ZigBee	1120 μs
N_{ZB}	Number of active ZigBee nodes	10

algorithm. The offered load of WLAN is set to 1.2 since the WLAN fully utilizes its allocated time resource when the offered load of WLAN becomes more than 1. Since the actual channel occupancy time of both systems varies, the throughput of the fully utilized system (i.e., WLAN) increases even if its offered load is fixed. If one system fully utilizes its given time and the other does not, then the algorithm allocates the remaining time resource of the less active system to the fully utilized system. Through this operation, the time resource can be more efficiently utilized than the original time-based fairness scheme.

Another key point of the result is that the ZigBee network with the *network-centric inter-system interference mediation algorithm* can achieve a significant amount of throughput while the throughput of conventional ZigBee network is nearly zero when the offered load of WLAN is 1.2. It is noted that the *network-centric inter-system interference mediation algorithm* can effectively mediate the inter-system interference and guarantee the ZigBee communications with fairness perspective.

So far, we investigated the coexistence algorithms between the ZigBee network and other networks such as WLAN in this section. In short, the coexistence algorithms presented in this section and their advantages and disadvantages are listed in Table 7.8.

Table 7.8: Coexistence Algorithms: Strong and Weak Points

Algorithm		Strong	Weak
Self-Interference Avoidance Algorithm	HBSIAA[a] [2]	simple, fast setup, self-algorithm	support of only single link, possible under available channels
	AIAMCCA[b] [17]	support of multiple nodes, self-algorithm	static clustering, possible under available channels
	ADIACA[c] [18]	support of multiple nodes, dynamic clustering, self-algorithm	possible under available channels
Network-Aided Interference Mitigation Algorithm	PDACA[d] [19]	support of multiple nodes, possible under fully utilized channels	need a portable device, performance loss in WLAN
	NCISIMA[e] [20]	support of multiple nodes, possible under fully utilized channels	need an IM node, performance loss in WLAN

[a]Hopping-baed interference avoidance algorithm
[b]Adaptive Interference-Aware Multi-Channel Clustering Algorithm
[c]Adaptive and Dynamic Interference-Aware Clustering Algorithm
[d]Portable Device Aided Coexistence Algorithm
[e]Network-Centric Inter-System Interference Mediation Algorithm

7.4 Discussion on Cognitive Radio as an Interference Mitigation Technology

Cognitive radio has been highlighted as a candidate technology by which secondary users can efficiently utilize temporarily unused spectrum of primary users. The cognitive radio can be considered as an interference mitigation technology in the viewpoint that the secondary users sense the shared channel, before transmitting their frames, to mitigate the collisions or interference to the primary users. The RawPEACH protocol is proposed for the cognitive radio concept to be applied as an interference mitigation technology in real network environments [24].

7.4.1 Cognitive Radio for Interference Mitigation: RawPEACH Protocol

In the RawPEACH protocol, the network model is considered as shown in Fig. 7.34. An access point (AP), primary users, and secondary users are considered. While the primary users require the strict guarantee of quality of service (QoS), the secondary

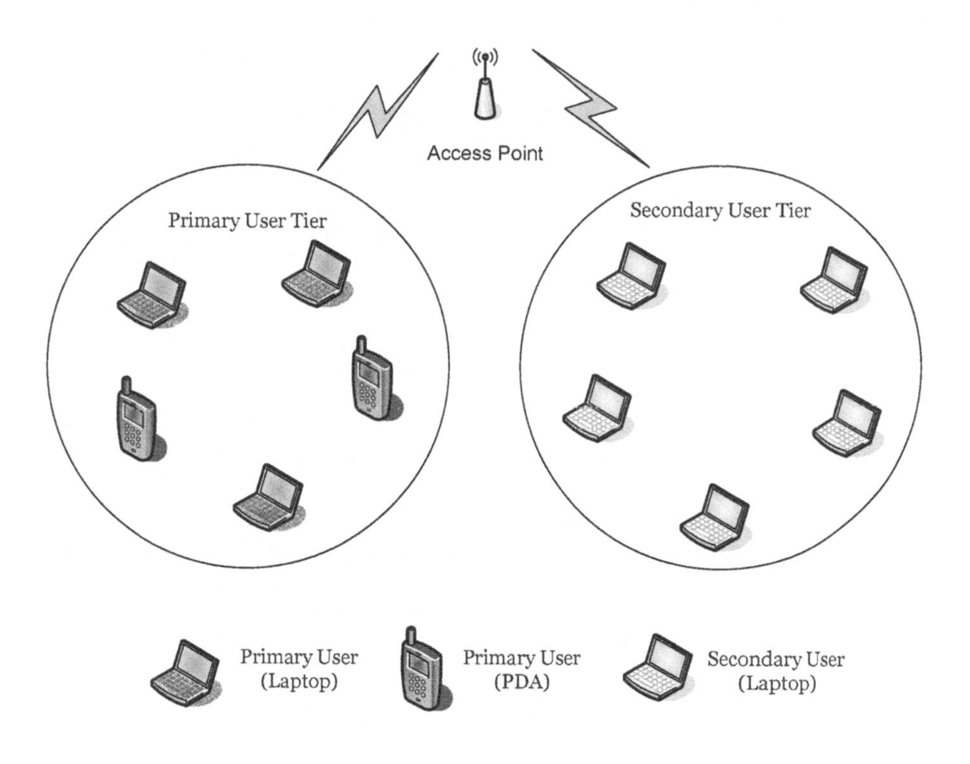

Figure 7.34: Network model in RawPEACH protocol.

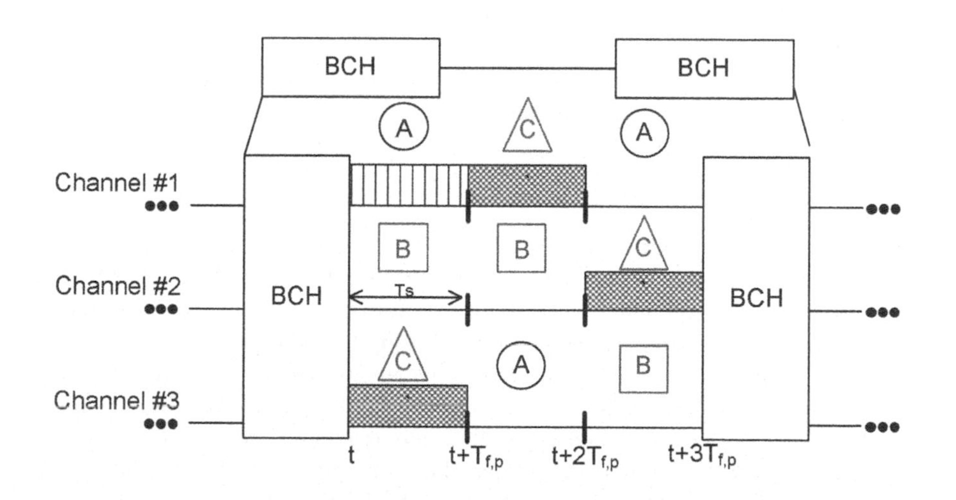

Figure 7.35: Operation of primary users in RawPEACH protocol.

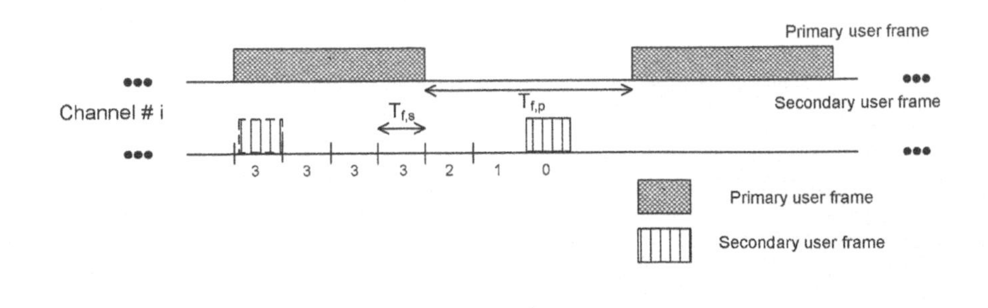

Figure 7.36: Operation of secondary users in RawPEACH protocol.

users just require best-effort services. The RawPEACH protocol can classify the user type in the association procedure.

The AP provides primary users with channelization as shown in Fig. 7.35. A contention-free channel is assigned to each primary user. For example, a channel defined as an orthogonal sequence $\{B_1, B_2, \cdots, B_n, \cdots\}$ with other primary users is allotted to one primary user. On the channel B_i, a primary user transmits its data frame during $[t + iT_{f,p}, t + (i+1)T_{f,p}]$, where $T_{f,p}$ is the frame interval of primary users.

The operation of the secondary users in RawPEACH is started by channel sensing; all the secondary users sense all the frequency channels at the beginning of every secondary time slot with length $T_{f,s}$ as shown in Fig. 7.36. The secondary frame length $T_{f,s}$ is required to be smaller than or equal to the frame length $T_{f,p}$ of primary users to prevent potential interference by secondary users. If the sensed channel is determined to be busy, secondary users do not transmit their frames to mitigate interference from/to primary users. The performance of secondary users can be managed by control parameters such as arbitrary interframe spaces (AIFSs), contention window (CW) size, and the frame lengths of primary and secondary users.

7.4.2 Performance Evaluation

Here, we evaluate the performance of RawPEACH protocol as a function of the number of orthogonal primary channels, the primary user activity and the number of secondary users. The secondary users are assumed to always have frames to transmit. Simulation and analysis results are obtained using OPNET simulation and [24], respectively. Figs. 7.37 and 7.38 show the normalized throughput S for varying the primary user activity P_p when the number N_{CH} of frequency channel is 1 and 3, respectively and the number of secondary users is increased from 2 to 12 in both scenarios. As expected, the throughput decreases as the primary user activity increases. When NCH is equal to one, the throughput decreases deeply while when NCH is equal to three an increase in the primary user activity slightly decreases the throughput.

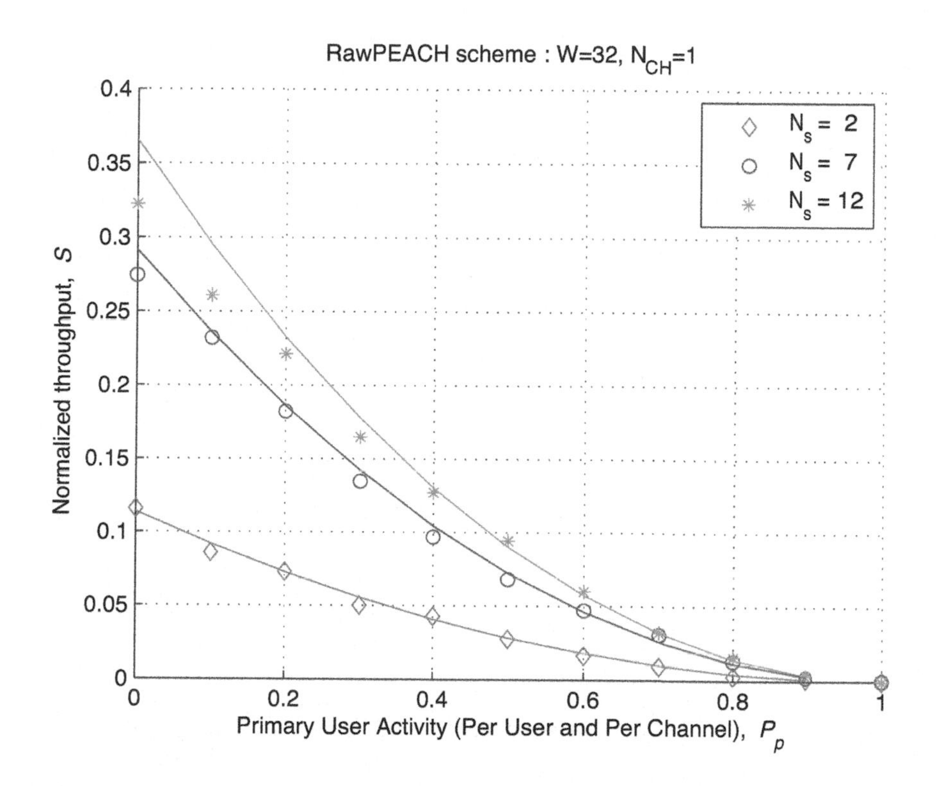

Figure 7.37: Throughput performance of RawPEACH protocol $N_{CH} = 1$.

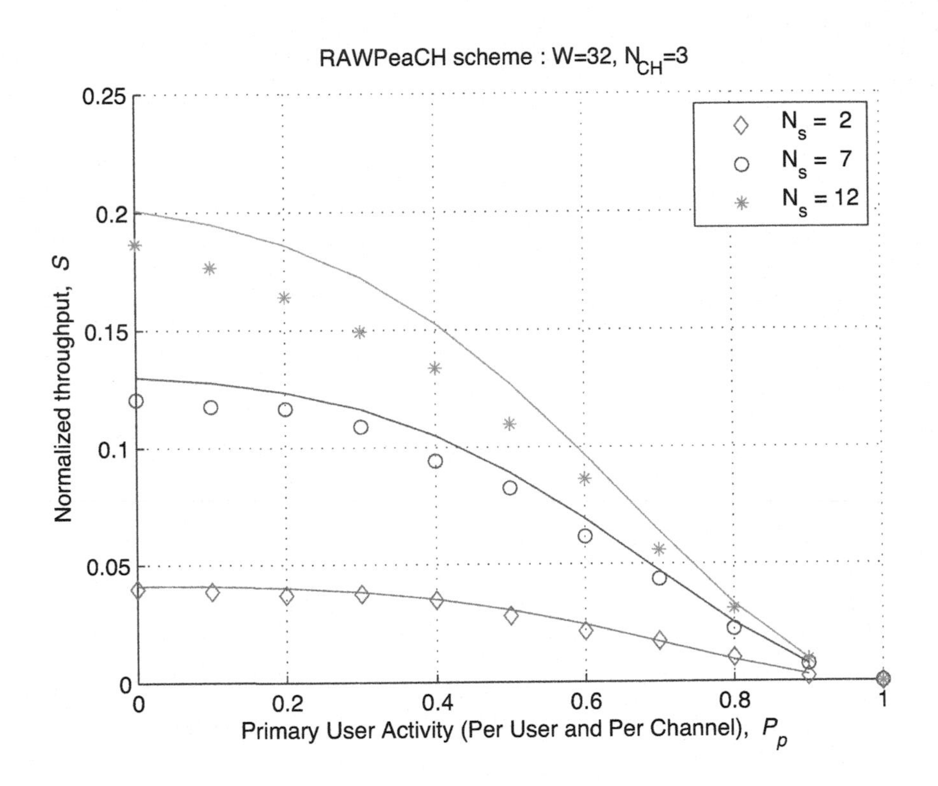

Figure 7.38: Throughput performance of RawPEACH protocol $N_{CH} = 3$.

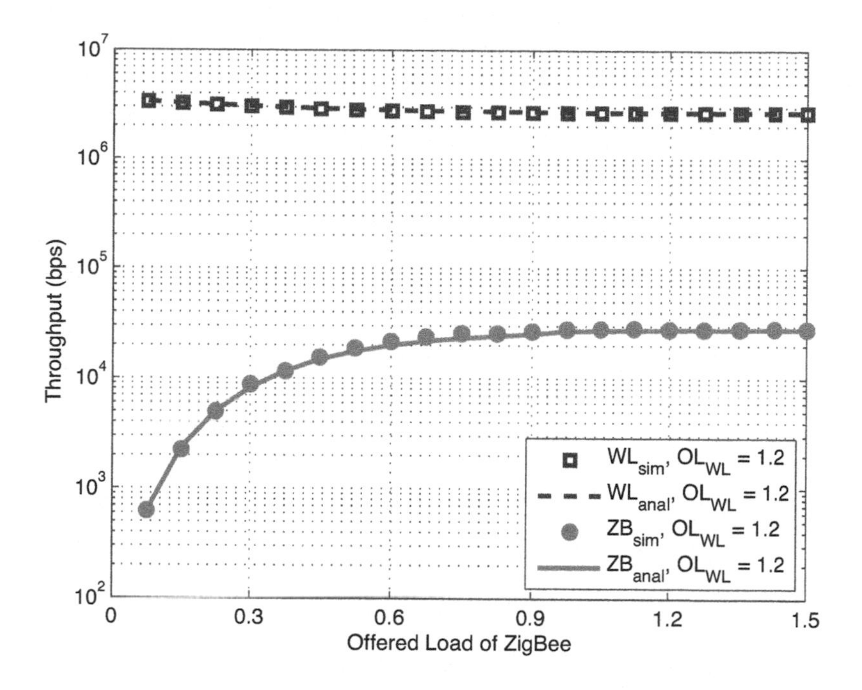

Figure 7.39: Throughput of WLAN and ZigBee networks for varying offered loads of ZigBee network in the network-centric inter-system interference mediation algorithm.

From these results, the application of the cognitive radio concept as an interference mitigation technology can be effective if the interference level is not severe.

7.5 Summary and Conclusion

This chapter presented several coexistence issues between ZigBee and other networks in 2.4GHz ISM band. First, the performance of the ZigBee network in the presence of external networks, such as WLAN, Bluetooth, and microwave ovens was evaluated through experiments and mathematical modeling and simulation. Analyses based on measurements and mathematical modeling show that the ZigBee network is prone to errors and its throughput is significantly degraded in the presence of interference since its transmission power emission is much lower than other systems in the 2.4 GHz ISM band. In order to mitigate this problem, two types of interference avoidance/mitigation algorithms have been proposed and their performance has been evaluated through experiments in a real testbed environment. If there are some interference-free channels, the self-interference avoidance algorithm effectively enhances the performance of ZigBee networks. The network-aided interference mitigation algorithm guarantees the ZigBee data transmissions even if the ZigBee network suffers from inevitable interference situations.

References

[1] IEEE 802.15.2 WG, *Part 15.2: Coexistence of Wireless Personal Area Networks with Other Wireless Devices Operating in Unlicensed Frequency Bands*, August 2003.

[2] S. M. Kim, J. W. Chong, C. Y. Jung, T. H. Jeon, J. H. Park, Y. J. Kang, S. H. Jeong, M. J. Kim, and D. K. Sung, "Experiments on Interference and Coexistence between ZigBee and WLAN Devices Operating in the 2.4 GHz ISM Band," *The Journal of Korean Institute of Next Generation PC*, Vol. 1, No. 2, pp. 24-33, December 2005.

[3] A. Sikora and V. F. Groza, "Coexistence of IEEE802.15.4 with other Systems in the 2.4GHz-ISM-Band," in Proceedings of *IEEE Instrumentation and Measurement Technology Conference*, May 2005.

[4] N. Golmie, D. Cypher and O. Rebala, "Performance analysis of low rate wireless technologies for medical applications," *Computer Communications*, Vol. 28, No. 10, pp. 1266-1275, June 2005.

[5] K. Shuaib, M. Boulmalf, F. Sallabi, and A. Lakas, "Co-exsitence of ZigBee and WLAN, A Performance Study," in Proceedings of *IEEE Wireless Telecommunications Symposium*, April 2006.

[6] J. W. Chong, H. Y. Hwang, C. Y. Jung and D. K. Sung, "Analysis of throughput and energy consumption in a ZigBee network under the presence of Bluetooth interference," in Proceedings of *IEEE Global Communications Conference (GLOBECOM)*, November 2007.

[7] J. W. Chong, H. Y. Hwang, C. Y. Jung and D. K. Sung, "Analysis of throughput in a ZigBee network under the presence of WLAN interference," In Proceedings of *IEEE International Symposium on Communications and Information Technologies (ISCIT)*, October 2007.

[8] D. K. Yoon, S. Y. Shin and W. H. Kwon, "Packet error rate analysis of IEEE 802.11b under IEEE 802.15.4 interference," In Proceedings of *IEEE Vehicular Technology Conference (VTC)*, May 2006.

[9] I. Howitt, and J. A. Gutierrez, "IEEE 802.15.4 Low Rate - Wireless Personal Area Network Coexistence Issues," in Proceedings of*IEEE Wireless Communication Networking Conference (WCNC)*, March 2003.

[10] S. Y Shin, H. S. Park, S. Choi, and W. H. Kwon, "Packet Error Rate Analysis of ZigBee Under WLAN and Bluetooth Interferences," *IEEE Transactions on Wireless Communications*, Vol. 6, No. 8, pp. 2825-2830, August 2007.

[11] M. Petrova, J. Riihijärvi, P. Mähönen, and S. Labella, "Performance Study of IEEE 802.15.4 Using Measurements and Simulations," in Proceedings of *IEEE Wireless Communications and Networking Conference (WCNC)*, May 2006.

[12] IEEE 802.15.1 Specification, *Wireless Medium Access Control (MAC) and Physical Layer (PHY) Specifications for Wireless Personal Area Networks (WPANs)*, June 2002.

[13] IEEE 802.15.4 Specification, *Wireless Medium Access Control (MAC) and Physical Layer (PHY) Specifications for Low Rate Wireless Personal Area Networks (LR-WPANs)*, October 2003.

[14] IEEE 802.11b Specification, *Wireless LAN Medium Access Control (MAC) and Physical Layer (PHY) Specifications: High-Speed Physical Layer Extension in the 2.4 GHz Band*, September 1999.

[15] IEEE 802.11g Specification, *Wireless LAN Medium Access Control (MAC) and Physical Layer (PHY) Specifications Amendment 4: Further Higher Data Rate Extension in the 2.4 GHz Band*, June 2003.

[16] T. R. Park, T. H. Kim, J. Y. Choi, S. Choi and W. H. Kwon, "Throughput and energy consumption analysis of IEEE 802.15.4 slotted CSMA/CA," *IEE Electronics Letters*, Vol. 14, No. 18, pp. 1017-1019, September 2005.

[17] M. S. Kang, J. W. Chong, H. Hyun, S. M. Kim, B. H. Jung, and D. K. Sung, "Adaptive Interference-Aware Multi-Channel Clusturing Algorithm in a ZigBee Network in the Presence of WLAN Interference," In Proceedings of *IEEE International Symposium on Wireless Pervasive Computing (ISWPC)*, February 2007.

[18] D. N. Ko, J. W. Chong, B. H. Jung, and D. K. Sung, "Dynamic Clustering Schemes of ZigBee Networks in the Presence of WLAN Interference," In Proceedings of *IEEE Military Communications Conference (MILCOM)*, November 2008.

[19] B. H. Jung, J. W. Chong, S. H. Jeong, H. Y. Hwang, S. M. Kim, M. S. Kang, and D. K. Sung, "Ubiquitous Wearable Computer (UWC)-Aided Coexistence Algorithm in an Overlaid Network Environment of WLAN and ZigBee Networks," In Proceedings of *IEEE International Symposium on Wireless Pervasive Computing (ISWPC)*, February 2007.

[20] B. H. Jung, J. W. Chong, C. Y. Jung, S. M. Kim, and D. K. Sung, "Interference Mediation for Coexistence of WLAN and ZigBee Networks," In Proceedings of *IEEE Personal, Indoor and Mobile Radio Communications Symposium (PIMRC)*, September 2008.

[21] A. V. Babu and L. Jacob, "Performance Analysis of IEEE 802.11 Multirate WLANs: Time Based Fairness versus Throughput Based Fairness," In Proceedings of *IEEE International Wireless Communications and Mobile Computing Conference (IWCMC)*, June 2005.

[22] G. Tan and J. Guttag, "Time-based Fairness Improves Performance in Multi-rate WLANs," In Proceedings of *USENIX Annual Technical Conference*, June 2004.

[23] B. H. Jung, S. J. Kim, H. Jin, H. Y. Hwang, J. W. Chong, and D. K. Sung, "Performance Improvement of Error-Prone Multi-Rate WLANs through Payload Size Adjustment," In Proceedings of *IEEE International Conference on Communications (ICC)*, June 2009.

[24] J. W. Chong, Y. Sung, and D. K. Sung, "RawPEACH: Multiband CSMA/CA-Based Cognitive Radio Networks," *Journal of Communications and Networks*, Vol. 11, No. 2, pp. 174-185, April 2009.

Chapter 8

ZigBee Profile

Pan Zhou, Tao Jiang, Chonggang Wang, Qian Zhang

CONTENTS

ZigBee is a specification for a suite of high level communication protocols used to create personal area networks built from small, low-power digital radios. ZigBee is a low-cost, low-power, wireless mesh network standard that is based on an IEEE 802 standard. Though low-powered, ZigBee devices often transmit data over longer distances by passing data through intermediate devices to reach more distant ones, creating a mesh network; i.e., a network with no centralized control or high-power transmitter/receiver able to reach all of the networked devices. The decentralized nature of such wireless ad-hoc networks make them suitable for applications where a central node can't be relied upon.

8.1 ZigBee Overview

ZigBee is used in applications that require a low data rate, long battery life, and secure networking. ZigBee has a defined rate of 250 kbit/s, best suited for periodic or intermittent data or a single signal transmission from a sensor or input device. Applications include wireless light switches, electrical meters with in-home-displays, traffic management systems, and other consumer and industrial equipment that require short-range wireless transfer of data at relatively low rates. The technology defined by the ZigBee specification is intended to be simpler and less expensive than other WPANs, such as Bluetooth or Wi-Fi. ZigBee networks are secured by 128 bit symmetric encryption keys. In home automation applications, transmission distances range from 10 to 100 meters line-of-sight, depending on power output and environmental characteristics[1].

ZigBee was conceived in 1998, standardized in 2003 and revised in 2006. The name refers to the waggle dance of honey bees after their return to the beehive[2].

The low cost allows the technology to be widely deployed in wireless control and monitoring applications. Low power-usage allows longer life with smaller batteries. Mesh networking provides high reliability and more extensive range. ZigBee chip vendors typically sell integrated radios and microcontrollers with between 60 KB and 256 KB flash memory.

ZigBee operates in the industrial, scientific and medical (ISM) radio bands; 868 MHz in Europe, 915 MHz in the USA and Australia and 2.4 GHz in most jurisdictions worldwide. Data transmission rates vary from 20 kilobits/second in the 868 MHz frequency band to 250 kilobits/second in the 2.4 GHz frequency band[3].

The ZigBee network layer natively supports both star and tree typical networks, and generic mesh networks. Every network must have one coordinator device, tasked with its creation, the control of its parameters and basic maintenance. Within star networks, the coordinator must be the central node. Both trees and meshes allow the use of ZigBee routers to extend communication at the network level.

ZigBee builds upon the physical layer and media access control defined in IEEE standard 802.15.4 (2003 version) for low-rate WPANs. The specification goes on to complete the standard by adding four main components: network layer, application layer, ZigBee device objects (ZDOs) and manufacturer-defined application objects which allow for customization and favor total integration. Besides adding two high-level network layers to the underlying structure, the most significant improvement is the introduction of ZDOs. These are responsible for a number of tasks, which include keeping of device roles, management of requests to join a network, device discovery and security.

ZigBee is not intended to support powerline networking but to interface with it at least for smart metering and smart appliance purposes. Because ZigBee nodes can go from sleep to active mode in 30 ms or less, the latency can be low and devices can be responsive, particularly compared to Bluetooth wake-up delays, which are typically around three seconds[4]. Because ZigBee nodes can sleep most of the time, average power consumption can be low, resulting in long battery life. A Bluetooth SMART device can, when advertising is pushed to maximum connect, exchange data and disconnect in 3 ms. This significantly enhances the experiences for HID devices.

8.1.1 History

ZigBee-style networks began to be conceived around 1999, when many installers realized that both Wi-Fi and Bluetooth were going to be unsuitable for many applications. In particular, many engineers saw a need for self-organizing ad-hoc digital radio networks. The real need for mesh has been cast in doubt since that, in particular as mesh is largely absent in the market.

The IEEE 802.15.4-2003 standard was completed in May 2003 and has been superseded by the publication of IEEE 802.15.4-2006[5]. In the summer of 2003, Philips Semiconductors, a major mesh network supporter, ceased the investment. Philips Lighting has, however, continued Philips' participation, and Philips remains a promoter member on the ZigBee Alliance Board of Directors.

The ZigBee Alliance announced in October 2004 that the membership had more than doubled in the preceding year and had grown to more than 100 member companies, in 22 countries. By April 2005 membership had grown to more than 150 companies, and by December 2005 membership had passed 200 companies.

The ZigBee specifications were ratified on 14 December 2004[6]. The ZigBee Alliance announced availability of Specification 1.0 on 13 June 2005, known as ZigBee 2004 Specification. In September 2006, ZigBee 2006 Specification was announced. In 2007, ZigBee PRO, the enhanced ZigBee specification was finalized.

The first stack release is now called ZigBee 2004. The second stack release is called ZigBee 2006, and mainly replaces the MSG/KVP (Message/Key Value Pair) structure used in 2004 with a "cluster library". The 2004 stack is now more or less obsolete.

ZigBee 2007, now the current stack release, contains two stack profiles, stack profile 1 (simply called ZigBee), for home and light commercial use, and stack profile 2 (called ZigBee PRO). ZigBee PRO offers more features, such as multi-casting, many-to-one routing and high security with Symmetric-Key Key Exchange (SKKE), while ZigBee (stack profile 1) offers a smaller footprint in RAM and flash. Both offer full mesh networking and work with all ZigBee application profiles.

ZigBee 2007 is fully backward compatible with ZigBee 2006 devices: A ZigBee 2007 device may join and operate on a ZigBee 2006 network and vice versa. Due to differences in routing options, ZigBee PRO devices must become non-routing ZigBee End-Devices (ZEDs) on a ZigBee 2006 network, the same as for ZigBee 2006 devices on a ZigBee 2007 network must become ZEDs on a ZigBee PRO network. The applications running on those devices work the same, regardless of the stack profile beneath them.

The ZigBee 1.0 specification was ratified on 14 December 2004 and is available to members of the ZigBee Alliance. Most recently, the ZigBee 2007 specification was posted on 30 October 2007. The first ZigBee Application Profile, Home Automation, was announced 2 November 2007.

8.1.2 Standard and Profiles

The ZigBee Alliance is a group of companies that maintain and publish the ZigBee standard[10]. The term ZigBee is a registered trademark of this group, not a single technical standard. The Alliance publishes application profiles that allow multiple OEM vendors to create interoperable products. The relationship between IEEE 802.15.4 and ZigBee[11] is similar to that between IEEE 802.11 and the Wi-Fi Alliance.

8.1.3 License

For non-commercial purposes, the ZigBee specification is available free to the general public[12]. An entry level membership in the ZigBee Alliance, called Adopter, provides access to the as-yet unpublished specifications and permission to create products for market using the specifications. The requirements for membership in the ZigBee Alliance causes problems for Free Software developers because the annual fee conflicts with the GNU General Public Licence[13]. The requirement for the developer to join the ZigBee Alliance similarly conflicts with most other free software licenses[14]. ZigBee Alliance board has been asked to make their license compatible with GPL, but the ZigBee board refused. The refusal came, even though Bluetooth had already changed its license to make it compatible with GPL. Linux developers seem ready to abandon ZigBee, and use TCP/IP instead[15][16].

8.1.4 Application Profiles

ZigBee protocols are intended for embedded applications requiring low data rates and low power consumption. The resulting network will use very small amounts of power: individual devices must have a battery life of at least two years to pass ZigBee certification[7].

Typical application areas include[8]:

- Home Entertainment and Control: Home automation, smart lighting[9], advanced temperature control, safety and security, movies and music
- Wireless sensor networks: Starting with individual sensors like Telosb/Tmote and Iris from Memsic
- Industrial control
- Embedded sensing
- Medical data collection
- Smoke and intruder warning
- Building automation

The current list of application profiles either published, or in development are:

- Released specifications
  - ZigBee Home Automation
  - ZigBee Smart Energy 1.0
  - ZigBee Telecommunication Services
  - ZigBee Health Care
  - ZigBee RF4CE C Remote Control
  - ZigBee RF4CE C Input Device
  - ZigBee Light Link
- Specifications under development
  - ZigBee Smart Energy 2.0
  - ZigBee Building Automation
  - ZigBee Retail Services

The ZigBee Smart Energy V2.0 specifications define an IP-based protocol to monitor, control, inform and automate the delivery and use of energy and water. It is an enhancement of the ZigBee Smart Energy version 1 specifications[17], adding services for plug-in electric vehicle (PEV) charging, installation, configuration and firmware download, prepay services, user information and messaging, load control, demand response and common information and application profile interfaces for wired and wireless networks. It is being developed by partners including:

- HomeGrid Forum responsible for marketing and certifying ITU-T G.hn technology and products
- HomePlug Powerline Alliance
- International Society of Automotive Engineers SAE International
- IPSO Alliance
- SunSpec Alliance
- Wi-Fi Alliance.

In 2009 the RF4CE (Radio Frequency for Consumer Electronics) Consortium and ZigBee Alliance agreed to jointly deliver a standard for radio frequency remote controls. ZigBee RF4CE is designed for a wide range of consumer electronics products, such as TVs and set-top boxes. It promises many advantages over existing remote control solutions, including richer communication and increased reliability, enhanced features and flexibility, interoperability, and no line-of-sight barrier[18]. The ZigBee RF4CE specification lifts off some networking weight and does not support all the mesh features, which is traded for smaller memory configurations for lower cost devices, such as remote control of consumer electronics.

With the introduction of the second ZigBee RF4CE application profile in 2012 and increased momentum in MSO market, the ZigBee RF4CE team provides an overview on current status of the standard, applications, and future of the technology[19][20].

8.1.5 ZigBee Radio Hardware

The radio design used by ZigBee has been carefully optimized for low cost in large scale production. It has few analog stages and uses digital circuits wherever possible. Though the radios themselves are inexpensive, the ZigBee Qualification Process involves a full validation of the requirements of the physical layer. All radios derived from the same validated semiconductor mask set would enjoy the same RF characteristics. An uncertified physical layer that malfunctions could cripple the battery lifespan of other devices on a ZigBee network. ZigBee radios have very tight constraints on power and bandwidth. Thus, radios are tested with guidance given by Clause 6 of the 802.15.4-2006 Standard. Most vendors plan to integrate the radio and microcontroller onto a single chip[21] getting smaller devices[22]. This standard specifies operation in the unlicensed 2.4 GHz (worldwide), 915 MHz (Americas and Australia) and 868 MHz (Europe) ISM bands. Sixteen channels are allocated in the 2.4 GHz band, with each channel requiring 5 MHz of bandwidth. The radios use direct-sequence spread spectrum coding, which is managed by the digital stream into the modulator. Binary phase-shift keying (BPSK) is used in the 868 and 915 MHz bands, and offset quadrature phase-shift keying (OQPSK) that transmits two bits per symbol is used in the 2.4 GHz band. The raw, over-the-air data rate is 250 kbit/s per channel in the 2.4 GHz band, 40 kbit/s per channel in the 915 MHz band, and 20 kbit/s in the 868 MHz band. The actual data throughput will be less than the maximum specified bit rate due to the packet overhead and processing delays. For indoor applications at 2.4 GHz transmission distance may be 10-20 m, depending on the construction materials, the number of walls to be penetrated and the output power permitted in that geographical location[23]. Outdoors with line-of-sight, range may be up to 1500 m depending on power output and environmental characteristics[1]. The output power of the radios is generally 0-20 dBm (1-100 mW).

8.1.6 Device Types and Operating Modes

ZigBee devices are of three types:

- ZigBee Coordinator (ZC): The most capable device, the Coordinator forms the root of the network tree and might bridge other networks. There is exactly one ZigBee Coordinator in each network since it is the device that started the network originally (the ZigBee LightLink specification also allows operation without a ZigBee Coordinator, making it more usable for over-the-shelf home products). It stores information about the network, including acting as the Trust Center & repository for security keys[24][25].

- ZigBee Router (ZR): As well as running an application function, a Router can act as an intermediate router, passing on data from other devices.

- ZigBee End Device (ZED): Contains just enough functionality to talk to the parent node (either the Coordinator or a Router); it cannot relay data from other devices. This relationship allows the node to be asleep a significant amount of the time thereby giving long battery life. A ZED requires the least amount of memory, and therefore can be less expensive to manufacture than a ZR or ZC.

The current ZigBee protocols support beacon and non-beacon enabled networks. In non-beacon-enabled networks, an unslotted CSMA/CA channel access mechanism is used. In this type of network, ZigBee Routers typically have their receivers continuously active, requiring a more robust power supply. However, this allows for heterogeneous networks in which some devices receive continuously, while others only transmit when an external stimulus is detected. The typical example of a heterogeneous network is a wireless light switch: The ZigBee node at the lamp may receive constantly, since it is connected to the main supply, while a battery-powered light switch would remain asleep until the switch is thrown. The switch then wakes up, sends a command to the lamp, receives an acknowledgment, and returns to sleep. In such a network the lamp node will be at least a ZigBee Router, if not the ZigBee Coordinator; the switch node is typically a ZigBee End Device.

In beacon-enabled networks, the special network nodes called ZigBee Routers transmit periodic beacons to confirm their presence to other network nodes. Nodes may sleep between beacons, thus lowering their duty cycle and extending their battery life. Beacon intervals depend on data rate; they may range from 15.36 milliseconds to 251.65824 seconds at 250 kbit/s, from 24 milliseconds to 393.216 seconds at 40 kbit/s and from 48 milliseconds to 786.432 seconds at 20 kbit/s. However, low duty cycle operation with long beacon intervals requires precise timing, which can conflict with the need for low product cost. In general, the ZigBee protocols minimize the time the radio is on, so as to reduce power use. In beaconing networks, nodes only need to be active while a beacon is being transmitted. In non-beacon-enabled networks, power consumption is decidedly asymmetrical: some devices are always active, while others spend most of their time sleeping.

Except for the Smart Energy Profile 2.0, ZigBee devices are required to conform to the IEEE 802.15.4-2003 Low-Rate Wireless Personal Area Network (LR-WPAN) standard. The standard specifies the lower protocol layers: the physical layer (PHY), and the media access control portion of the data link layer (DLL). The basic channel access mode is "carrier sense, multiple access/collision avoidance" (CSMA/CA). That is, the nodes talk in the same way that humans converse; they briefly check to see that no one is talking before they start, with three notable exceptions. Beacons are sent on a fixed timing schedule, and do not use CSMA. Message acknowledgments also do not use CSMA. Finally, devices in beacon-enabled networks that have low latency real-time requirements may also use Guaranteed Time Slots (GTS), which by definition do not use CSMA.

8.1.7 Network Layer

The main functions of the network layer are to enable the correct use of the MAC sublayer and provide a suitable interface for use by the next upper layer, namely the application layer. Its capabilities and structure are those typically associated to such network layers, including routing.

On the one hand, the data entity creates and manages network layer data units from the payload of the application layer and performs routing according to the current topology. On the other hand, there is the layer control, which is used to handle configuration of new devices and establish new networks: it can determine whether a neighboring device belongs to the network and discovers new neighbors and routers. The control can also detect the presence of a receiver, which allows direct communication and MAC synchronization.

The routing protocol used by the Network layer is AODV. In order to find the destination device, it broadcasts out a route request to all of its neighbors. The neighbors then broadcast the request to their neighbors, etc., until the destination is reached. Once the destination is reached, it sends its route reply via unicast transmission following the lowest cost path back to the source. Once the source receives the reply, it will update its routing table for the destination address with the next hop in the path and the path cost.

8.1.8 Application Layer

The application layer is the highest-level layer defined by the specification, and is the effective interface of the ZigBee system to its end users. It is comprised of the majority of components added by the ZigBee specification: both ZDO and its management procedures, together with application objects defined by the manufacturer, are considered part of this layer.

8.1.9 Main Components

The ZDO is responsible for defining the role of a device as either coordinator or end device, as mentioned above, but also for the discovery of new (one-hop) devices on the network and the identification of their offered services. It may then go on to establish secure links with external devices and reply to binding requests accordingly.

The application support sublayer (APS) is the other main standard component of the layer, and as such it offers a well-defined interface and control services. It works as a bridge between the network layer and the other components of the application layer: it keeps up-to-date binding tables in the form of a database, which can be used to find appropriate devices depending on the services that are needed and those that the different devices offer. As the union between both specified layers, it also routes messages across the layers of the protocol stack.

8.1.10 Communication Models

An application may consist of communicating objects which cooperate to carry out the desired tasks. The focus of ZigBee is to distribute work among many different devices which reside within individual ZigBee nodes which in turn form a network (said work will typically be largely local to each device, for instance the control of each individual household appliance).

The collection of objects that form the network communicate using the facilities provided by APS, supervised by ZDO interfaces. The application layer data service follows a typical request-confirm/indication-response structure. Within a single device, up to 240 application objects can exist, numbered in the range 1-240. 0 is reserved for the ZDO data interface and 255 for broadcast; the 241-254 range is not currently in use but may be in the future. Two services are available for application objects to use (in ZigBee 1.0):

- The key-value pair (KVP) service is meant for configuration purposes. It enables description, request and modification of object attributes through a simple interface based on get/set and event primitives, some allowing a request for response. Configuration uses compressed XML (full XML can be used) to provide an adaptable and elegant solution.

- The message service is designed to offer a general approach to information treatment, avoiding the necessity to adapt application protocols and potential overhead incurred by KVP. It allows arbitrary payloads to be transmitted over APS frames.

Addressing is also part of the application layer. A network node consists of an 802.15.4-conformant radio transceiver and one or more device descriptions (basically collections of attributes which can be polled or set, or which can be monitored through events). The transceiver is the base for addressing, and devices within a node are specified by an endpoint identifier in the range 1-240.

8.1.11 Communication and Device Discovery

In order for applications to communicate, their comprising devices must use a common application protocol (types of messages, formats and so on); these sets of conventions are grouped in profiles. Furthermore, binding is decided upon by matching input and output cluster identifiers, unique within the context of a given profile and associated to an incoming or outgoing data flow in a device. Binding tables contain source and destination pairs.

Depending on the available information, device discovery may follow different methods. When the network address is known, the IEEE address can be requested using unicast communication. When it is not, petitions are broadcast (the IEEE address being part of the response payload). End devices will simply respond with the requested address, while a network coordinator or a router will also send the addresses of all the devices associated with it.

This extended discovery protocol permits external devices to find out about devices in a network and the services that they offer, which endpoints can report when queried by the discovering device (which has previously obtained their addresses). Matching services can also be used.

The use of cluster identifiers enforces the binding of complementary entities by means of the binding tables, which are maintained by ZigBee coordinators, as the table must be always available within a network and coordinators are most likely to have a permanent power supply. Backups, managed by higher-level layers, may be needed by some applications. Binding requires an established communication link; after it exists, whether to add a new node to the network is decided, according to the application and security policies.

Communication can happen right after the association. Direct addressing uses both radio address and endpoint identifier, whereas indirect addressing uses every relevant field (address, endpoint, cluster and attribute) and requires that they be sent to the network coordinator, which maintains associations and translates requests for communication. Indirect addressing is particularly useful to keep some devices very simple and minimize their need for storage. Besides these two methods, broadcast to all endpoints in a device is available, and group addressing is used to communicate with groups of endpoints belonging to a set of devices.

8.1.12 Security Services

As one of its defining features, ZigBee provides facilities for carrying out secure communications, protecting establishment and transport of cryptographic keys, cyphering frames and controlling devices. It builds on the basic security framework defined in IEEE 802.15.4. This part of the architecture relies on the correct management of symmetric keys and the correct implementation of methods and security policies.

8.1.13 Basic Security Model

The basic mechanism to ensure confidentiality is the adequate protection of all keying material. Trust must be assumed in the initial installation of the keys, as well as in the processing of security information. In order for an implementation to globally work, its general conformance to specified behaviors is assumed.

Keys are the cornerstone of the security architecture; as such their protection is of paramount importance, and keys are never supposed to be transported through an insecure channel. A momentary exception to this rule occurs during the initial phase of the addition to the network of a previously unconfigured device. The ZigBee network model must take particular care of security considerations, as ad hoc networks may be within the protocol stack, different network layers are not cryptographically separated, so access policies are needed and correct design assumed. The open trust model within a device allows for key sharing, which notably decreases potential cost. Nevertheless, the layer which creates a frame is responsible for its security. If malicious devices may exist, every network layer payload must be ciphered, so unauthorized traffic can be immediately cut off. The exception, again, is the transmission of the network key, which confers a unified security layer to the network, to a new connecting device.

8.1.14 Security Architecture

ZigBee uses 128-bit keys to implement its security mechanisms. A key can be associated either to a network, being usable by both ZigBee layers and the MAC sublayer, or to a link, acquired through pre-installation, agreement or transport. Establishment of link keys is based on a master key which controls link key correspondence. Ultimately, at least the initial master key must be obtained through a secure medium (transport or pre-installation), as the security of the whole network depends on it. Link and master keys are only visible to the application layer. Different services use different one-way variations of the link key in order to avoid leaks and security risks.

Key distribution is one of the most important security functions of the network. A secure network will designate one special device which other devices trust for the distribution of security keys: the trust center. Ideally, devices will have the trust center address and initial master key preloaded; if a momentary vulnerability is allowed, it will be sent as described above. Typical applications without special security needs will use a network key provided by the trust center (through the initially insecure channel) to communicate.

Thus, the trust center maintains both the network key and provides point-to-point security. Devices will only accept communications originating from a key provided by the trust center, except for the initial master key. The security architecture is distributed among the network layers as follows:

- The MAC sublayer is capable of single-hop reliable communications. As a rule, the security level it is to use is specified by the upper layers.
- The network layer manages routing, processing received messages and being capable of broadcasting requests. Outgoing frames will use the adequate link key according to the routing, if it is available; otherwise, the network key will be used to protect the payload from external devices.
- The application layer offers key establishment and transport services to both ZDO and applications. It is also responsible for the propagation across the network of changes in devices within it, which may originate in the devices themselves (for instance, a simple status change) or in the trust manager (which may inform the network that a certain device is to be eliminated from it). It also routes requests from devices to the trust center and network key renewals from the trust center to all devices. Besides this, the ZDO maintains the security policies of the device.

The security levels infrastructure is based on CCM, which adds encryption- and integrity-only features to CCM.

8.1.15 Simulation of ZigBee Networks

Network simulators, like OPNET, NetSim and NS2, can be used to simulate IEEE 802.15.4 ZigBee networks. These simulators come with open source C or C++ libraries for users to modify. This way users can check out the validity of new algorithms prior to hardware implementation.

8.2 Latest Situation of ZigBee Specification Overview

The latest ZigBee specification, officially named ZigBee 2012, offers full wireless mesh networking capable of supporting more than 64,000 devices on a single network. Its designed to connect the widest range of devices, in any industry, into a single control network. ZigBee supports the largest number of interoperable standards including ZigBee Building Automation, ZigBee Health Care, ZigBee Home

Automation, ZigBee Light Link, ZigBee Smart Energy, ZigBee Telecom Services, and the forthcoming ZigBee Retail Services.

The ZigBee 2012 specification has two implementation options or Feature Sets: ZigBee and ZigBee PRO. The ZigBee Feature Set is designed to support smaller networks with hundreds of devices in a single network. The ZigBee PRO Feature Set is the most popular choice of developers and the specification used for most Alliance developed standards. It maximizes all the capabilities of the ZigBee Feature Set, plus facilitates ease-of-use and advanced support for larger networks comprised of thousands of devices ZigBee PRO now offers an optional new and innovative feature, Green Power to connect energy harvesting or self-powered devices into ZigBee PRO networks. Both Feature Sets are designed to interoperate with each other, ensuring long-term use and stability.

The ZigBee specification enhances the IEEE 802.15.4 standard by adding network and security layers and an application framework. From this foundation, Alliance developed standards can be used to create a multi-vendor interoperable solutions. For custom application where interoperability is not required, manufacturers can create their own manufacturer specific standards.

Some of the characteristics of ZigBee include:

- Global operation in the 2.4GHz frequency band according to IEEE 802.15.4
- Regional operation in the 915Mhz (Americas) and 868Mhz (Europe)
- Frequency agile solution operating over 16 channels in the 2.4GHz frequency
- Incorporates power saving mechanisms for all device classes, plus support for battery-less devices
- Discovery mechanism with full application confirmation
- Pairing mechanism with full application confirmation
- Multiple star topology and inter-personal area network (PAN) communication
- Various transmission options including broadcast
- Security key generation mechanism
- Utilizes the industry standard AES-128 security scheme
- Supports Alliance standards or manufacturer specific innovations

A complete list of Features and Benefits for the ZigBee specification can be seen in [27]. Here we discuss the following features that get the most attention for ZigBee in near future applications: green power feature, RF4CE control feature, smart energy feature, smart energy profile 2 and IP support feature.

8.3 ZigBee PRO Green Power Feature

With the enhanced ZigBee 2012 specification, the ZigBee PRO feature set gains an new optional feature: Green Power. The ZigBee PRO Green Power feature allows battery-less devices to securely join ZigBee PRO networks. It is the most eco-friendly way to power ZigBee products such as sensors, switches, dimmers and many other devices. These devices can now be powered just by using widely available, but often missed sources of energy like motion, light, vibration, to name a few. The energy used to flip a typical light switch via common energy harvesting techniques, is powerful enough to generate and send commands through a ZigBee PRO 2012 network.

A ZigBee PRO 2012 network is Green Power-ready if two of its devices implement the Green Power Proxy and Green Power Sink functionalities. The Proxy interfaces with the Green Power device and is typically part of an always powered on device that serves as a ZigBee PRO router. The Sink function (lamp) is typically found in the target device to be controlled and handles the interpretation of the Green Power device (energy harvesting light switch) commands. Both Proxy and Sink functionalities need to be present in a ZigBee PRO 2012 network and they can even be implemented in the same device. While ZigBee PRO networks have always been efficient power users, the ZigBee PRO Green Power feature significantly expands the capabilities of ZigBee PRO. It strengthens its leadership position as the global standard for wireless sensor and control networks and the Internet of Things. Product manufacturers can implement ZigBee into more products with confidence, knowing ZigBee is backed by a thriving, innovative and competitive ecosystem versus proprietary and single-source technologies.

With ZigBee PRO Green Power products, consumers and businesses will appreciate the install-it and forget it simplicity. They can add ZigBee devices to more areas with greater ease, including locations where power is unavailable, not allowed for safety reasons or for historical preservation purposes. Plus, these devices can join larger ZigBee networks and deliver more control than ever before.

8.4 The ZigBee RF4CE Specification

8.4.1 ZigBee RF4CE Overview

The ZigBee RF4CE specification offers an immediate, low-cost, easy-to-implement solution for control of products and opportunity for a couple of standards including ZigBee Remote Control and ZigBee Input Device. The ZigBee RF4CE specification is designed to control a wide range of products including home entertainment devices, garage door openers, keyless entry systems and many more.

The ZigBee RF4CE specification defines a remote control (RC) network that defines a simple, robust and low-cost communication network allowing wireless connectivity in applications for consumer electronic (CE) devices.

The ZigBee RF4CE specification enhances the IEEE 802.15.4 standard by providing a simple networking layer and standard application layer that can be used to create a multi-vendor interoperable solution for use within the home.

Some of the characteristics of ZigBee RF4CE include:

- Operation in the 2.4GHz frequency band according to IEEE 802.15.4
- Frequency agile solution operating over three channels
- Incorporates power saving mechanisms for all device classes
- Discovery mechanism with full application confirmation
- Pairing mechanism with full application confirmation
- Multiple star topology with inter-personal area network (PAN) communication
- Various transmission options including broadcast
- Security key generation mechanism
- Utilizes the industry standard AES-128 security scheme
- Specifies a simple RC control profile for CE products
- Support Alliance developed standards or manufacturer specific profiles

8.4.2 ZigBee RF4CE Network Topology

The ZigBee RF4CE network is composed of two types of devices: a target node and a controller node. A target node has full PAN coordinator capabilities and can start a network in its own right. A controller node can join networks started by target nodes by pairing with the target. Multiple remote control (RC) PANs form an RC network and nodes in the network can communicate between RC PANs.

In order to communicate with a target node, a controller node first switches to the channel and assumes the PAN identifier of the destination RC PAN. It then uses the network address, allocated through the pairing procedure, to identify itself on the RC PAN and thus communicate with the desired target node.

This graphic illustrates an example ZigBee RF4CE topology that includes three target nodes: a TV, a DVD player and a Set-Top Box (STB) with each target node creating its own RC PAN. There's also an optional gateway providing access to the Internet to expand control options. The TV, DVD and STB also have dedicated RCs that are paired to each appropriate target node. A multifunction RC, capable of controlling all three target nodes itself, is added to the network by successively pairing to the desired target nodes. The ZigBee RF4CE Network Topology is shown in Figure 8.1.

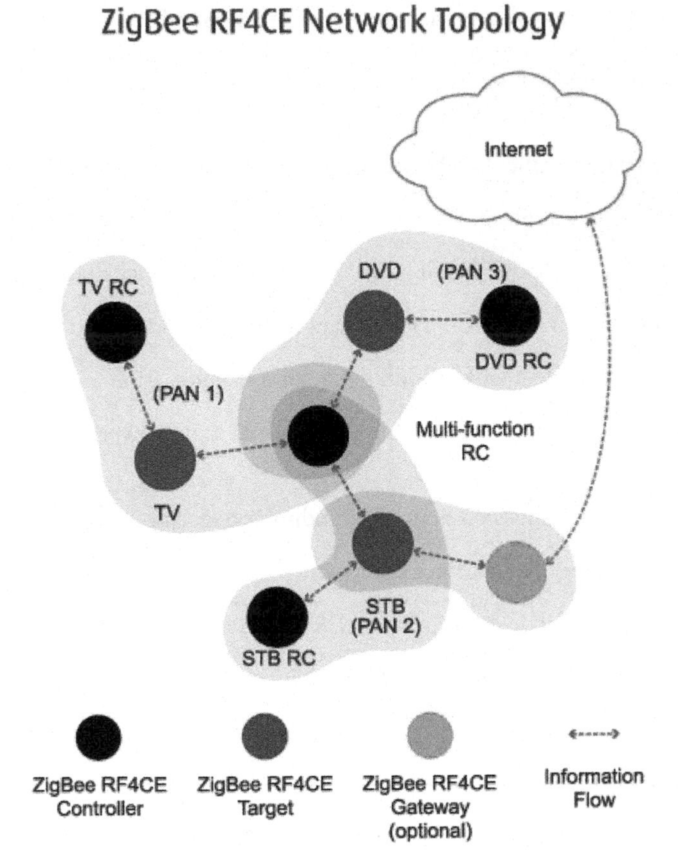

Figure 8.1: ZigBee RF4CE network topology. (From www.zigbee.org)

8.5 ZigBee Smart Energy Standard Overview

ZigBee Smart Energy version 1.1, the newest version for product development, adds several important features including dynamic pricing enhancements, tunneling of other protocols, prepayment features, over-the-air updates and guaranteed backwards compatibility with certified ZigBee Smart Energy products version 1.0. You can also watch our webinar from April, 2012[6] to learn even more about how companies are using it to deliver energy management to consumers today. This standard supports the diverse needs of a global ecosystem of utilities, product manufacturers and government groups as they plan to meet future energy and water needs. This is shown in Figure 8.2.

All ZigBee Smart Energy products are ZigBee Certified to perform regardless of manufacturer, allowing utilities and consumers to purchase with confidence. Every product needed to implement a robust ZigBee Smart Energy home area network

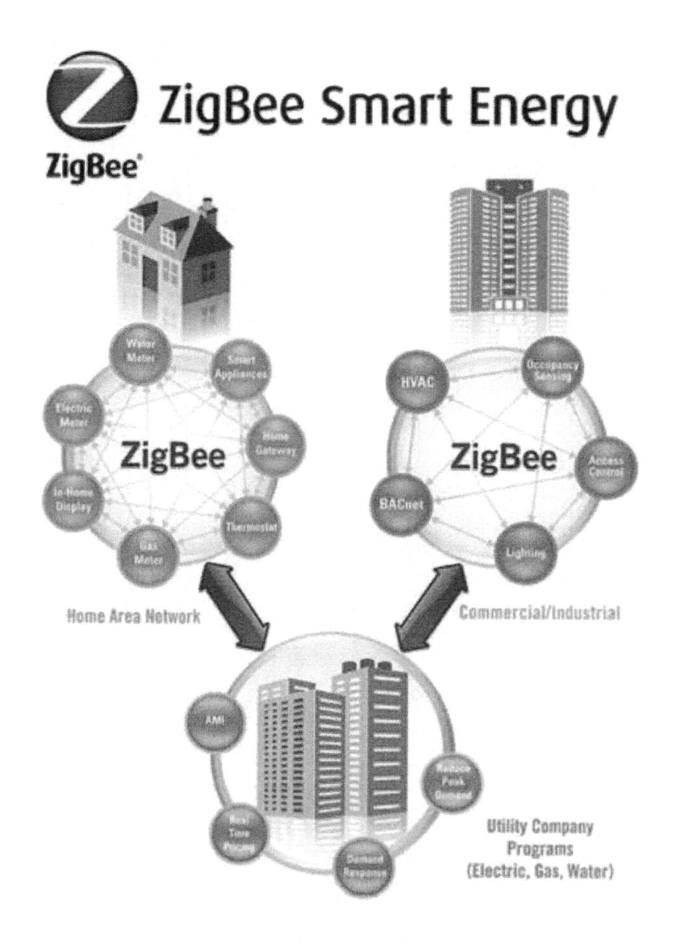

Figure 8.2: ZigBee smart energy standard overview. (From www.zigbee.org)

(HAN) is available. These products make it easy for utilities and governments to deploy smart grid solutions that are secure, easy to install and consumer-friendly.

Some of the world's leading utilities, energy service providers, product manufacturers and technology companies are supporting the development of ZigBee Smart Energy. Several other standards groups are also involved with extending the reach of ZigBee Smart Energy to more homes around the world. ZigBee Smart Energy is in active deployment in many areas.

8.5.1 ZigBee Smart Energy Benefits

ZigBee Smart Energy Benefits is listed in Table 8.1.

Table 8.1: ZigBee Smart Energy Benefits

	UTILITIES	OEMS	END USERS (Residential)	END USER (Commercial)
AFFORDABLE	■ Open standard supports competitive marketplace of multiple vendors, which lowers cost	■ Open standard supports competitive marketplace of multiple platform vendors offering lower cost	■ Open standard supports competitive marketplace of multiple products that lowers cost through competition ■ Lowers energy costs	■ Open standard supports competitive marketplace of multiple vendors that lowers cost through competition ■ Lowers energy costs
EASY TO USE	■ Standard technology, certified interoperability, global 2.4GHz spectrum eases adoption and customer service	■ Standardized technology provides fast access to new markets ■ Reduced product research and development cycle	■ Wireless technology eliminates cost and hassle of running wires ■ Certified interoperability, global 2.4GHz spectrum simplifies installation and operation ■ Automatic features simplify use ■ Internet connection for greater access, control	
REDUCES ENERGY CONSUMPTION	■ Improves energy availability, reliability ■ Improves regulatory compliance	■ Lowers customers' cost to operate	■ Immediate usage information drives decisions to conserve ■ Improves compliance with conservation programs	
REDUCES ENVIRONMENTAL IMPACT	■ Improves regulatory compliance ■ Increased efficiency reduces need for additional generation plants	■ Improves regulatory compliance ■ Allows entry into markets with regulatory requirements	■ Reduces impact on environment ■ Improves emissions "footprint"	■ Reduces impact on environment ■ Improves compliance with emission standards

8.6 Smart Energy Profile 2 (SEP 2.0)

The Smart-Energy Profile 2.0 is being developed to create a standard and interoperable protocol that connects smart energy devices in the home to the Smart Grid. While the original work for SEP 2.0 was done via a joint liaison agreement between the ZigBee Alliance and the HomePlug Alliance, the standard itself is designed to run over Transmission Control Protocol / Internet Protocol (TCP / IP) and is therefore media access control (MAC) and physical layer (PHY) agnostic. A new coalition of Alliances had been formed (composed of Wi-Fi Alliance, ZigBee Alliance, HomePlug Alliance and HomeGrid Alliance) with the intent of developing SEP 2.0 application level interoperability testing and certification program. Even though the standard is still under development, it is expected to be widely adopted by Electric Utilities to implement their consumer facing programs.

In addition to providing connectivity to the Smart Grid, it is also envisioned that self-contained implementations of Home Energy Management systems, which leverage SEP 2.0 application connectivity, will be deployed in the marketplace. Furthermore, SEP 2.0 enables the development and deployment of smart-appliances and services that not only provide energy related services, but also provide the ability for device OEMs and service providers to innovate and offer consumers enhanced user-experiences and value-added services.

Smart Energy Profile 2 (SEP 2) offers a global standard for IP-based control, both wired and wireless, for energy management in Home Area Networks (HANs). SEP 2 is an evolution of ZigBee Smart Energy 1.x and provides new capabilities such as control of plug-in hybrid electric vehicles (PHEVs) charging, HAN deployments in multi-dwelling units such as apartment buildings, support for multiple energy service interfaces into a single premises and support for any transport based on IETF IP compliant protocols such as ZigBee IP specification.

SEP 2 is supported by many stakeholders across the energy supply ecosystem including manufacturers of smart meters, appliances, programmable thermostats and other devices in homes, utilities, energy service providers as well as various government and standards organizations around the world. Smart Energy Profile 2 does not replace ZigBee Smart Energy 1.x. Instead, it offers utilities and energy service providers another choice in the creation of Home Area Networks.

8.6.1 SEP 2.0 Functionality

The SEP 2.0 protocol is built around the notion of function-sets and each function-set represents a minimum set of device behaviors required to deliver a particular functionality. Some of the core function sets defined in the specification include metering, pricing and demand-response load control (DRLC). These function-sets enable informed consumer participation in managing energy consumption in home and participating in the Utilities efforts to manage peak-demand loads in the Grid. Additionally, function-sets are being defined for more advanced use-cases such as Plug-in Electric Vehicles and Distributed Renewable Energy Management.

The metering function set allows devices to get usage-information from the smart-meter or any other device that has metering capability (e.g., a smart-appliance or a sub-meter). Thus, an SEP 2.0 compliant application running on a smartphone, tablet or a dedicated display device, could present to users real-time energy consumption information for the whole home. This information would be delivered directly from the smart-meter / indirectly from a cloud-based server, or from any smart-appliance or smart-energy device supporting metering function.

The pricing function set allows a Utility to send pricing signals to smart-energy devices such as smart-appliances and programmable thermostats. These devices can then take actions to reduce or shift usage when the energy price is high (based on pre-set user-preferences). Alternatively, the pricing signals may go to an energy management service, to store user-preferences and intelligently manage the energy consumption across multiple user-devices. Finally, an energy management service might seek user-input to initiate actions through any method that the user might want, such as email, short message service (SMS), voice alerts and others.

The DRLC function-set provides a more active means for Utilities to implement demand response programs where the DRLC events are targeted to specific devices and have expectations of specific load-curtailment (for example, set point offset for thermostats or duty-cycling for other loads). These programs are used when a Utility expects to hit peak demand, typically offering financial incentives to users to participate in these programs.

8.6.2 SEP 2.0 Architecture and Technologies

The SEP 2.0 application protocol is built on a representational state transfer (REST) architecture that is used widely to deploy Webservices over Hypertext Transfer Protocol (HTTP). A REST architecture is based on a client-server model in which servers contain and perform operations on resources. Servers expose resource representations to clients and clients make requests to access representations of resources on the servers such as read, write, create and delete. In SEP 2.0, resources represent things like meter reading, pricing tariffs, demand response events, etc.

SEP 2.0 resource representations are built to be compatible with the International Electrotechnical Commissions Common Information Model (CIM). The result is an Extensible Markup Language C-based (XML) protocol developed on a REST architecture utilizing HTTP for transport. In addition, the protocol uses other commonly used Standards. For example, it uses Multicast Doman Name System (mDNS) and DNS-Service Discovery to enable SEP 2.0 devices to be discovered on a local network implementing the service discovery method, popularized by Apple, through its Bonjour protocol and is now widely available on all major Operating System platforms from desktops to tablets to phones. Also, SEP 2.0 makes use of Transport Layer Security (TLS) to secure communications between devices, thus ensuring that the protocol meets rigorous security requirements needed to protect sensitive consumer information and to ensure integrity of Smart Grid transactions.

The SEP 2.0 architecture and the technologies used by the protocol standard are the same technologies that are used to implement the rich ecosystem of applications running on smartphones, tablets and browsers communicating with Web-based services. Thus, there is a broad-based developer community, know-how and tools to innovate around SEP 2.0 enabled devices. Similar to recent innovation in the mobile Internet world, SEP 2.0 enables technology that can quickly build and deploy cloud-based services. The only missing ingredient is the lack of widely available smart-energy devices implementing SEP 2.0 based communications.

8.6.3 Enabling Smart-Energy Devices and Services

A standard like SEP 2.0 only addresses part of the challenge as it still requires solutions to be built that overcome adoption barriers, engaging users and motivating participants. A prominent barrier for consumers is the availability of affordable solutions.

The easiest way to save energy is to not use it unless it is required. When users are away from their home, a significant amount of energy can be saved by turning off any electric/electronic devices that do not need to stay powered on and by changing the setpoints for thermostat and water heaters to energy-saving away mode. Heating/cooling and water-heater consume about 50 percent of energy in a typical home which can amount to significant savings for consumers.

While these actions can be taken with the existing mass-deployed devices, few users engage in these behaviors as they are inconvenient and compromise user comfort. For example, a user leaving home can set the thermostat to away mode, but will end up returning to a home that it either too hot or too cold for their comfort. Setting the hot-water heater to away mode would typically require a trip to the basement and it becomes a challenge to motivate users to do this in their busy lives.

Smart-energy devices that make it convenient for users to save energy without sacrificing comfort are a key ingredient to engaging consumers in managing their energy consumption. Consider a smartphone application that enables a user to set his/her thermostat and/or hot-water heater to away mode when they leave home and to remotely set it back to comfort mode so they can ensure that their home is at a comfortable temperature when they get back home. The simple, reliable, and affordable solutions are likely to engage users and are critical to mass-market adoption.

8.6.4 Offering Consumer Value beyond Energy Savings

The core technology to build a SEP 2.0-compliant appliance gives device manufacturers the opportunity to offer more consumer benefits, not just the prospect of saving energy costs. This is critical to mass adoption as it will provide consumers with other motivations to purchase smart-appliances and thermostats. The first consumer benefit occurs by simply allowing applications to be developed on devices like smartphones and tablets that can interact with a smart-thermostat or appliance. Appliance manufacturers are under competitive pressures to provide more user-friendly interfaces, especially as they add more features to their appliances. This quickly gets expensive as technologies such as touch screen interfaces (which users are quickly becoming accustomed to) are expensive. Connectivity offers a solution to the appliance manufacturers as it can be used to provide consumers with rich user-interfaces on devices that they already own and provide a rich user-interface platform.

Beyond that, connected appliances enable an appliance manufacturer to offer a whole range of value-added services to consumers by having the device communicate with a cloud-based services using the technologies already present in a SEP 2.0 compliant software platform (HTTP and TLS client). For example, a water heater might be able to detect that it is about to leak and a service could send an alert to the user before the damage. Moreover, the owner of a smart-refrigerator might get a notification that the refrigerator door was left open, or the water filter needs changing. Cloud services could gather usage data from an appliance and present it to users through dashboards. They could analyze the data and present guidance to users on how to more effectively use their appliance, and when to upgrade. There are endless possibilities of value-added services and it is imperative that we leverage the SEP 2.0 protocol to ensure mass deployment of smart-energy devices.

8.7 ZigBee IP Specification

8.7.1 ZigBee IP Compliant Platform

The ZigBee IP specification is an IPv6-based full wireless mesh networking solution and provides seamless Internet connections to control low-power, low-cost devices. ZigBee IP was designed to specifically support the forthcoming ZigBee Smart Energy version 2 standard.

The ZigBee Compliant Platforms found below are the foundation of ZigBee products. Each platform is comprised of a 2.4 GHz radio and a microprocessor with storage running ZigBee IP firmware.

The following companies have successfully completed the rigorous ZigBee Certified testing program to ensure their platforms meet strict performance criteria established by the Alliance.

8.7.2 ZigBee IP: The First Open Standard for IPv6-Based Wireless Mesh Networks

New specification brings end-to-end IPv6 control to low-power, low-cost devices SAN RAMON, Calif. C March 27, 2013 C The ZigBee Alliance[26], a global ecosystem of companies creating wireless solutions for use in energy management, commercial and consumer applications, today announced the completion and public availability of its third specification, ZigBee IP. ZigBee IP is the first open standard for an IPv6-based full wireless mesh networking solution and provides seamless Internet connections to control low-power, low-cost devices. A free public webinar hosted by industry experts will discuss the unique capabilities of ZigBee IP.

ZigBee IP specification enriches the IEEE 802.15.4 standard by adding network and security layers and an application framework. ZigBee IP offers a scalable architecture with end-to-end IPv6 networking, laying the foundation for an Internet of Things without the need for intermediate gateways. It offers a cost-effective and energy-efficient wireless mesh network based on standard Internet protocols, such as 6LoWPAN, IPv6, PANA, RPL, TCP, TLS and UDP. It also features proven, end-to-end security using TLS1.2 protocol, link layer frame security based on AES-128-CCM algorithm and support for public key infrastructure using standard X.509 v3 certificates and ECC-256 cipher suite.

ZigBee IP offers a significant step forward in the expansion of IP-based control, said Mark Grazier, marketing manager and ZigBee board member, Wireless Connectivity Solutions, Texas Instruments Incorporated (TI). "Having low-power, low-cost wireless mesh devices that connect to a variety of Smart Grid IPv6-based protocols will further expand the Internet of Things."

ZigBee IP continues the ZigBcc tradition of self-organizing and self-healing mesh networking to enable robust communications over the globally available 2.4 GHz frequency as well as over the 868/915/920 MHz frequencies in select countries. For a complete list of features and benefits, visit www.ZigBee.org/IP ZigBee. IP provides another option for product manufacturers to add mesh networking to their products, said Tobin Richardson, chairman and CEO of the ZigBee Alliance. "We expect to see a wave of ZigBee Certified products based on ZigBee IP when Smart Energy Profile 2 is completed later this year since ZigBee IP was designed specifically to meet SEP2 requirements for the Smart Grid."

Exegin, Silicon Labs and Texas Instruments provided Golden Units against which all future ZigBee Certified products using the ZigBee IP specification will be tested. This testing process ensures compliance with the specification so that manufacturers can be assured of consistent communications. Grid2Home and Sensinode achieved ZigBee Compliant platform status. Testing services were provided by NTS, TRaC and TV Rheinland.

8.7.3 ZigBee IP Specification Overview

ZigBee IP is the first open standard for an IPv6-based full wireless mesh networking solution and provides seamless Internet connections to control low-power, low-cost devices. It connects dozens of the different devices into a single control network. ZigBee IP was designed to specifically support the forthcoming ZigBee Smart Energy version 2 standard and was made available to members and the public in 2013. The ZigBee IP specification enriches the IEEE 802.15.4 standard by adding network and security layers and an application framework. ZigBee IP offers a scalable architecture with end-to-end IPv6 networking, laying the foundation for an Internet of Things without the need for intermediate gateways. It offers a cost-effective and energy-efficient wireless mesh network based on standard Internet protocols, such as 6LoWPAN, IPv6, PANA, RPL, TCP, TLS and UDP. It also features proven, end-to-end security using TLS1.2 protocol, link layer frame security based on AES-128-CCM algorithm and support for public key infrastructure using standard X.509 v3 certificates and ECC-256 cipher suite. ZigBee IP enables low-power devices to participate natively with other IPv6-enabled Ethernet, Wi-Fi and, HomePlug devices.

From this foundation, product manufacturers can use the ZigBee Smart Energy version 2 standard to create multi-vendor interoperable solutions. As with any ZigBee Alliance specification, custom applications, known as manufacturer specific profiles, can be developed without multi-vendor interoperability. Characteristics of ZigBee IP include:

- Global operation in the 2.4GHz frequency band according to IEEE 802.15.4
- Regional operation in the 915Mhz (Americas), 868Mhz (Europe) and 920 MHz (Japan)
- Incorporates power saving mechanisms for all device classes
- Supports development of discovery mechanisms with full application confirmation
- Supports development of pairing mechanisms with full application confirmation
- Multiple star topology and inter-personal area network (PAN) communication
- Unicast and multi-cast transmission options
- Security key update mechanism
- Utilizes the industry standard AES-128-CCM security scheme
- Supports Alliance standards or manufacturer specific innovations

A complete list of Features and Benefits for the ZigBee IP specification can be seen in[26].

8.7.4 ZigBee IP Specification Network Topology

ZigBee IP networks are composed of several device types: ZigBee IP Coordinator, ZigBee IP Routers and ZigBee IP Hosts, as shown in Figure 8.3. Coordinators control the formation and security of networks. Routers extend the range of networks. Hosts perform specific sensing or control functions. Manufacturers often create devices that perform multiple functions, for example a programmable communicating thermostat may also route messages to the rest of the network. This graphic illustrates an example ZigBee IP topology that includes one coordinator, a border router for Internet control access, five routing devices and, two end devices creating a control network. An example network in a smart home, the coordinator may be a programmable communicating thermostat with advanced support for an in-home display. Devices such as smart plugs, thermostats and smart appliances could be configured as routing devices. Simple devices such as smart appliances and temperature sensors could be end devices.

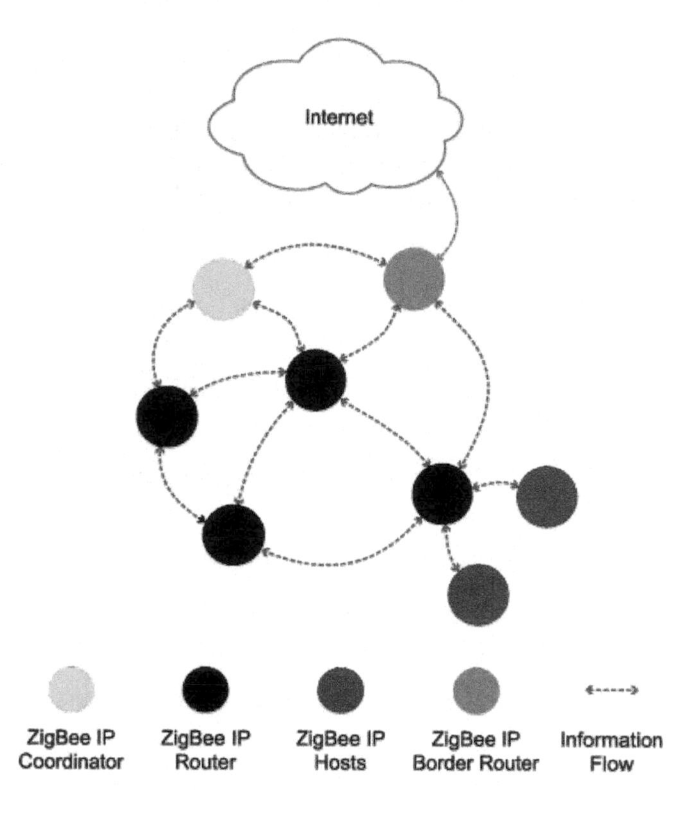

Figure 8.3: ZigBee IP network topology. (From www.zigbee.org)

References

[1] "ZigBee Specification FAQ". ZigBee Alliance. ZigBee Alliance. Retrieved 14 June 2013.

[2] "ZigBee Wireless Networking", Drew Gislason (via EETimes)

[3] "Frequently Asked Questions". ZigBee Alliance. Retrieved March 23, 2013.

[4] "ZigBee: Wireless Technology for Low-Power Sensor Networks". Commsdesign.com. Retrieved 2012-10-18.

[5] "IEEE 802.15.4". Ieee802.org. Retrieved 2012-10-18.

[6] ZigBee Document 053474r06, Version 1.0, ZigBee Specification. ZigBee Alliance. 2004.

[7] ZigBee IP Specification, "The ZigBee Smart Energy ". http://www.zigbee.org/Standards/ZigBeeSmartEnergy/Features.aspx

[8] "What's so good about ZigBee networks?" Daintree Networks. Retrieved 2007-01-19.

[9] Bellido-Outeirino, Francisco J. (February 2012). "Building lighting automation through the integration of DALI with wireless sensor networks". IEEE Transactions on Consumer Electronics 58 (1): 47C52. doi:10.1109/TCE.2012.6170054.

[10] "The ZigBee Alliance". ZigBee.org. Retrieved 2012-10-18.

[11] "Wireless Sensor Networks Research Group". Sensor-networks.org. 2008-11-17. Retrieved 2012-10-18.

[12] "ZigBee Cluster Library Specification Download Request". zigbee.org. Retrieved 2010-04-10.

[13] "ZigBee is only royalty-free if not used for commercial purposes", FAQ for BEN WPAN

[14] "ZigBee, Linux, and the GPL". freaklabs.org. Retrieved 2009-06-14.

[15] Jon Smirl (7 January 2011). "6lowpan/ROLL support on 802.15.4 radios". eLinux.org. Retrieved 2013-06-14.

[16] Bob Gohn (May 30, 2011). "The ZigBee IP-ification Wars". Navigant. Retrieved 2013-06-14.

[17] "ZigBee Smart Energy Overview". ZigBee.org. Retrieved 2012-10-18.

[18] "Introducing ZigBee RF4CE". Daintree Networks. Retrieved 2009-05-04.

[19] "ZigBee RF4CE: A Quiet Revolution is Underway (December 2012)". ZigBee Alliance. Retrieved 2012-12-06.

[20] "ZigBee RF4CE Webinar: A Quiet Revolution is Underway (December 2012)". ZigBee Alliance. Retrieved 2012-12-06.

[21] "Zigbit Modules MCU Wireless- Atmel Corporation". Atmel.com. Retrieved 2012-10-18.

[22] "n-Core Platform & Polaris Real-Time Locating System C Wireless Sensor Networks C ZigBee". N-core.info. Retrieved 2012-10-18.

[23] David Egan, "ZigBee Propagation for Smart Metering Networks", Electric Light & Power vol. 17 issue12

[24] "Wireless Sensor Networks Research Group". Sensor-networks.org. 2010-04-15. Retrieved 2012-10-18.

[25] "Wireless Sensor Networks Research Group". Sensor-networks.org. 2009-02-05. Retrieved 2012-10-18.

[26] ZigBee IP Specification, "The ZigBee Alliance". http://www.zigbee.org/Specifications /ZigBeeIP/Overview.aspx

[27] ZigBee Specifications, "The ZigBee Alliance". http://www.zigbee.org/Specifications/ZigBee

Chapter 9

Developing ZigBee Certified Products

Huasong Cao, Sergio González-Valenzuela, Victor C. M. Leung

CONTENTS

Recent years have seen a myriad research efforts aimed at conserving energy in hardware-constrained devices equipped with a low-power radio interface employed to retrieve environmental data. Clusters of these devices, referred to as wireless sensor networks (WSNs) [7] must implement the necessary algorithms and communications protocols to achieve this energy efficiency objective efficiently. As of late, the applicability boundaries of WSNs have been pushed to the limit by employing networked sensors to retrieve bodily signals and readings that provide information about the state of health in people [8]. A number of challenges remain to be solved before these wireless body area sensor networks (WBASNs) become pervasive [11]. In particular, ZigBee technology [5] aims at providing a standardized platform upon which this kind of technology can rely to create reliable commercial products. In this chapter, we explore the steps involved in the development of WBASN products for certification by the ZigBee Alliance, and describe important challenges, procedures, and a walk-through example that users venturing into this endeavour can reference.

9.1 ZigBee Application Layer

ZigBee is a standard for data communications between low-power devices that are normally associated with sensor reading tasks in a variety of scenarios, including home automation, smart home and industrial appliances, and even human health monitoring [24]-[30]. It is based on IEEE 802.15.4, which specifies the physical and media access layer standards [6]. This relationship is similar to the relationship between the Wi-Fi Alliance [9] and the IEEE 802.11 standard [10]. To this regard, a conglomerate of companies form the ZigBee Alliance, which issues the communications protocol standards that ZigBee-certified products adhere to.

The ZigBee stack architecture is a simplified implementation of the Open System Interconnection reference model (OSI) [13]. The application layer (APL) in ZigBee consists of the Application support Sub-layer (APS), and Application Objects [1].

APS in ZigBee is a counterpart of the transport layer in OSI, and application objects in ZigBee are equivalent to protocols in the application layer of OSI. We recall that the main idea of the OSI model, or more practically the current Internet, is to have a sandglass-like architecture that enables data packet routing over heterogeneous networks using a unique address, while providing support for multiple end-to-end data streams through "ports". ZigBee applies the same approach, whereby the "endpoint" abstraction replaces that of ports [14] [15]. In the following sections, we introduce the new terms and concepts that appear in the ZigBee APL specification, which are essential to developing ZigBee-certified end-products. These terms include: APS, application profile, device description, cluster and attribute, application object, ZigBee Device Objects (ZDO), and binding.

9.1.1 Application Support Sub-layer

Producing devices for ZigBee certification requires a good understanding of its Application Support Layer. The APS resides within the APL, and services application objects and ZDO through endpoints. Each endpoint is described by an 8-bit identifier (ID), and up to 241 endpoints can be allocated (endpoint 0 is reserved for ZDO). Together, they enable multiplexing of data streams coming from different application objects. However, unlike the Internet's protocol stack, the current ZigBee stack is intended for use only in networks employing the IEEE 802.15.4 at the lower layers, in which hop-by-hop acknowledgements can be enabled to ensure reliability. Nevertheless, APS enhances reliability through an optional end-to-end acknowledgement mechanism. A data frame from an application object usually contains the information of a destination address and a destination endpoint. It is optional but beneficial to enable binding in APS, so that these two fields need not be specified each time the application object passes data to APS. We refer to binding as a mechanism that links the endpoint and cluster identifier of the source to the network address and endpoint of the destination(s). A binding table is optionally maintained at APS of each device.

9.1.2 Profile

A profile (or application profile) in ZigBee is a specification that defines the message types in a target application, the message formats, and the processing actions of ZigBee devices. It is equivalent to an application layer protocol in the Internet (e.g., File Transfer Protocol (FTP), Hypertext Transfer Protocol (HTTP), or Domain Name System (DNS)), but based on the ZigBee application framework. Two profile classes exist: public and manufacturer-specific. A public profile is defined by the ZigBee Alliance as one aimed at providing agreements and cooperation between different vendors. ZigBee devices certified to comply with a particular public profile (e.g., Smart Energy public profile [2]) can interoperate with each other in order to facilitate the targeted application. Taking the Smart Energy profile as an example, devices shall communicate based on the profile specification and exchange metering and control information to build an energy-aware home/factory.

The manufacturer-specific profile is defined by a company with the intent of providing agreements and cooperation between its own designed devices. This class of profile complements existing public profiles with a broader area of applications, e.g., data/file transfer between personal computer (PC) and its peripherals.

ZigBee defines a 16-bit long Profile identifier that can be used to enumerate up to 65,536 different items. Because of the relatively limited number, every company that wishes to design a new profile must request for allocation of a profile identifier from the ZigBee Alliance. However, a new profile identifier is not needed for cases where manufacturer-specific extensions are supported through a public profile, e.g., making use of the reserved fields. The granting of a unique profile identifier allows the requester to define the device descriptions and cluster identifiers within the scope of that profile, as described next.

9.1.3 Device Description

A device description contains the device's name and its identifier. The name usually indicates the role of the device. For example, in the Smart Energy profile, a device's name could be "energy service portal", "metering device", "in-premise display", "wirelessly programmable thermostat", "load control device", "smart appliance", or "pre-payment terminal". It is therefore straightforward to distinguish the functionality of the corresponding device by looking at its description. In most Internet application protocols, however, there are only two abstract roles: a server and a client.

9.1.4 Cluster and Attributes

A cluster is a collection of attributes and its corresponding access commands. More specifically, an attribute is a data entity that represents a physical quantity or abstract state, which can be retrieved by reading the sensor or register associated with it. The corresponding command is used to set/get the attribute, which provides the necessary interface information between devices in a client/server approach. Revisiting our Smart Energy profile example, clusters can be formed by associating information related to price, demand response and load control, simple metering messages, smart energy tunnelling, and pre-payment. Attributes can be used to describe information related to the utility enrollment group and device's class value. Similarly, commands to retrieve events' status and schedules can be employed. The criterion for designing clusters and attributes should ensure that each cluster covers an appropriate range of attributes that describe the monitoring/control environment. There is also a ZigBee cluster library (ZCL) [4] designed for promoting re-using of profile identifiers. In addition, cluster and attribute definitions can be shared across application profiles.

9.1.5 Application Objects

An application object is an implementation of a certain ZigBee profile. It resides within the application framework and communicates with APS through a destined endpoint. As explained earlier, an endpoint in ZigBee is the counterpart of a port in the transport layer of the Internet protocol stack. Up to 240 endpoints (from endpoint 1 to 240) can be used to accommodate 240 application objects, as mentioned before. When a data frame is delivered from APS to the application framework, it is matched to a specific end point with a specific profile. Upon successful matching, the payload is passed onto the application object. Otherwise, it is discarded.

9.1.6 ZigBee Device Objects

A ZigBee Device Object (ZDO) is an implementation of a special public profile named the ZigBee Device Profile (ZDP). The profile differs from other public and manufacturer-specific profiles in the following aspects:

- Its implementation is mandatory in each ZigBee device. As a result, it defines capabilities supported in all devices.
- It utilizes only one device description, which is enough for its unique role in every device.
- Its clusters do not employ attributes, as it is designed only for network management purposes rather than sensor monitoring.

ZDO is served on endpoint 0 in each device. However, it contains six objects: device and service discovery, network manager, binding manager, security manager, node manager, and group manager. Each of these objects specifies mandatory and optional clusters and attributes defined in ZDP.

9.2 ZigBee Certified Products

A variety of ZigBee-enabled products are being sold worldwide. These products can be categorized as belonging to any of these four classes: ZigBee Certified Products, Manufacturer Specific Certified Products, ZigBee Compliant Platforms, and Designed for ZigBee Products. The differences between them lie in whether or not they have passed certification and compliance tests as required by the ZigBee Alliance, and which application profile they conform to, i.e., public or manufacturer-specific. In the following sections, we compare these products by referring to off-the-shelf products as examples, and explain the procedures followed during the certification process by ZigBee Alliance when a product is ready.

9.2.1 ZigBee Certified Products

ZigBee Certified Products are those end-products that conform to a public application profile and have been certified by the ZigBee Alliance. They are interoperable between vendors within the scope of that particular profile. Currently, there are two profiles publicly available. They are the ZigBee Smart Energy Public Application Profile [2] and the ZigBee Home Automation Public Application Profile [3]. Smart Energy realizes energy management and bill control functionalities in a wireless context, with the goal of enabling a smart grid. Off-the-shelf products are mostly metering devices, e.g., gas/electricity consumption meters and thermostats. Home automation products complement the operation of these in a home environment with various applications to ensure security and remote control. Devices can be as diverse as: monitors of room/corridor motion, monitors of window movement, wireless doorbell/panic button, door lock remote control, light switch and media management.

Although only two profiles have been published on the Alliance's website, ZigBee does target a much wider market in the area of building automation, heathcare, telecommunications services and remote control. The corresponding public profiles are being drafted. The corresponding products are either certified as manufacturer-specific, or listed as designed-for-ZigBee.

9.2.2 Manufacturer Specific Certified Products

Manufacturer Specific Certified Products are also end-products, but conform to the manufacturer-specific application profile instead. This provides manufacturers with flexibility in designing new products targeting new applications, while sacrificing interoperability. As a result, products in this class enjoy a wider range of applications than ZigBee Certified Products, including: connecting PC/mobile phone to peripherals, serial-to-ZigBee converters, industry monitoring and control systems employing sensors and actuators, human presence management solutions, and fire extinguisher monitoring.

9.2.3 ZigBee Compliant Platforms

ZigBee compliant platforms are building blocks used for developing end-products. The main difference here is that the company that develops the platform (not the ZigBee Alliance) ensures the compliance of the platform. A platform normally has an IEEE 802.15.4 radio, a low-power Micro-Controller Unit (MCU), a ZigBee protocol stack, Application Programming Interfaces (APIs), and possibly sensors. More details of platform architecture are given in Section 9.3. The end-product designer finds the best combination of these components for the targeted application and starts writing application programs using APIs, as discussed in Section 9.4.

9.2.4 Designed for ZigBee Products

Designed-for-ZigBee Products are end-products, or original equipment manufacturer (OEM) platforms that have not yet passed ZigBee certification and compliance tests. In this class of products, we see more novel use-cases as well as testing equipments in addition to traditional ZigBee applications. More and more ZigBee solutions are providing remote access to the monitoring/control environment through the Internet.

End users can then get real-time data or issue a command through a mobile phone. An important question that arises here is: how do Designed for ZigBee Products become certified? The answer to this question will be addressed shortly.

9.2.5 Certification Process

Firstly, the certification process requires that the product manufacturer becomes a ZigBee Alliance member. After that, the product is submitted to one of three Test Providers, who in turn ensures that it meets ZigBee specifications. The Test Provider informs the ZigBee Alliance of the outcome upon successful completion of the testing phase. Then, after some administrative processing tasks, the submitting company obtains the logo usage rights, as well as a certificate. The product is then entered into the registry of certified products. The details of the actual certification process are outside the scope of this work.

9.3 OEM Platforms

As mentioned before, a ZigBee OEM platform consists of an IEEE 802.15.4 radio, a low-power MCU, a ZigBee protocol stack, APIs, and sensors. The radio and MCU can be integrated into one chip utilizing the popular system-on-chip (SoC) technology (e.g., EM351 chip made by Ember [16]). It is often desirable that the MCU supports low-power sleep mode so that a given battery can last longer. Existing ZigBee protocol stack implementations take on the form of firmware (e.g., eZeeNet stack by Atmel [17]), or hardware (e.g., the CC2480 chip by Texas Instruments - TI [18]) based on compliant platforms. In either case, APIs are provided for utilizing APS and ZDO when programming application objects. There are also open firmware stacks developed by communities of software programmers. For example, the TinyOS ZigBee Working Group [19] is developing the protocol stack based on its proven architecture that is working in tens of platforms. In the following subsections, we discuss the selection criterions of hardware and firmware environments as core steps in developing a ZigBee certified product.

9.3.1 SoC, ZigBee Processor, and Software Stack

A SoC incorporates an IEEE 802.15.4 radio, a ZigBee upper layer stack, and user-developed application objects together. On the platform, no other MCU is needed in addition to the SoC chip. See Figure 9.1. This arrangement yields a simpler layout of the printed circuit board (PCB), and eliminates possible errors in software design for interfacing the MCU to the radio chip. Ideally, the compound system also consumes less power. On the other hand, computing and memory resources on the SoC are comparatively more restricted, leading to the support of simpler application objects that are able to run on the platform. An example platform with ZigBee SoC chip is EM35x listed in Table 9.1.

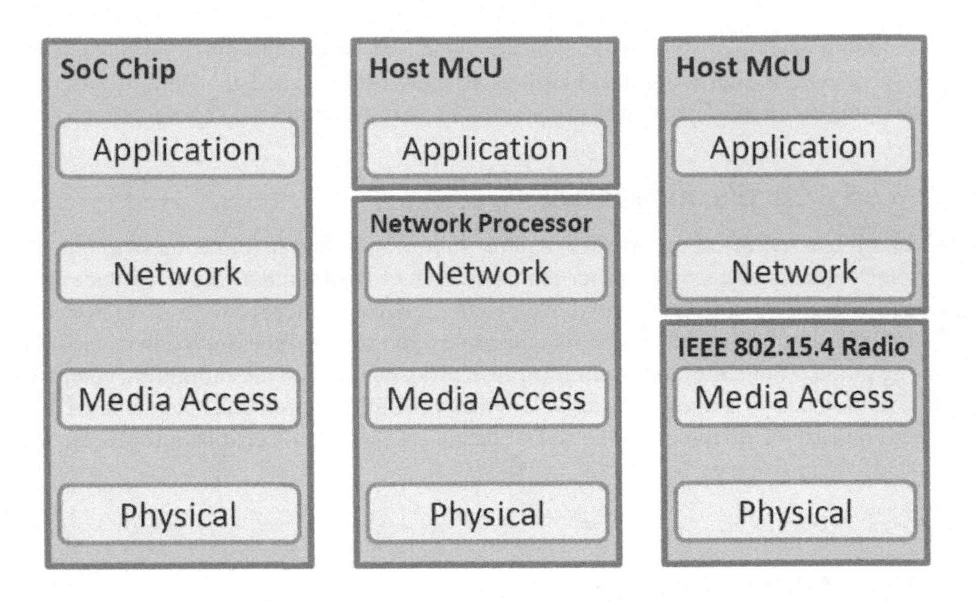

Figure 9.1: Comparisons of three platforms.

A ZigBee Processor is a hardware implementation of the ZigBee stack, excluding application objects. The application objects run on a standalone MCU, which interfaces with the ZigBee Processor and other peripherals. The PCB is relatively more complex since more than one chip and associated components are accommodated. In addition, the separation of ZigBee networking functionalities and application objects enables the ZigBee Processor to periodically enter the sleep mode. This approach leverages power savings for applications that require intense local computation, but less frequent radio communications. In addition, a standalone MCU provides better programming, interfacing, and computing capabilities to support complex application objects. Refer to eZ430-RF2480 in Table 9.1 as an example of this configuration.

Table 9.1: OEM Platform Examples

Platforms	MCU and Radio	OS	Flash	Feature
EM35x	Ember SoC	Proprietory	128 KB	Prototyping area available
eZ430-RF2480	MSP430 + CC2480	TI's Z-Accel	32 KB	Small in size
MicaZ	ATmega128L + CC2420			Sensor boards available

The software stack that defines the upper ZigBee layers can be implemented into firmware that runs on a MCU, or possibly a PC, together with the user-developed application objects that interface with the corresponding IEEE 802.15.4 radio chip and peripherals. This kind of platform is inferior to the above two in terms of ease of development, and does not take advantage of the simplicity of the SoC approach, or power of the ZigBee Processor approach. However, when developing applications, such as ZigBee testing equipment that requires additional processing capabilities of lower layer packets, this approach provides the most flexible choice. Particularly, this advantage comes from the assumption that access to lower layers is available through APIs. Otherwise, another implementation of the ZigBee software stack is needed to allow these privileges. MicaZ in Table 9.1 is an example of this configuration.

9.3.1.1 ZigBee in TinyOS

TinyOS [31] is an embedded operating system, that is open and widely used in the community of WSN researchers. It provides support of drivers for tens of hardware platforms, and can be easily ported to new custom-made ones. It prevails in the research community also because numerous protocols, from media access methods to applications, have been implemented in TinyOS. TinyOS ZigBee Working Group [19] is currently working on an implementation of the ZigBee stack in TinyOS based on existing work by TinyOS 802.15.4 Working Group. The most recent release (TinyOS 2.1.1) does not fully support ZigBee standard-compliant networking and application protocols yet.

9.3.2 Free versus Open

Most commercial ZigBee platforms are distributed with a license-free, manufacturer-implemented stack that is realized either as firmware or hardware that yields a shorter development life cycle. In this regard, the manufacturer ensures that its stack strictly conforms to ZigBee specifications, and that it provides the corresponding APIs that application-layer developers can employ. The particular APIs offered by the manufacturer may vary between products, depending on the type of ZigBee specification they are conforming to, which can either be ZigBee Feature Set, or ZigBee Pro. In addition, some APIs provide support for a public and manufacturer-specific cluster library, which helps shorten development time. However, this kind of platform provides limited insight into the ZigBee stack itself because developers are restricted from modifying the implementation (e.g., to migrate the code to a new platform, or to enhance the support of new protocols).

Other community-contributed platforms [20] have produced open-source implementations of the ZigBee stack. Compared to their freeware competitors, these platforms benefit from wider use in academic research. The focus of these platforms is shifted from development life cycle to openness and system performance. However, there is no assurance that the stack fully conforms to ZigBee specifications. From the perspective of an application developer and a system developer, it is expected that the specific architecture of the stack implementation is well studied before programming

Bits: 8	0/16	8	8	Variable
Frame Control	Manufacturer Code	Sequence Number	Command Identifier	Frame Payload
Header				Payload

Figure 9.2: Command packet format.

the application object. APIs implementing APS and ZDO may not be fully available, as delivering end products is not the target of these platforms. However, accessibility and API support for research purposes are readily available.

9.3.3 Sensors and Actuators

Sensors and actuators may be the key components in a ZigBee network. In fact, many platforms come with sensors embedded for easy use, such as the TI's EZ430-RF2480 kit that has temperature and light sensors. Other sensors and actuators for specific applications can be readily developed, revealing the need for having readily available interfaces on the chosen platform. To this end, sensor readings are normally fed into an analog-to-digital converter, a serial interface, or an integrated circuit. Actuators can be enabled using the same set of serial interfaces, but using instead a digital-to-analog converter. When choosing a platform, it is important to evaluate what sensors/actuators are already available, and what interfacing capabilities are supported on the platform.

9.4 Developing ZigBee Certified Products: An Example

In this section, the reader will find a practical example of how to develop a ZigBee product. In particular, we target a ZigBee health care application focused on monitoring important physiological signals and parameters of the human, i.e., heart rate, temperature, and measurement of oxygen saturation in the blood. See Figure 9.3. Because the ZigBee health care public profile has not been published yet, we walk the readers through the process of designing a manufacturer-specific profile. The process of applying for a profile identifier and certification from ZigBee Alliance is out of the scope of this section, but has been briefly introduced in previous sections.

9.4.1 Challenges

The objective to achieve energy efficiency in WSNs applies when developing ZigBee products. However, there is little room left for ZigBee Certified Product developers to modify lower-layer protocols in order to conserve energy. The developers

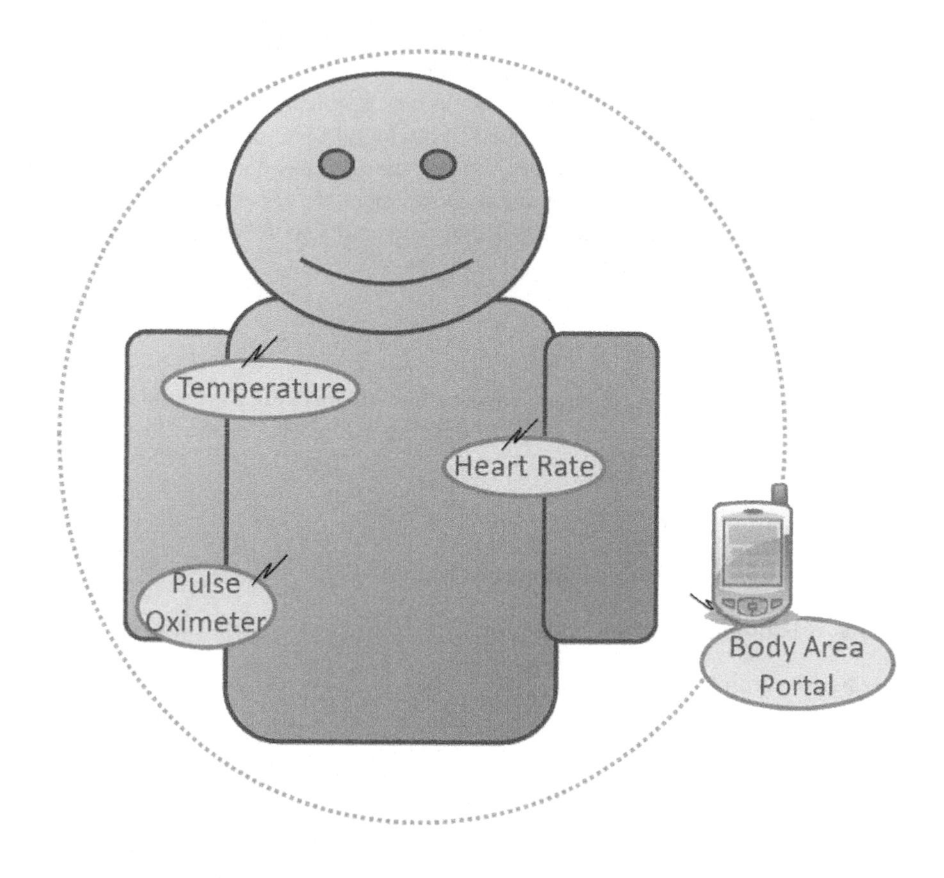

Figure 9.3: Body area sensors for healthcare.

are required to achieve this goal at the system level, by choosing an appropriate network topology, designing a compact application profile and implementing an efficient power management scheme in software.

Currently, a ZigBee network supports three topology configurations: star, tree and mesh. From star topology to mesh topology, the complexity of communications between devices is increasing, but not necessarily the power consumption. The developers are expected to match the topology configuration to the logical connections of targeted networked devices.

As aforementioned, an application profile is the same as an application layer protocol that defines the message types, packet fields, and actions upon receiving a message. The problems are how many states a sensor/actuator has, what is the resolution of the sensor/actuator, and how to convey the sensor/actuator information efficiently and accurately. On one hand, we require more states to better describe the sensor/actuator, larger resolution to enhance end-user experience; on the other hand, simplified message structure will help improve energy efficiency. The developers are expected to make decisions on these tradeoffs.

Besides software implementation of drivers, application profile and user interface, power management schemes need to be devised and realized in software too. Such a scheme sits on a system level, which manipulates hardware functionalities, and adapts itself to end-user's practice. Developers are expected to study not only the application scenarios of their ZigBee devices, but also the targeted users to better develop the adaptable scheme.

9.4.2 Development Steps

With these challenges in mind, we now see how the decisions are made during system realization. The necessary steps to develop a ZigBee Certified Product can be summarized as follows (Figure 9.6.):

- Specify the requirements of the application;
- Design the application profile;
- Choose the functional platform;
- Develop firmware;
- Test the system;
- Initiate certification procedure.

Advanced knowledge of the target application aids in the design of the profile and the selection of the OEM platform. This entails a thorough thinking to reach a resolution of a number of issues. For instance: What is the development life cycle, and what is the monetary budget? These are always the first two questions to be asked before technical questions: What are the required types of devices (sensors/actuators) that need to be used? How are they powered? What are their physical dimensions? What command/messages do they need to exchange, and how frequently are they exchanged? With this knowledge, the developers are then able to determine the network topology that suits the specific sensors/actuators, the message formats, packet fields and actions upon receiving messages.

When choosing an OEM platform, the development life cycle and budget again play an important part. Other tradeoffs have been discussed in Section 9.3. In essence, the decision shall be made between system transparency and development flexibility. After the design of application profile and selection of OEM platform, we come to the implementation and testing stages. Developers are expected to realize the previous design decisions on the chosen hardware-constrained platform.

The following explains the steps by walking the readers through an example of designing ZigBee Certified products for a health care application.

9.4.3 Specify the Requirements of the Application

Our aim is to develop a sample ZigBee network of sensor nodes, each of which monitors a particular physiological signal in order to enable ubiquitous health care. With this goal in mind, we devise the network to have four types of devices, namely: a body area network portal, a heart rate monitor, a temperature sensor, and pulse-oximeter in accordance to the parameters and signals we intend to monitor. The role of the body area network portal can be assigned to a smart phone, which serves as the controller and provides a data display. Considering that all of these devices are portable, they shall be configured as end-devices, indicating that they are battery powered. We also assume that a user would wear these devices for a long time each day, and so comfort would be of prime importance during the design process. While actual sensor design is outside the scope of our discussion, we take into consideration their corresponding form/shape factor when selecting the platform for prototyping.

Knowing the four types of devices and their functionalities in the ZigBee network, we can further devise command/message formats, which are known as clusters and attributes, as explained before. For the sake of simplicity, we assume that each of the three sensor devices (heart rate monitor, temperature sensor, and pulse oximeter) possesses similar outputs and inputs. Here, the outputs from one sensor device are fed to the data consumer (the body area network portal) that pools sensor readings information, whereas the inputs carrying the configuration information from the body area portal are sent to the corresponding sensor device, e.g., the sampling frequency and reporting frequency. These parameters are specified in the clusters. Furthermore, we also consider that these readings span different ranges for different sensors. For instance, heart rate samples ranges from 50 to 150 beats-per-minute, whereas temperature ranges from 20°C to 45°C. Therefore, two simple attributes are added to restrict sensor reading values on different sensor devices.

9.4.4 Design the Application Profile

First, we assign to our sample application profile the name "*Body Area Healthcare Profile*". As introduced in Section 9.1, the profile specifications shall consist of at least the device descriptions, clusters and attributes. The designed profile is illustrated in Figure 9.4.

9.4.4.1 Device Descriptions

The four types of devices previously described for our sample application are listed in Table 9.2, each possessing a unique identifier, which will be used when programming the devices. Each device is also listed with the clusters it supports, which will be explained in detail below.

9.4.4.2 Cluster and Attributes

The desired network functionalities can be abstracted into two clusters. We refer to the first as *Sensor Control*, and the second one as *Monitoring*. The Sensor Control

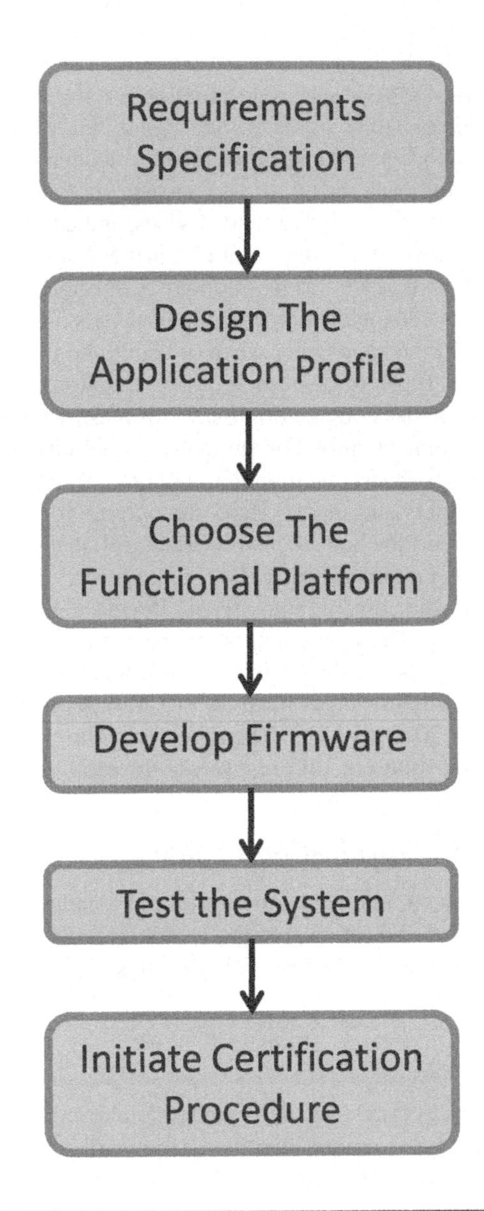

Figure 9.4: Steps of ZigBee product development.

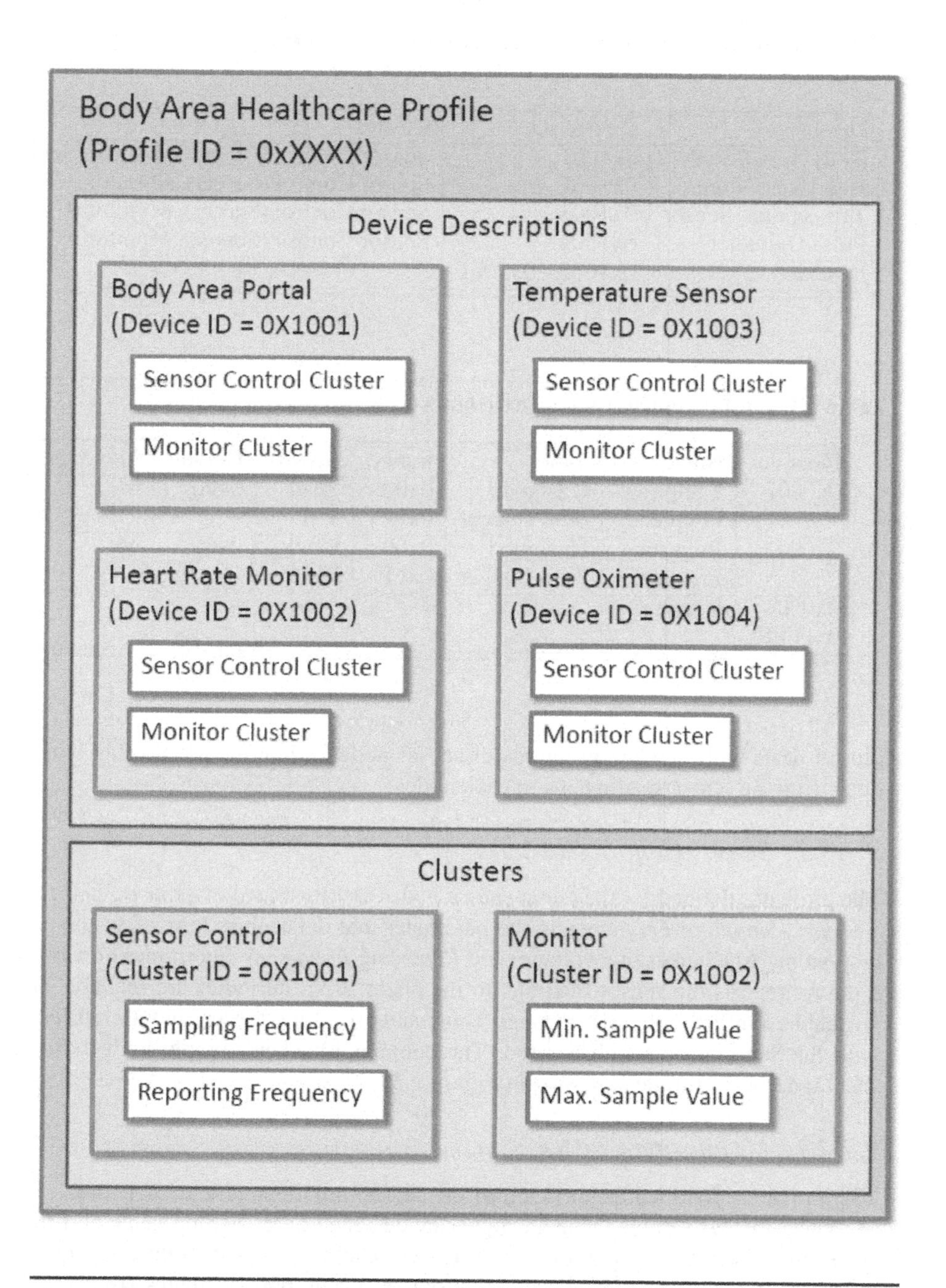

Figure 9.5: Body area healthcare profile.

Table 9.2: Device Descriptions in Body Area Healthcare Profile

Device	Device ID	Cluster
Body Area Portal	0x1001	Sensor Control (Client), Monitor (Server)
Heart Rate Monitor	0x1002	Sensor Control (Server), Monitor (Client)
Temperature Sensor	0x1003	Sensor Control (Server), Monitor (Client)
Pulse Oximeter	0x1004	Sensor Control (Server), Monitor (Client)
Reserved	0x1005 - 0x10FF	

Table 9.3: Sensor Control Cluster Attributes

Identifier	Name	Type	Range	Access	Default	Man.Opt.
0x0001	Sampling Frequency	Unsigned 8-bit Integer	0x00 - 0xFF	Read/ Write	0x00	M
0x0002	Reporting Frequency	Unsigned 8-bit Integer	0x00 - 0xFF	Read/ Write	0x00	M
0x0003 - 0xFFFF	Reserved					

cluster deals with the configurations of sensor nodes, while the Monitoring cluster deals with the sensor readings from each node.

9.4.4.3 Sensor Control Cluster

The attributes defined for the client sensor nodes are illustrated in Table 9.3.

Here, *Sampling Frequency* is the parameter that determines how often the corresponding ADC takes in a voltage, and *Reporting Frequency* determines how often a device reports the sensor readings to the client. Both attributes are readable and writeable, can be configured through commands from the server, and are affiliated with the Sensor Control cluster too. The command packets comply with those in ZCL and share a format as shown in Figure 9.2.

9.4.4.4 Monitoring Cluster

The attributes identified for the Monitoring Cluster are illustrated in Table 9.4.

The Min/Max sample value attributes are designed to restrict the range of different sensor readings. As mentioned before, the reading of a temperature sensor has a different range as that of a heart rate monitor. Again, both attributes are readable and writeable, and can be configured through commands. However, in the Monitoring Cluster, sensor devices are seen as servers, and as a result, those configuration commands come from the client, i.e., the Body Area Network Portal in our example.

Table 9.4: Sensor Control Cluster Attributes

Identifier	Name	Type	Range	Access	Default	Man.Opt.
0x0001	Min. Sample Value	Unsigned 8-bit Integer	0x00 - 0xFF	Read/ Write	0x00	M
0x0002	Max. Sample Value	Unsigned 8-bit Integer	0x00 - 0xFF	Read/ Write	0xFF	M
0x0003 - 0xFFFF	Reserved					

9.4.5 Choose the Functional Platform

Considering the need for easily interfacing with sensors, and the need for a platform with a small size, we choose TI's kit eZ430-RF2480 as an example for furthering our explanation. This kit comes with three CC2480 boards, three battery boards and one USB emulator board. Each CC2480 board has a MCU and a ZigBee network processor on it that is connected to a sensor output, all of which comprise a full sensor node in the network. The USB emulator board interfaces the target board to a PC, and is used for flashing binary codes, or for debugging purposes [21]. TI's ZigBee stack is implemented by the CC2480 chip, which handles time-critical and process-intensive ZigBee protocol tasks. Access to functionalities is provided through APIs when the CC2480 is interfaced with a MCU using the SPI/UART port.

9.4.6 Develop Firmware

Implementing the profile in firmware is a relatively straightforward task. However, care shall be taken when realizing the following routines.

Firstly, after booting the MCU and the network processor, the sensors' configuration is expected to complete before reading in data [22]. The MCU shall register its application with the network processor in order to inform it about the endpoint, profile ID, device description ID, device version, and cluster IDs. Without this knowledge, the network process cannot initiate or join a ZigBee network by itself. For the case of the CC2480 chip, this is done by sending an *AF_REGISTER* command to it from the MCU [23].

Before commencing the network's operation, the network processor shall also be configured with the logical device type, i.e., coordinator, router or end device, the personal area network ID (known as PAN ID), and the physical channels to scan. Having additional configurations at hand is always useful to better realize control functionalities of network processors when needed; however, it is not a requirement since network processors can work with default parameters. In the CC2480, this can be done by issuing multiple *ZB_WRITE_CONFIGURATION* commands by the MCU, each time with the corresponding configuration identifier specified.

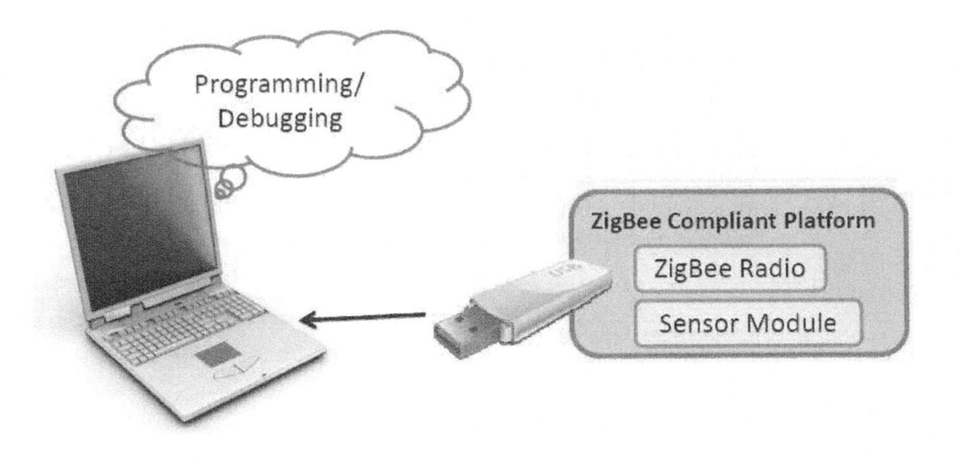

Figure 9.6: Debugging.

After the network is initialized, devices begin to join in. At this stage, the MCU needs to have little intervention since the tasks are handled solely by the network processor. It is also necessary for devices to identify others as they join the network because sources need to know the network addresses of their destinations. Otherwise it is impossible for the network to forward the packet. Without prior knowledge of the destination address, a binding request is flooded in the network, and matching devices will respond with the address information. This can be also achieved through an external commissioning tool, which is outside the scope of our discussions. In the CC2480, binding is created through the *ZB_BIND_DEVICE* command.

After the implementation of the devised profile, we come to the debugging stage. As devices normally have limited resources, it is always desirable to see indications through LEDs on board that serve as an indication of the current system state. Also, some platforms provide JTAG or some other proprietary interface for real-time debugging. In the case of the CC2480 board and the USB emulator board, the latter provides for the former a serial-to-USB converter to connect to a PC.

9.4.7 Test the System

A thorough debugging of the software implementation guarantees a fully functional network. However, it is desirable to test the whole system from each of the following perspectives to evaluate the overall system performance. These include physical layer characteristics, timing of packets, power consumption, and possibly the portability of devices.

The tests of physical layer characteristics, e.g., power spectrum density and adjacent channel interference, require professional equipment to evaluate. However, timing information can be obtained by employing a "packet sniffer" in close proximity to the network devices to display every packet it receives with a timestamp in the tester. Normally, the sniffer is an IEEE 802.15.4 radio interfaced to PC through a USB port. The forwarding of packets from a radio chip to a PC may take some time, but can be considered to be the same for each packet. As a result, the timestamp for each packet can be a delayed version of the actual time, from which the information regarding the time delay between packets can be readily obtained. Evaluating both the timing and relevant fields of packets in all the layers and from all devices provides the tester with a reliable picture of the network's current operating state.

Since power consumption gives an estimate of the lifetime of a device, it is sensible to register the power consumption curve over time and evaluate the outcome. A straightforward approach is placing a small resistor in series with the platform and measure the voltage difference at two ends of the resistor. The current can then be calculated and when multiplied by the voltage supply and an estimate of the power consumption can be obtained.

Portability is supported in ZigBee protocols through orphan scan and network rejoining processes. When mobility of devices in a network is required (e.g., a ubiquitous health care application mentioned above), mobility support of devices in a ZigBee network shall be tested as well [12]. It is necessary to first identify what the relative movement will be, its distance, and its speed. Then, tests shall be carried out to evaluate the effectiveness and efficiency of ZigBee's support for this mobility.

9.4.8 Initiate Certification Procedure

After ensuring conformance to the specifications and devised profile, a newly developed ZigBee product can be submitted to the corresponding certification stage that follows the procedures established by the alliance.

9.5 Conclusion

We have explored the challenges and described the procedures involved in the development of a product targeted for certification by the ZigBee Alliance. We have also observed that this process requires a significant amount of planning and considerations to ensure that the proposed application meets with a number of requirements. In particular, we have explored a sample application targeted at WBASN product development. This example reveals the importance of taking into account all efficiency-related factors that make the difference between a poorly-designed device, and one that fulfills the necessary requirements for a quality product that can be used to monitor the health of people. Such healthcare applications have become the cornerstone of numerous research efforts, and is expected to yield significant advancements in this area for the years to come. The design and development of a product for ZigBee certification is a demanding task, but is also one that ensures interoperability with similar products depending on the profile being employed. In the near future, we are poised to see an even greater variety of products being developed as the ZigBee standard profiles mature and low-power, mobile devices become pervasive.

References

[1] I. F. Akyildiz, W. Su, Y. Sankarasubramaniam and E. Cayirci, "Wireless Sensor Networks: A Survey", Computer Networks: The International Journal of Computer and Telecommunications Networking, Volume 38, Issue 4, March 2002.

[2] IEEE 802.15 WPAN Task Group 6 (TG6) Body Area Networks, http://www.ieee802.org/15/pub/TG6.html.

[3] H. Cao, C. Chow, H. Chan and V. C. M. Leung, "Enabling Technologies for Wireless Body Area Networks: A Survey and Outlook", IEEE Communications Magazine, Volume 47, Issue. 12, pp. 84-93, Dec. 2009.

[4] ZigBee Alliance, http://www.zigbee.org/.

[5] K. Gill, S. Yang, F. Yao and X. Lu, "A zigbee-based home automation system", Consumer Electronics, IEEE Transactions on, Volume 55, Issue 2, pp.422-430, May 2009.

[6] J. Su, C. Lee and W. Wu, "The design and implementation of a low-cost and programmable home automation module", Consumer Electronics, IEEE Transactions on, Volume 52, Issue 4, pp.1239-1244, Nov. 2006.

[7] W. S. Lee and S. H. Hong, "KNX - ZigBee gateway for home automation", Automation Science and Engineering, 2008, IEEE International Conference Aug. 23-26, 2008.

[8] F. L. Zucatto, C. A. Biscassi, F. Monsignore, F. Fidelix, S. Coutinho and M. L. Rocha, "ZigBee for building control wireless sensor networks", Microwave and Optoelectronics Conference, 2007, SBMO/IEEE MTT-S International, pp.511-515, Oct. 29 - Nov. 1, 2007.

[9] K. F. Tsang, W. C. Lee, K. L. Lam, H. Y. Tung and K. Xuan, "An integrated ZigBee automation system: An energy saving solution", Mechatronics and Machine Vision in Practice, 2007, 14th International Conference on, pp.252-258, Dec. 4-6, 2007.

[10] J. Y. Jung and J. W. Lee, "ZigBee Device Access Control and Reliable Data Transmission in ZigBee Based Health Monitoring System", Advanced Communication Technology, 2008, 10th International Conference on, pp.795-797, Feb. 17-20, 2008.

[11] W. W. Lin and Y. H. Sheng, "Using OSGi UPnP and ZigBee to Provide a Wireless Ubiquitous Home Healthcare Environment", Mobile Ubiquitous Computing, Systems, Services and Technologies, 2008, The Second International Conference on, pp.268-273, Sept. 29 Oct. 4, 2008.

[12] IEEE Computer Society, Part 15.4: Wireless Medium Access Control and Physical Layer Specifications for Low-Rate Wireless Personal Area Networks. NY, 200.

[13] Wi-Fi Alliance, http://www.wi-fi.org/.

[14] IEEE 802.11 Wireless Local Area Networks, http://www.ieee802.org/11/.

[15] J. F. Kurose and K. W. Ross, Computer Networking: A Top-Down Approach, 5th edition. Addison Wesley, NY, 2009.

[16] ZigBee Alliance, ZigBee Specification, ZigBee Document 053474r17. CA, 2008.

[17] Daintree Networks, Getting Started with ZigBee and IEEE 802.15.4. 2008.

[18] Daintree Networks, Understanding the ZigBee Applications Framework. 2006.

[19] ZigBee Alliance, ZigBee Smart Energy Profile Specification, ZigBee Document 075356r15. CA, 2008.

[20] ZigBee Alliance, ZigBee Cluster Library Specification, ZigBee Document 075123r01ZB. CA, 2007.

[21] ZigBee Alliance, ZigBee Home Automation Public Application Profile, ZigBee Document 053520r25. CA, 2007.

[22] Ember: http://www.ember.com/.

[23] Atmel: http://www.atmel.com/.

[24] Texas Instruments: http://www.ti.com/.

[25] TinyOS ZigBee Working Group: http://www.hurray.isep.ipp.pt/activities/ZigBee_WG/.

[26] TinyOS: http://www.tinyos.net/.

[27] Crossbow Technology: http://www.xbow.com/.

[28] Texas Instruments, eZ430-RF2480 User's Guide, SWRU151A. 2008.

[29] Texas Instruments, CC2480 Developer's Guide, SWRU176. 2008.

[30] Texas Instruments, CC2480 Interface Specification, SWRU175A. 2008.

[31] H. Cao, X. Liang, I. Balasingham and V.C.M. Leung, "Performance Analysis of ZigBee Technology for Wireless Body Area Sensor Networks", ASIT 2009, Niagara Falls, ON, Sept. 2009.

Chapter 10

Monitoring the Efficiency of Workflow

Vincent Tam, Johnny Yeung

CONTENTS

10.1 Introduction

Wireless sensors [1, 2, 5, 6, 7, 8, 11, 13, 14, 16] have received much attention in both research and system development in recent years. With a wealth of exciting research results obtained over the past decade, they have been continuously fueling significant impacts and reshaping many possible areas of real-life applications including localization [6, 11, 14, 16] or object tracking [14] for military applications, effective event storage and distribution schemes [13, 15] in wireless sensor networks

(WSNs) for monitoring or surveillance, deploying wireless sensors to monitor the changes in temperature, pressure or other factors in chemical plants, refinery [17] or power lines [1] for real-time and intelligent control systems, using ultra-sound and/or other wireless sensors as parts of embedded systems for robot navigation [2] or control, incorporating the multi-axis accelerometer or intrinsic motion sensors relative to gravity into personal electronic devices such as the Apple iPhone® [18], the Nokia N96® mobile phones [3] or Nintendo's Wii Remote® controller [2] for mobile gaming, or even as digital compasses to work with 3G or global positioning system (GPS) receivers [20] on mobile devices [8, 22, 23, 25] for car or personal navigation. In many such innovative applications of wireless sensors, each sensor/node in the system/network is employed to detect some specific event such as changes of positions or temperature, and quickly sends the collected information to the nearest server or sink node for further processing or concatenation of information. In many industrial WSN applications, the automatic position determination, namely the localization, of sensors is important since a sensor's position must be known for the data to become meaningful, especially in the location-aware sensor network communication protocols such as sensing coverage [3]. Besides, the constrained resources of individual sensors including the maximally supported data rate, the limited battery lifetime, and possibly the high failure rate of date transmission due to some noisy terrains, together with the scalability, mobility or other specific requirements of the underlying applications all pose difficult challenges to the latest research in localization or WSN applications in general. Nevertheless, the innovative and suitable uses of wireless sensors would be expected to open up many potential markets of emergent technologies to assure a high quality of living in the future. A prominent example is the possible integration of wireless sensors into the next-generation intelligent buildings that can automatically find out the shortest path, open the involved exit doors and guides the tenants to safe places in extreme cases of fires or natural disasters. After a wireless sensor detects any interested event, it has to use a wireless network such as BlueTooth™ [9], the IEEE 802.11 or commonly named WiFi™ for Wireless Fidelity [6], or the ZigBee™ [4, 9, 21, 27, 29, 32] network. Mobile gaming devices like the Nintendo's Wii Remote® controller may use BlueTooth™ network for convenient and short-range communication in personal area networks (PANs) while popular mobile phones integrated with motion sensors, like the Apple iPhones®, may provide both connection functions of BlueTooth™ and WiFi for users to choose. In general, the ZigBee™ networks as defined by the IEEE 802.15.4 specification are intended to be simpler and less expensive than BlueTooth™ or IEEE 802.11 networks. This is mainly because the ZigBee™ networks can be activated (that is going from sleep to active mode) in 15 milliseconds or less whereas the wake-up delays in BlueTooth™ networks are typically around 3 seconds [7]. With the very low latency, devices connected to a ZigBee™ network can be very responsive. Besides, since ZigBee™ devices can sleep most of the time, the average power consumption can be very low, thus resulting in a relatively longer battery life. Overall speaking, the ZigBee™ technology is a low-cost, low-power, wireless mesh networking proprietary standard. The low cost allows the technology itself to be widely deployed in wireless control and monitoring applications. The low power-usage enables a longer life of the concerned devices with smaller batteries while the mesh networking provides high reliability and larger range for many industrial applications such as commercial building automation [7], telecommunication applications [9], or even personal or hospital care systems [9]. In this chapter, our focus is on investigating the integration of the latest wireless ad-hoc networks and sensor technology to monitor the efficiency of workflow in manufacturing plants. With the high competitiveness due to globalization, and the fast development of many developing countries, especially in Asia or Africa, it is extremely important for manufacturing plants like those in the Southern part of PRC to have high productivity, and more importantly with the availability of real-time data and analysis report generated to streamline the whole operation and efficiency of modern supply-chain management (SCM). With this critical mission in mind, our study is aimed at utilizing the latest sensor and mesh network technologies to build a working prototype with lowest possible costs so as to demonstrate how these technologies can be combined to facilitate the analysis of the workflow efficiency in industrial applications, with the real-time data being captured through numerous low-cost and customizable sensors installed at different machines/checkpoints of a manufacturing plant. It is worth noting that there are many commercially available wireless sensors, such as those produced by the Crossbow or FreeScale Semiconductor Inc., that readily support the ZigBee^{TM} wireless networks. However, the lower costs in designing and producing our own sensors that are highly adaptive and customizable for the specific tasks of monitoring the machine movement or item flow in manufacturing plants may provide extra benefits to our adopted approach. As thousands of sensors may be required for a typical small-to-medium sized manufacturing plants in PRC, the relatively lower production costs of our sensors can be a critical factor to consider, especially for the developing countries. In addition, with the real-time data or analysis report available in electronic forms on a local server PC, a manager/supervisor can use his/her own WiFi-enabled portable devices such as smart-phones or pocket PCs to quickly access such real-time data or analysis about the efficiency of any specific machine, pipeline or checkpoint in his/her manufacturing plant anytime and anywhere, possibly through the nearest WLAN access points available. Accordingly, we will review the design and implementation of our low-cost and highly customizable wireless sensors that can be easily integrated into realistic ZigBee^{TM} networks for uses in many industrial applications. To demonstrate the feasibility of our proposal, we have tested our working prototype in a real manufacturing plant in Guangdong, the southern part of PRC. The preliminary evaluation results of our prototype are encouraging, and prompt further investigation. All in all, this research has produced significant results that clearly demonstrate the expected performance and limitations of our integrative approach for monitoring workflow efficiency, and more importantly its untapped potential for the future development. This chapter is organized as follows. Section 10.2 reviews some previous works relevant to our proposal including the wireless sensors and their applications, and the ZigBee^{TM} networks. Section 10.3 details the system architecture design of our wireless sensors integrated with the ZigBee^{TM} networks for monitoring the workflow efficiency in modern manufacturing plants. We give an empirical evaluation of our working prototype in a real industrial environment in Section 10.4.

Lastly, we summarize our work and shed lights on many possible future directions in Section 10.5.

10.2 Related Works

In this section, we will consider some related works including the use of wireless sensors for real-life applications, and also the low-cost and low-power ZigBee networks widely adopted as mesh networks for industrial applications.

10.2.1 Wireless Sensors and Their Applications

Autonomous and small-sized sensors can collaborate through a spatially distributed and wireless sensor network (WSN) [5, 6, 7, 8, 11, 13, 14, 16] to monitor many environmental or physical conditions, including pollutants, pressure, motion, sound, temperature or vibration. In the past, the development of wireless sensors was largely motivated by military or robotic application such as battlefield surveillance or robot navigation. However, they are used in many industrial application areas nowadays, such as environment and habitat monitoring, industrial process monitoring and control, home automation, healthcare applications, machine health monitoring, and traffic control.

Besides sensing capabilities, each node in a wireless sensor network is often equipped with a wireless transceiver or other wireless telecommunication device, a small microcontroller to preprocess or compress the collected data, and an energy source like a battery. Accordingly, each sensor node, commonly known as a "mote" in North America, is capable of performing processing, collecting sensory information and communicating with other connected nodes in the network. Figure 10.1 shows the typical architecture of a sensor node. Since the micro-controller module may not have sufficient capacity of random-access memory (RAM) or on-board cache to store the user program or real-time data, it will often interface with an external memory unit to extend its capacity. Moreover, in cases where the sensors can only produce analog signals, an analog-to-digital converter (ADC) is required to convert the analog signals to digital ones for further processing by the micro-controller board.

Basically, a wireless sensor module can vary in size from that of a box to the size of a grain of dust. Similarly, the cost of a sensor node is a variable that can range from a few hundred U.S. dollars, equivalent to thousands of Hong Kong dollars or Chinese yuans, to a few U.S. cents as dependent on the sizes of the wireless sensors and their required complexity to achieve the tasks at hand. As reported in many industrial applications and research study, size and cost constraints on sensor nodes result in the corresponding constraints on resources including energy, memory, computational speed and bandwidth.

As clearly explained, wireless sensors usually constitute a wireless ad-hoc network to collaboratively sense real-time data and distribute the "interested" events

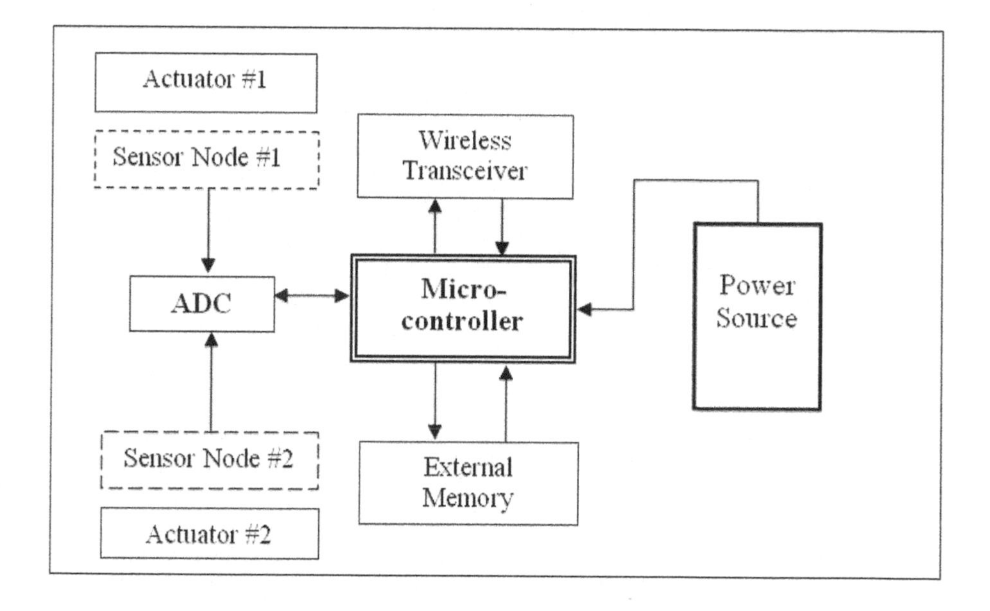

Figure 10.1: The typical architecture of a sensor node in a wireless network.

over the network in order to monitor the targeted machines or environment in practical applications. In many circumstances, a data packet sent by any particular sensor will need to go through multiple hops (or nodes) before being forwarded to the base stations or cluster-heads for further processing. Figure 10.2 shows the typical multi-hop message routing mechanism used in a WSN to effectively collect and distribute the sensory information for process control or monitoring.

10.2.2 The ZigBee Networks

As defined by the IEEE 802.15.4 specification [8], the ZigBee^{TM} networks are designed to be a simpler, low-cost and low-power de facto wireless standard for industrial applications. When compared to other wireless personal area networks (WPANs) inlcuding BlueTooth or IEEE 802.11 networks for important real-world applications [5, 9, 31, 32], ZigBee^{TM} networks will have their unique strength that can be detailed as follows:

- they can be activated in a relatively short period of time like 15 milliseconds or less. With respect to the typical wake-up delays of 3 seconds for BlueTooth^{TM} networks [9], the wake-up time for ZigBee^{TM} networks is only 1/200 of that required by BlueTooth^{TM};
- with this much lower latency of a ZigBee^{TM} network, any sensor node or device connecting to the network can be more responsive to any sudden change in the underlying environment being monitored;

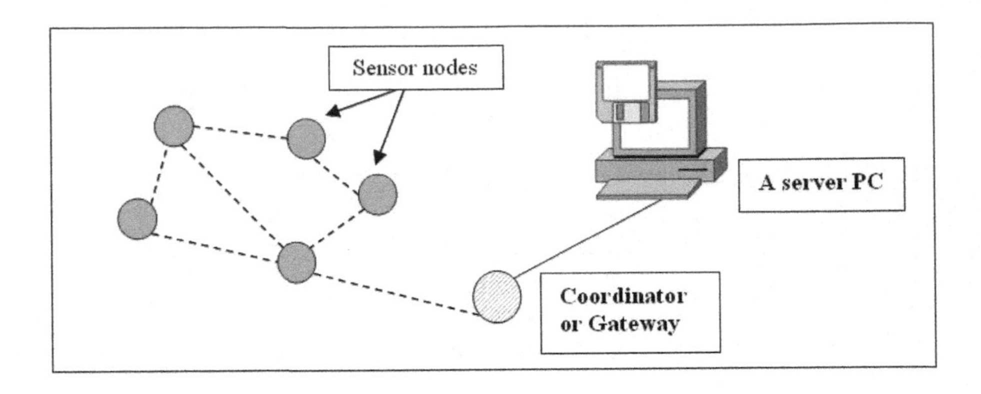

Figure 10.2: The multi-hop routing mechanism used in a wireless sensor network.

- since ZigBee™ devices can sleep most of the time, the average power consumption can be very low, thus resulting in a relatively longer battery life.

All in all, the low installation cost of a ZigBee™ network enables the technology itself to be widely deployed in numerous wireless control and monitoring applications while the low power-usage promotes a longer life of the concerned devices with smaller batteries. Therefore, the overall mesh networking technique provides high reliability and larger range for many industrial applications such as commercial building automation [7], telecommunication applications [9], or even healthcare systems [9].

10.3 The Architecture of Our Monitoring System

Our proposed system of wireless sensors and ZigBee™ network to monitor the efficiency of workflow in modern manufacturing plants is designed to process real-time data on multiple channels available in the ZigBee™ network. Each wireless channel is uniquely assigned to a production line or machine being monitored. Essentially, there is one sensor module allocated to each wireless channel so as to detect the interested event such as the motion of a specific machine, the movement of a finished or unfinished item on a production line, or the temperature of a machine, and then communicate to the server PC through the assigned wireless channel. Since each wireless channel is pre-assigned to monitoring a particular production line or machine, any real-time data received from the pre-assigned channel can be quickly associated to that of the specific machine. In fact, certain detected parameters associated with a machine such as the number of repetitive motions performed by the specific machine may also imply the rate of the goods production on the corresponding production line. However, it can be difficult to determine whether the produced item is of acceptable quality or not unless a high-resolution and high-speed video/image camera

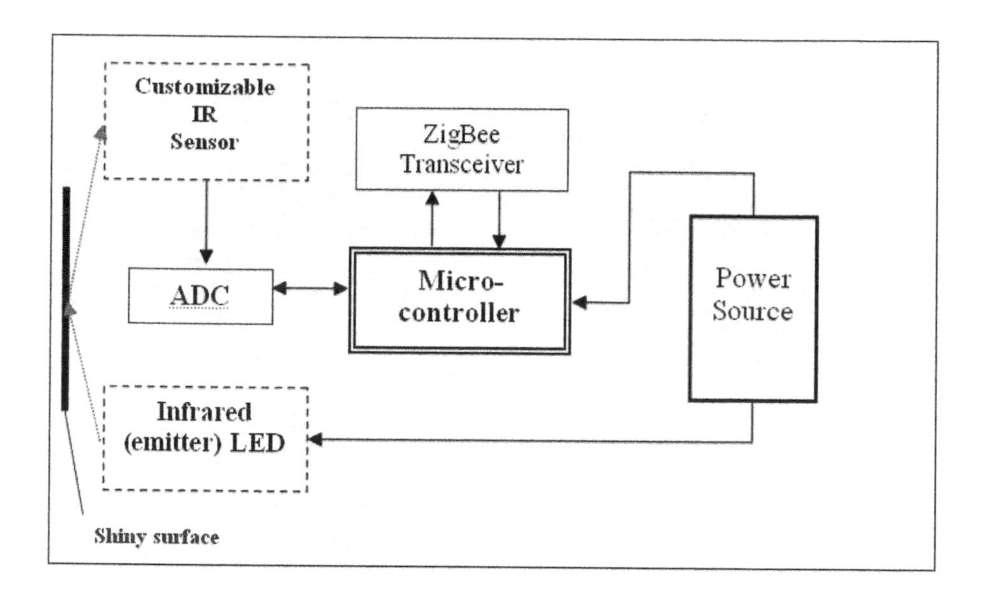

Figure 10.3: The system architecture of our low-cost and highly customizable sensor node for use in a ZigBee network to monitor the efficiency of workflow in any manufacturing plant.

is employed with an intelligent image processing algorithm. Nevertheless, Figure 10.3 gives the architecture of our low-cost and highly customizable sensor module whereas Figure 10.4 shows a working prototype of our sensor module embedded in a metal box for compactness and ease of attachment to any machine or production line in a manufacturing plant.

Essentially, the sensor module will detect the infrared signal emitted by an infrared light-emitting diode (LED) implanted on one side of the metal box. Based on the IR signal as reflected from a fixed clean and shiny surface on the moving part of a machine being monitored, the sensor module will generate the digital pulses for preprocessing by the micro-controller module which compress the sensory data into data packets for transmission to the server PC via the allocated wireless channel in the ZigBee network. In case no infrared signal is received by the sensor module, there will be no digital pulse generated. Accordingly, the micro-controller will assume that there is no machine movement. Ultimately, the compressed data packet will be received on the server machine. Installed with a noise filtering algorithm, which can work effectively after some fine tuning, the server PC will be able to remove very tiny movements caused by certain shaky/damping motions during the initialization or stopping of a machine from the collected data set. More importantly, with the rectified figures, the server PC can quickly generate a detailed report with various charts on Web pages for further analysis on the actual efficiency of workflow. The Web report may contain statistics as those required by certain Engineering Management standard such as the Overall Equipment Effectiveness (OEE) [4] as defined by the

Figure 10.4: Our working prototype of the low-cost and highly customizable IR-based sensor node that integrates a ZigBee transceiver, micro-controller module, IR LED and IR receiver with adjustable sensitivity.

world-wide organization of Six Sigma. For the proposed architecture, the network topology is hierarchical and multi-hop, with the top and middle layers of the network hierarchy as the coordinator nodes and the bottom layer (i.e., the leaf nodes) as the sensor module. Initially, the network hierarchy is pre-configured with a unique sensor node identification number (ID#) pre-assigned to each sensor module by the corresponding micro-controller module. The rate of machine movement as the raw data detected in each sensor module is then transmitted to the micro-controller module that filters out the noisy data and then combines with the corresponding sensor node ID# before being sent out to the ZigBee module for transmission to the coordinator node and ultimately to the server PC.

Essentially, the system is mainly developed to provide the essential functions facilitating the tasks of a manager/supervisor of any modern manufacturing plant in the following ways. For ease of management and improved security, the proposed system can continuously update the sensory data and information of the system such as the working processes and machines that are functioning normally, how all the processes are working and more importantly performance measures of each process/machine being monitored. The administrator can simply monitor the whole place in detail while sitting in his/her own office or even in countries far away from the manufacturing plant during an overseas visit. The efficiency of each process and also the roughly estimated working performance of each worker can be easily monitored anytime and anywhere with his/her smart-phone or portable device accessing the company server through any wireless or phone network. Besides overseeing the routine operations, our proposed system can also help to improve the security of the underlying manufacturing plant or safety of the machines being monitored. When there is any abnormal activity of the machines including their irregular movement possibly due to burglary, theft or machine breakdown, especially after the regular working hours, alerts will be sent to the concerned administrator so that the appropriate actions may be taken. Figure 10.5 illustrates the overall design of our proposed system integrated with our highly customizable IR-based wireless sensors and ZigBee networks to monitor the efficiency of workflow.

For a more accurate data recording, the manufacturing plants should be able to obtain specific sensory information such as accurate time, fast motions of the machines and their performance. However in many real-life situations, many manufacturing plants fail to obtain such accurate and real-time data for monitoring and control. For this particular reason, our proposed system can help to record the sensory information more accurately, as compared to some conventional manual counting methods still in use, so as to ultimately obtain much more accurate estimation on the working efficiency of the company. This is especially important when the supervisors/manager aims at optimizing the efficiency of the overall workflow, the supervisors/manager should know every single detail of the efficiency of involved machines/production lines. Otherwise, the supervisors/manager cannot further improve on the work flow and other factors based on the existing statistics in order to make the best use of all the available facilities and resources within the manufacturing plants to maximize on the potential profits. Figure 10.6 shows an ideal model of the modern manufacturing plant facilitated by our sensor and ZigBee based monitoring system to optimize on the efficiency of the workflow. Basically, in addition to the sensor modules installed on each individual machine, extra sensor modules can be installed along the conveyor belts to count the actual number of items produced/bypassed in certain section(s) of the manufacturing pipelines since some defected items may be manually picked up by some quality control workers or automatically rejected by the checking/monitoring devices. In this way, the whole sensor based monitoring system will have a more realistic and real-time figure about the productivity of certain section in a manufacturing plant.

Lastly, other details of our proposed system including the channel assignments, mesh network configuration and system security are carefully considered as follows.

- Channel Assignments: in our proposed system, data collection and processing on multiple wireless channels are possible. Therefore, the server PC has to distinguish whether the received signals are from the same sensor or from different sensors. In doing so, we have two different methods, the first one is to use the 8 channels which is from the standard of ZigBee networks (while there are some ZigBee networks with 16 channels). Nevertheless, the ZigBee networks can typically be used to transmit signals that are simple and have similar frame structure. For those signals denoting more complicated information, we have to use the micro-controller module/chip to encrypt the signal so as to distinguish the concerned signal while reserving the number of possible channels that can still be used in our proposed workflow monitoring

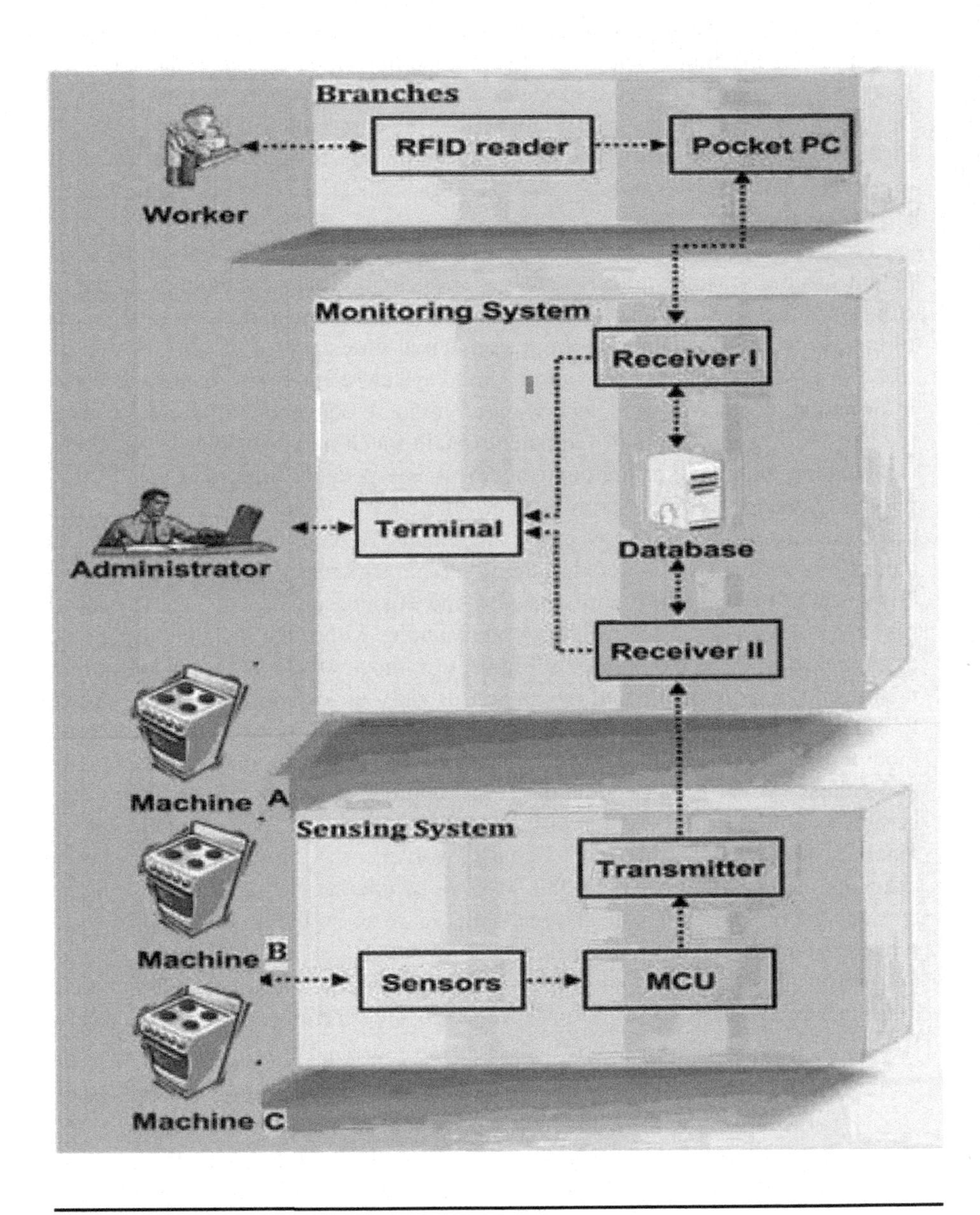

Figure 10.5: The overall system design of our highly customizable IR-based sensor nodes integrated ZigBee networks.

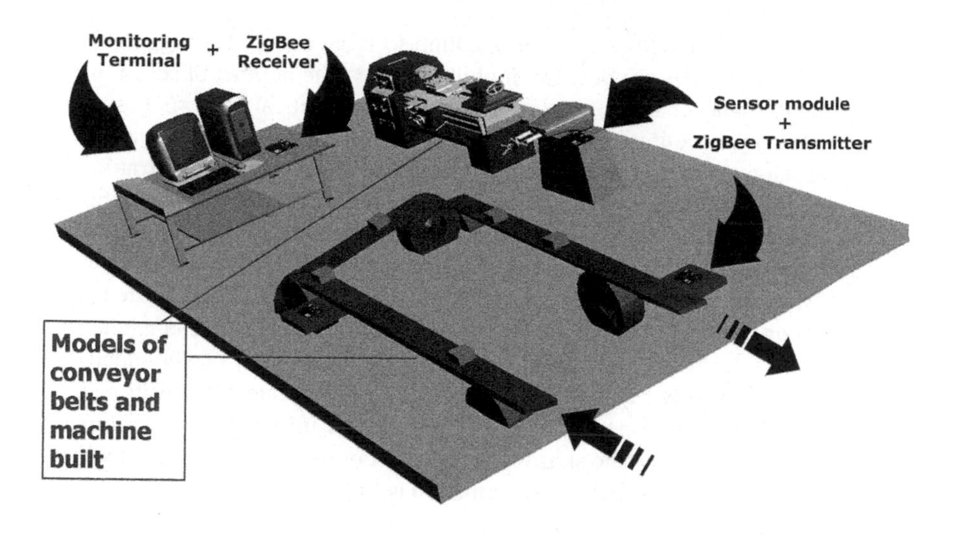

Figure 10.6: An ideal model of the modern manufacturing plant facilitated by our sensor and ZigBee based monitoring system.

system. Accordingly, there are two possible schemes for channel assignments as this workflow monitoring application is concerned.

1. Using default channel - the default channel that can be used is designed for distinguishing between different signals and recipients. A possible advantage of this scheme is to allow the use of simple software and hardware configuration, and also reducing the installation cost required. However, the signals with a complicated format are not easy enough to be handled by this method.
2. Using micro-controllers - more complicated signals are not simple enough to use the default channels only to distinguish different signals. So a microcontroller module is added to the sensor modules to encrypt the corresponding signals before being sent through the underlying ZigBee network.

In view of the complexity of the signals involved, we adopted the latter approach for channel assignment in our proposed monitoring system.

- Mesh network configuration: in a sizeable manufacturing plant containing hundreds or thousands of machines, the wireless communication network will cover large areas and the wireless routing modules are always highly mobile, so direct transmission of data is nearly impossible to meet the requirements. Mesh network is one of the possible options to solve the problem and therefore adopted to build such communication networks. In our proposed system, we aimed to use it in medium-to-large manufacturing plants where most of the sensor nodes are mobile. Besides, the number of sensor nodes to be used may change over time, so mobility is a critical concern. For the hardware, we have implemented the mesh network configuration since its transmission can cover large areas. Moreover, to enable the administrators to add the sensor nodes in a more convenient way, our network configuration software is built so that it can automatically detect any newly added sensor nodes by performing the required authentication with them. Once authenticated, the list of the sensor nodes stored inside the server's monitoring system will be automatically added to the new sensor node so that they can exchange the latest information with each other. As each existing or newly added sensor node is already pre-assigned to the nearest coordinator/gateway node with which it is connected, routing of the detected information is travelled unidirectionally from the sensor node to its coordinator node, and ultimately to the server PC in which the information is further processed before uploading the final report to the Web server.

- Security: security should be carefully considered in our application since the produced data is strictly confidential to the concerned company. In the past, wireless networks always lacked security due to the open transmission media such that others could easily get the information anywhere within the wireless network. With this concern, authentication is necessary in the concerned system so that only the authorized personnel with the assigned machines can get the information such that the provided personal identity and security key match. In our proposed system, the centralized server system keeps a list of authorized people that can collect the concerned data, and a security key for each of the assigned machines. Only when both items match, the authenticated personnel can start to retrieve the stored data of the assigned machine. Otherwise, no information retrieval is allowed. In other words, in our proposed system, just holding the security key is not good enough. This is very important since the wireless network is likely to be opened to the public, and anyone may have the chance to get the sensitive data by various means if the security is not sufficiently stringent. In addition, to increase the security in our system, all the signals are encrypted before they are being transmitted in the open media. The micro-controller module is used to encrypt the data in a different pattern before the signals are being sent out. On the other hand, the central monitoring process run on the server PC should have installed our provided software to decrypt the signals so as to understand their meaning. In this way, the sensory information is safer to be transmitted as encrypted data packets through the underlying ZigBee network.

10.4 Prototype Implementation and Evaluation

We have implemented a working prototype of our proposed sensor and ZigBee based monitoring system and firstly performed tests on them in our laboratory. Later, an empirical evaluation was carried out under the real industrial environment in Guangdong, the southern part of PRC. The micro-controller module is pre-programmed using a tailored assembly language for encrypting the data to be transmitted over the assigned wireless channel assigned to a particular machine/production line. Besides, the server application to decrypt the received data packets, filter out the possible noises (due to the damping/shaky motion of the machine) in the data, and lastly perform a detailed analysis on the data is implemented in a C# program of around 1,000 lines of code running on the Microsoft .NET platform of our server PC, that is essentially a Microsoft Windows XP based notebook PC in our case.

Firstly, we designed our own circuit for the sensor module using a trial-and-error approach. The major reason for us to build our own sensor module is to try to achieve a lower production cost of each sensor module yet being adaptive enough for the different manufacturing environments. All the involved components including the micro-controller chip, the infrared LED emitter and receiver were purchased from some local electronic shops in the Shum Shui Po area of Hong Kong. After constructing and fixing the sensor module, we proceeded to program the micro-controller chip using the C language and its associated development toolkit available on a CD-ROM. As we chose to use a fairly low-cost micro-controller chip that works with a specific Basic Input/Output System (BIOS), the commonly used tinyOS is not suitable for the development of our application. Overall speaking, during the design and development phase, we employed an iterative improvement approach in which vigorous testing will be performed for each fixed version of designs and prototypes until we obtain the final version in which all the sensor, micro-controller and ZigBee modules can be integrated together into a single metal box for compactness and portability in the modern manufacturing environment.

For the experiment, we have performed the test in two parts. For the first part, we have tested the performance of the sensor module for their sensitivity and the stability. We would also like to know if the prototype really works under the real factory environment. In particular, there are many stamping or other machines in operation that may create lots of electromagnetic noises or mechanical vibrations in the background that may interfere with the sensing capabilities of our sensor modules. To resolve this problem of fairly severe background noises, our team has arrived at the manufacturing plant much earlier before the actual testing to perform fine tuning of our sensor modules. Besides, we have used the data filtering function implemented in our server PC to filter out part of the potentially noisy data obtained from the background.

As for the empirical evaluation, we have tested the performance of the sensor module and the ZigBee components. The stability and the transmission range were tested using the real machines of the industries. During the evaluation, the whole system was set up with the real-time data collected and stored in the database of the server PC running our monitoring system. As shown in Figure 10.7, the data is mostly

(A)	(B)	(C)	(D)
2008-04-17	12:43:20:875	00:00:00.6718750	:3
2008-04-17	12:43:21:546	00:00:00.6718750	:3
2008-04-17	12:43:22:218	00:00:00.6718750	:3
2008-04-17	12:43:22:906	00:00:00.6875000	:3
2008-04-17	12:43:23:562	00:00:00.6562500	:3
2008-04-17	12:43:24:234	00:00:00.6718750	:3
2008-04-17	12:43:24:906	00:00:00.6718750	:3
2008-04-17	12:43:25:578	00:00:00.6718750	:3
2008-04-17	12:43:26:250	00:00:00.6718750	:3

Figure 10.7: A record of the real-time data collected by our sensor-based monitoring system in a real manufacturing plant in the southern part of PRC.

consistent and stable over a long period. The leftmost column (A) denotes the date when the test was performed whereas column (B) and (C) represent the specific time and duration of the observed event of the concerned machine. Lastly, column (D) depicts the number of motions occurred for the observed event. Basically, the rate of the process and work flow analysis is accurate. More importantly, it is obvious that our server program can clearly display various charts to show the activities of potentially multiple machines being monitored.

Figure 10.8 illustrates how our sensing module can be mounted on a stamping machine in the real manufacturing environment whereas Figure 10.9 shows a previous working prototype.

Figure 10.10 illustrates how the server PC integrated with the ZigBee receiver can instantly display the analysis result and continuously monitor on the efficiency of the stamping machine while Figure 10.11 shows a series of stamping machines that may potentially be installed with our highly customizable sensor nodes in the same manufacturing plant. Lastly, interested readers are welcome to visit our project website at: http://www.eee.hku.hk/~sennet/workflow_monitoring.html, where more pictures and video clips about our site testing are available. More updated test results will be posted onto our project website when we can pay more visits to the same or different manufacturing plants in PRC in the near future.

10.5 Concluding Remarks

Undoubtedly, our proposed wireless sensor and ZigBee based monitoring system will form an important part of the future real-time process monitoring and control system for modern manufacturing plants so as to increase their productivity and also competitiveness, especially due to the new challenges brought out by globalization. It is worth noting that our infra-red (IR) based wireless sensor is relatively much lower

Figure 10.8: Our highly customizable sensor node mounted on a stamping machine in the real manufacturing plant.

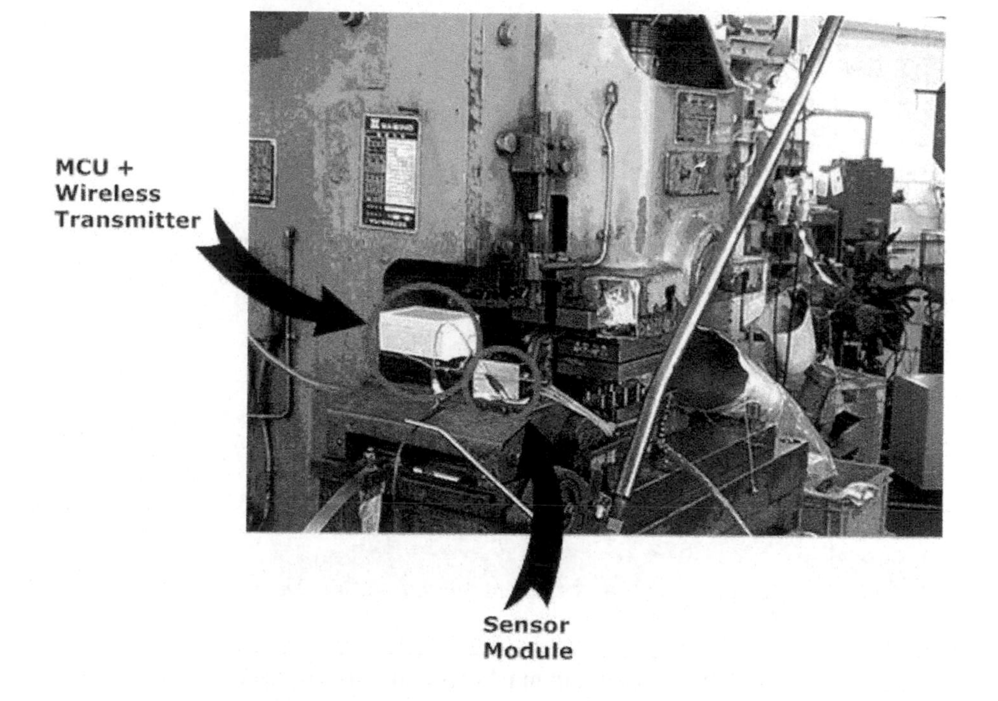

Figure 10.9: A previous working prototype of our sensor node with separate modules mounted on a stamping machine in the real manufacturing plant.

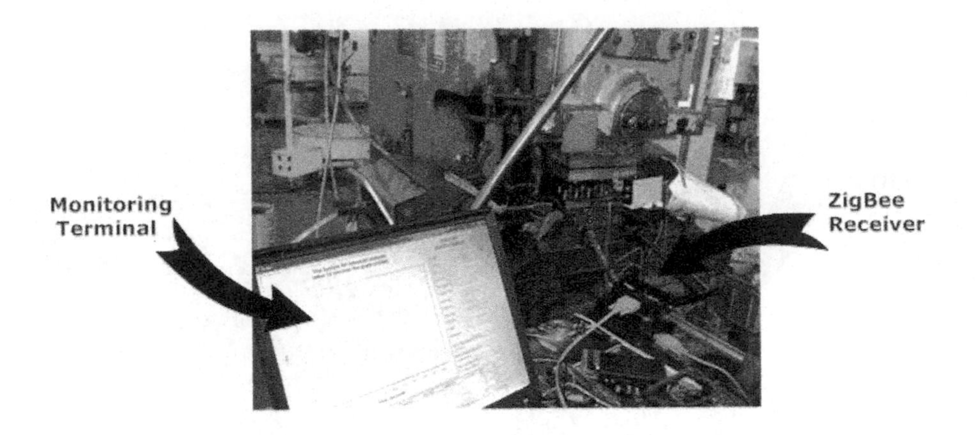

Figure 10.10: The server PC integrated with the ZigBee receiver to instantly collect and display the analysis report about the efficiency of the concerned stamping machine.

Figure 10.11: A series of stamping machines that may potentially be installed with our sensor nodes to concurrently monitor their activities.

in cost, just around several hundreds for each working prototype built as compared to the cost of commercially available products in the range of several thousand Hong Kong dollars. Besides, when producing in mass quantity, the manufacturing costs and resulting sizes of our wireless sensors will surely be further reduced, thus making our sensor nodes much affordable to those small-to-medium enterprises (SMEs) in the southern part of PRC. Besides reduced installation costs, our IR based wireless sensor is highly customizable, meaning that it can be easily adapted to various working environments for its sensitivity (as adjustable by a variable resistor) and range of sensed motion (typically in the range of several centimeters to tens of centimeters) after some tuning quickly done on the spot, as revealed by our previous experience in performing the empirical evaluation. There are several interesting directions for future investigation. First, the integration of other sensing devices such as temperature or pressure sensors with our existing monitoring system is worth investigating. Besides, a refinement of our server application to provide more detailed analysis report or various charts should be both interesting and useful. Lastly, the inclusion of other mechanisms such as the short message service (SMS) for the alerts whenever abnormal machine activity is detected should be thoroughly studied.

Acknowledgements

The authors wish to thank Professor Eryk Dukewit and Professor Yi Shang for their fruitful discussions. In addition, the authors are grateful to the generous supports by the Techcrystal Group and particularly their Marketing Manager, Mr. Clive Tsang, for allowing us to conduct on-site measurement for this research study. The authors would like to express their gratitude to the various supports provided by the Department of Electrical and Electronic Engineering, the University of Hong Kong.

References

[1] B. Alphenaar and D. Szucs, "Wireless Sensor Networks for Electric Transmission Line Monitoring", available at: http://www.netl.doe.gov/moderngrid/docs/Genscape_OE_DER-005.pdf, lastly visisted on: October 23, 2009.

[2] M. Batalin, M. Hattig and Gaurav S. Sukhatme, "Mobile Robot Navigation using a Sensor Network", In Proceedings of the IEEE International Conference on Robotics and Automation, pp. 636-642, 2003.

[3] M. Chan and V. Tam, "BlueGame A BlueTooth Enabled Multi-Player and Multi-Platform Game: An Experience Report", in Proceedings of the IEEE Consumer Communications and Networking Conference (CCNC06), Harrahs Las Vegas, Nevada, USA, January 8-10, 2006.

[4] F. Chen, N. Wang, R. German and F. Dressler, "Simulation Study of IEEE 802.11.4 LR-WPAN for Industrial Applications", in Wireless Communications and Mobile Computing, volume 10, Issue 5, pp. 609-621, John Wiley & Sons, January 2009.

[5] F. Chen, N. Wang, R. German and F. Dressler, "Real-time Enabled IEEE 802.15.4 Sensor Networks in Industrial Automation", in Proceedings of the 2009 IEEE International Symposium on Industrial Embedded Systems, pp. 136-139, IEEE, July 2009.

[6] K.Y. Cheng, K.S. Lui, Y.C. Wu and V. Tam, "A Distributed Multihop Time Synchronization Protocol for Wireless Sensor Networks using Pairwise Broadcast Synchronization", in the IEEE Transactions on Wireless Communications, Volume 8, Number 4, April 2009.

[7] Jose A. Gutierrez, "IEEE 802.15.4 Low-Rate Wireless Personal Area Networks: Enabling Wireless Sensor Networks", published by the IEEE, ISBN-10: 0738135577, November 2003.

[8] M. Kohvakka, M. Kuorilehto, M. Hännikäinen, and T.D. Hämäläinen, "Performance Analysis of IEEE 802.15.4 and ZigBee for Large-Scale Wireless Sensor Network Applications", in the 3rd ACM International Workshop on Performance Evaluation of Wireless Ad Hoc, Sensor, Ubiquitous Networks (ACM PE-WASUN 2006), pp. 48-57, Terromolinos, Spain, 2006.

[9] K. Koumpis, L. Hanna, M. Andersson and M. Johansson, "Wireless Industrial Control and Monitoring beyond Cable Replacement", in Proceedings of the PROFIBUS International Conference, pp. 1-7, Coombe Abbey, Warwickshire, U.K., June 2005.

[10] Z. Li, "ZigBee Wireless Sensor Network in Industrial Applications", in Proceedings of the International Joint Conference on SICE-ICASE, Busan, pp. 1067-1070, October 2006.

[11] H. Lim and J.C. Hou, "Distributed Localization for Anisotropic Sensor Networks", in the ACM Transactions on Sensor Networks, Volume 5, Number 2, Article 11, pp. 1-26, March 2009.

[12] F.G. Liu and Z. Miao, "The Applications of RFID Technology in Production Control in the Discrete Manufacturing Industry", in Proceedings of the IEEE AVSBS06, pp. 68, 2006.

[13] K.T. Ma, K.Y. Cheng, K. Lui and V. Tam, "Data Centric Storage with Diffuse Caching in Sensor Networks", In Special Issue on Distributed Systems of Sensors and Applications, Wireless Communications and Mobile Computing (WCMC) Journal, Volume 9, pp. 347- 356, Wiley InterScience, April 7, 2008.

[14] D. Niculescu and B. Nath, "Ad hoc positioning system (APS)", in Proceedings of the IEEE Globecom 2001, pp. 2926-2931, 2001.

[15] S. Shenker, S. Ratnasamy, B. Karp, R. Govindan, and D. Estrin. "Data-Centric Storage in Sensornets", ACM SIGCOMM Computer Communication Review, vol. 33, pp. 137-142, 2003.

[16] Y. Shang, W. Rumi, Y. Zhang and M. Fromherz, "Localization from mere connectivity", in Proceedings of the ACM MobiHoc 2003, pp. 201-212, 2003.

[17] C.J. Sreenan, U. Roedig, J. Brown, C.A. Boano, A. Dunkels, Z. He, T. V. Vassiliou, J. Sa Silva, L. Wolf, O. Wellnitz, R. Eiras, G. Hackenbroich, A. Klein, and D. Agrawal. "Performance Control in Wireless Sensor Networks", available at: http://www.carloalbertoboano.com/documents/ginseng09secon.pdf, last visited on October 23, 2009.

[18] The Apple iPhone Development Team, "Apple iPhone", available at: http://www.apple.com/iphone/, last visited on October 21, 2009.

[19] The BlueTooth Development Team, "BlueTooth website", available at: http://www.bluetooth.org/, last visited on January 21, 2009.

[20] The GPS Wing, "Global Positioning System", available at: http://www.gps.gov/, last visited on October 22, 2009.

[21] The IEEE Standing Committee on 802.15.4, "Wireless Personal Area Networks: Proposal for Factory Automation", the IEEE Proposed Standard for 802.15.4-15/09/057 lr0, August 2008.

[22] The Nintendo Wii Development Team, "Wii at Nintendo", available at: http://www.nintendo.com/wii, last visited on October 23, 2009.

[23] The Nokia N96 Development Team, "Nokia N96 Full phone specifications", available at: http://www.gsmarena.com/nokia_n96-2253.php, last visited on October 22, 2009.

[24] The OEE Team, "OEE Overall Equipment Effectiveness", available at: http://www.oee.com/, last visited on June 1, 2009.

[25] The Wikipedia Development Team, "Mobile Devices", available at: http://en.wikipedia.org/wiki/Mobile_device, last visited on October 22, 2009.

[26] The Wikipedia Development Team, "WiFi Wikipedia", available at: http://en.wikipedia.org/wiki/Wi-Fi, last visited on October 23, 2009.

[27] The Wikipedia Development Team, "ZigBee Wikipedia", available at: http://en.wikipedia.org/wiki/ZigBee, last visited on October 24, 2009.

[28] The Working Group Setting the Standards for Wireless LANs, "IEEE 802.11", available at: http://www.ieee802.org/11/, last visited on October 21, 2009.

[29] The ZigBee Alliance Team, "ZigBee Alliance", available at: http://www.zigbee.org/, last visited on October 22, 2009.

[30] H.W. Tsai, C.P. Chu and T.S. Chen, "Mobile object tracking in wireless sensor networks", Computer Communications, Volume 30, Issue 8, pp. 1811-1825, June 2007.

[31] Q. Wang, X, Liu, W. Chen, L. Sha and M. Caccamo, "Building Robust Wireless LAN for Industrial Control with the DSSS-CDMA Cell Phone Network Paradigm", IEEE Transactions on Mobile Computing, Vol. 6, Issue 6, pp. 706-719, June 2007.

[32] Y. Zhou, Y. Wang, J. Ma, J. Jia and F. Wang, "A Low-Latency GTS Strategy in IEEE 802.15.4 for Industrial Applications", in Proceedings of the 2008 2nd International Conference on Future Generation Communication and Networking, vol. 1, pp. 411-414, December 2008.

Chapter 11

ZigBee versus Other Protocols and Standards

Kumar Padmanabh, Sougata Sen, Sanjoy Paul

CONTENTS

11.1 Introduction

The business consumer drives the development of any new system. The development of any new system is based on a consumer's need. But sometimes the consumer's need is created artificially by business houses. However, industrial standards are created for different reasons. One of the corporate houses develops some product and hence has its monopoly in the market. The competitors of the said company feel left out in the race and hence create new competitive products. By this time end users or consumers are habituated with the older product. Replacement of older products by new products will become chaotic as the system may be entirely different. Therefore a standard system is created to make the system simple, interoperable and easy to use.

11.1.1 The Purpose of Standarization

The well defined standard brings simplicity and interoperability of the system; however, reasons of standardization are not limited to these facts. There are various other factors due to which standardization of the system takes place. Some of them are given below:

Business and user needs: As described above it is the user which drives the standardization process. Users want to replace the existing system with better functionalities, experience and for competitive cost, however they don't want to learn the operation of a new system. In some of the business needs, users want two or more devices of different vendors to communicate with each other. Unless or until all vendors follow the same standard this cannot be possible.

Competitiveness: Proprietary systems introduce a monopoly while standard systems bring competition among the vendors to reduce cost and add newer and improvised features in to the system.

Consistent Installation: In today's world system equipment comes from different vendors. Standards are needed for consistent installation of that equipment so that they can play their individual roles, while the system works seamlessly.

System Interoperability: Though two systems can be made interoperable by knowing the architecture and design, without a well defined standard it is almost impossible to make a third system interoperable with the first two. However, if any two systems are following the same standard, they can talk to each other.

Product Interchangeability: The end user may want to replace the older equipment with new equipment. However due to cost or performance point of view they may want to choose another vendor. This cannot be possible without all of the equipment from different vendors following the same standard.

Training and Maintenance: Equipment improves. Cost of training and maintenance will increase if all equipment do not follow the same standard. Training will be affected because for every equipment there will be a separate training.

In the next few sections we will study the various standards which is either competing with ZigBee or complementing it.

11.2 BACNet

We are living in an instrumented world. The place of residence, work and fun are completely instrumented to give comfort and security. The comfort is given in terms of air conditioning systems and security is given in the form of alerts for fire, intrusion, etc. The buildings of offices and residences use various automated facilities such as heating, ventilation and air-conditioning (HVAC), lighting, fire alarm, life safety, and security systems [1]. These building systems require automation and control systems for building facilities in real time. The demands of automated building facilities and services have increased which has resulted in the use of distributed, network-based control systems. This network based control provides real-time control and monitoring of building facilities. The data gathered by these systems can be used in analyzing, storing building related information and generating alarms and reports. Thus, network based system is the core of automation and control in modern day buildings.

As explained in the last section, the differences in the proprietary networking technologies of different vendors result in major challenges in integrating two proprietary technologies in monitoring and controlling a building. Moreover these proprietary technologies are not suitable for flexibilities and expendabilities of the systems. In order to solve these problems, the American Society of Heating, Refrigerating, and Air-Conditioning Engineers (ASHRAE) developed BACNet, the communication protocol standard specifically designed to meet the needs of building automation and control systems.

11.2.1 Motivation for BACnet Standards

The flexibilities and expendabilities of the building automation and control system motivated the technologies to design an interoperable device (see Figure 11.1). The building automation solutions which were proprietary started evolving into interoperable systems in the form of proprietary gateways and converters. In doing so, a particular manufacturer may have been able to read the code of other manufacturers and design a device that could communicate with this device. However this type of approach had specific restrictions. Firstly, the devices and gateways were very expensive and with emerging competition, the devices had a very short life span. Also, with multiple numbers of vendors in the market reading the code of a device of one vendor and subsequently developing new devices was not a scalable solution. This prevents new vendors from entering the business.

The first attempt to provide standard for communication protocol for building automation was in the form of the LonTalk protocol. This was based on an integrated

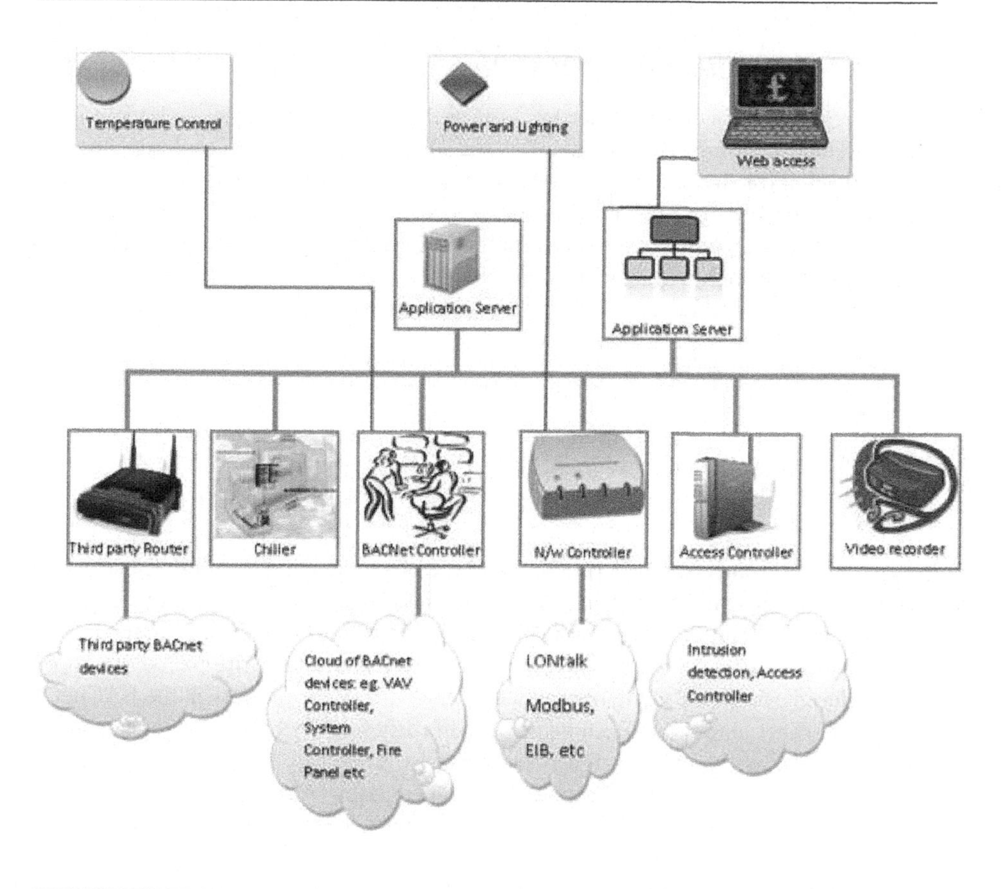

Figure 11.1: Devices in a typical BACNet system.

circuit (the Neuron Chip) with three, 8-bit processors onboard. These three processors were used individually to handle physical network connections, application, and hand-shaking between the other two.

Even though the building industry realized the need to create standard, the LON was not fully developed till 1992. Moreover it was proprietary until 1999. The use of neuron chips was made mandatory for all Lon works system. This didn't give openness and the entire standard was based on a proprietary single chip. Due to the proprietary nature of LON, the BACNet committee was formed in 1987 within ASHRAE and started developing a standard that the industry could adopt.

BACNet has ended the restrictions of proprietary systems which resulted in competitiveness of vendors to increase users' choices. The ultimate goal of BACNet was the complete interoperability among the ever increasing number of manufacturers' building automation control products. In reaching this goal, the BACNet Committee produced definition standards for BACNet data, control and communication functions.

11.2.2 Architecture of BACNet

The architecture of the BACnet has emerged from the OSI layer architecture with several modifications. BACnet emphasizes efficient, lightweight communication and optimization for small networks and short messages. Typically BACnet is for a wired network; however, the protocol is not limited for a wired network. Several implementations have been done for BACnet with end devices connected to the BACnet coordinator through wireless links.

The four layers from the OSI model within the BACnet architecture include:

- Application Layer
- Network Layer
- Data Link Layer
- Physical Layer

BACnet is a connection-less protocol and therefore the need for message segmentation and end-to-end error checking is much less than in a connection oriented network. Secondly the messages are very short in BACNet. In a conventional data network where messages are divided into several packets and entire packets cannot be reproduced even if only one packet is lost. In contrary the packets in the BACnet are of a very short length and each packet is self sufficient to convey the intended information. This has resulted in fewer error checks. Therefore a transport layer doesn't exist in the BACnet at all and some essential functions of transport layer have been delegated to the application layer. Due to self sufficient packet nature we also don't need to maintain the session layer. In addition, since BACnet uses a fixed encoding scheme and offloads security to the application layer, a separate presentation layer is not needed.

Table 11.1: Architecture of BACNet

<table>
<tr><td colspan="5">BACNet Application Layer</td><td></td><td>Application</td></tr>
<tr><td colspan="5">BACNet Network Layer</td><td></td><td>Network</td></tr>
<tr><td colspan="2">ISO8802-3, type-1</td><td>MS/TP</td><td>PTP</td><td></td><td></td><td>Data Link</td></tr>
<tr><td>ISO8802-3</td><td>ARCNET</td><td>EIA-485</td><td>EIA-232</td><td>LonTalk</td><td></td><td>Physical</td></tr>
</table>

11.2.3 Wireless Technologies in BACNet

Traditionally BACnet has been a wired communication technology. Converting BACnet into a wireless system will result in cost savings and other benefits for many facilities. ZigBee Alliance has been working with the BACNet project committee to add BACnet application profile in the ZigBee.

Table 11.2: Comparision of BACNet and ZigBee

	BACNet	ZigBee
Governance	ASHRAE	ZigBee Alliance
Standard Established	1995	2004
Markets	Commercial Buildings	Commercial and Residential Buildings
Object Model	Well-Defined	Work-in-Progress
Network	Work various wired media No wireless	Wireless 900 mHz and 2.4gHz
Power	Line Powered Devices	Battery powered, Line Powered
Availability	Devices Always On	Devices Always On or Sleeping
Data Rates	9.6 to 76.8 kbps (MS/TP) 9.6 to 56.0 kbps (PTP) 78.8 kbps (LonTalk) 156 kbps (Arcnet) 10 to 100+ mbps (BACnet IP)	256 kbps
Network Technology	Multiple	Mesh

Once a node is switched on and put into the ZigBee network it searches for the ZigBee coordinator. Only one coordinator is available in a particular personal area network at any moment of time. Coordinator and the new node need to recognize each other to establish a connection. Once the new node and a coordinator find and recognize each other, a connection is established. ZigBee is also capable of transmitting encrypted data for security [2] and [3].

The subset of Personal Area Network (PAN) is identified to form a BACnet network. BACnet can be formed by a broadcast, multicast, or unicast call. Each node in the PAN receives the broadcast and replies with whether or not it belongs to the appropriate BACnet network. Broadcast is simple to implement but it consumes a lot of bandwidth. A group of nodes is predefined for multicasting. Nodes not belonging to this predefined set can reject the call and they don't need to receive and interpret the data. Since the multicast process consumes less bandwidth it is preferable.

11.2.4 Physical and Data Link Layer in BACNet

There are various parameters which restrict the BACnet to choose the interfacing devices: namely speed, compatibility or chip for the protocols and cost. This has resulted in four types of LAN and point to point communication protocol. This has also resulted in optimized performance and the cost. The system designer can choose one of the four options to choose the physical interface of the BACnet device and the router. Large building automation systems frequently have multiple networks arranged in a hierarchical structure. Application specific controllers reside on a low-cost, low-speed LAN and are supervised by more sophisticated controllers that are interconnected by a high-speed LAN. The flexibility in the BACnet layer architecture has been provided to accommodate the future change in the technologies.

The ISO 8802-3, better known as "Ethernet" uses carrier sense multiple access with collision detection (CSMA/CD) and offers high speed (10 Mbps). It is widely popular in conventional data communication. This is the best communication channel to connect a BACnet device with the rest of the world. BACnet is compatible with several physical media which are defined for ISO 8802-3. Provision has been there to accommodate the future change in Ethernet protocol automatically.

The second option is ANSI 878.1 (ARCNET) which was developed by Datapoint Corporation. This is a token passing protocol and therefore deterministic in nature which gives the advantage of bounded transmission time. It is a lower-cost alternative but one needs to compromise with a respectable speed of 2.5 Mbps. The amendments and modification in the ARCNET standard are also supposed to be accommodated in BACnet. Thus, like Ethernet, any future enhancements automatically become part of BACnet.

The EIA-485 is one of the most commonly used physical layer technologies in building control systems, particularly for application-specific controller networks. This is the third networking possibility in BACnet. EIA-485 standard is a physical layer standard and doesn't provide any data link layer option within EIA485 standard. BACnet defines a Master-Slave/Token-Passing (MS/TP) protocol to provide this data link layer function. An MS/TP network has one or more master nodes that are peers on a logical token-passing ring. It may also have slave nodes that are unable to transmit messages until requested to do so by a master node. An MS/TP network can be made up entirely of master nodes forming a peer-to peer network. It can be a true master-slave network with a single master node and all other nodes as slaves. It can also be a combination of the two with multiple masters and slaves.

The fourth LAN protocol option in BACnet is LonTalk. LonTalk is a proprietary protocol based on a seven-layer architecture on a single chip called a neuron. One can choose any physical medium but compatible device is not defined in the LonTalk. The LonTalk protocol is low-cost and low-speed and has a very short message of 128 octed size. Since this is proprietary, therefore this is not interoperable. Interoperability can only be achieved by imposing external constraints.

The last Point-To-Point (PTP) protocol accesses the communication medium through full duplex EIA-232 interface. The typical application of EIA would be to connect a modem for dial-up access to a remote building automation system. This protocol doesn't specify the physical connection. But it defines the initialization, maintenance and termination of a physical connection. The system designer can choose any physical connection which is compatible.

11.2.5 Network Layer of BACNet

The performance and cost are the two parameters which define the network layer of BACnet. BACnet provides a wide range of options. The most important part of the

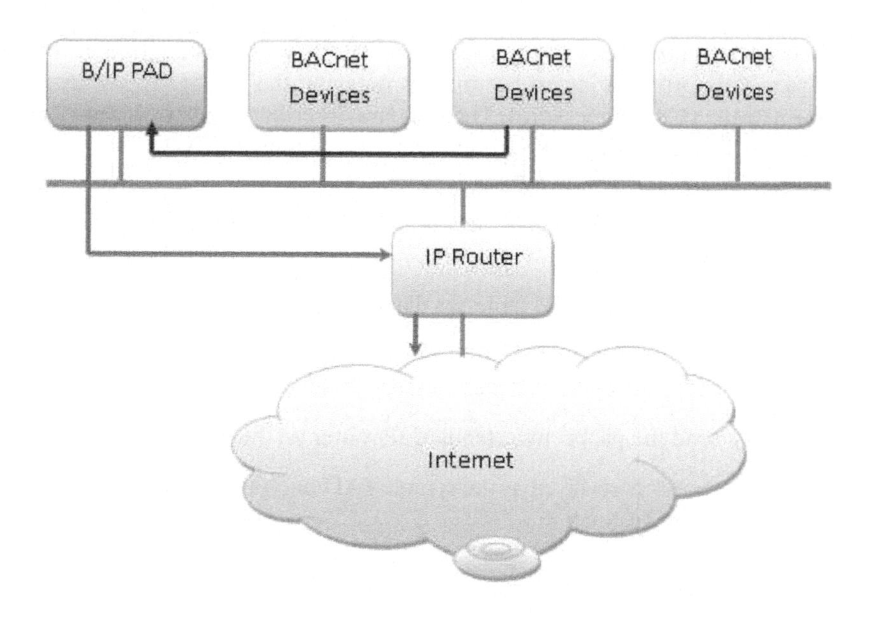

Figure 11.2: A typical BACNet system.

BACnet protocol is that it can use either one network layer in the entire network or it can use a combination of multiple network layer protocol. Ultimately the purpose of the network layer is to route packet from one BACnet device to other. There can be several routes from one device to other. However BACnet specifies only one active device. As described earlier, generally messages in BACnet are very short and therefore, most of the information is accommodated in a single packet and therefore message segmentation is not required at the network layer. The maximum length of the packet is not supposed to exceed the capability of any data link technology encountered along the path from source to destination. Messages longer than this can still be conveyed, but they must be segmented and reassembled at the application layer.

The BACnet message at the network layer contains a version number of one octet, one octet of control for indication of presence or absence of other network layer information. To address the device in the same network, no additional network layer information is needed. However, to address a device in the remote network, destination network number and MAC address of the destination device is used. The provision is given such that a device doesn't need to know its own network number. The router of the local network maintains the addressing information.

There are services defined in the BACnet which can configure the tables of the routers and the protocol also allow routers to search for the path to a destination network and manage temporary connections to remote networks. Routers can also indicate to client devices that a path to the destination device cannot be located if it fails to find a path to reach a destination.

BACnet also provides a way to route messages through existing Internet Protocol networks. This protocol is capable of re-processing the packets to encapsulate or decapsulate the BACnet messages. The standard describes the procedures in terms of devices called BACnet/IP Packet-Assembler-Disassemblers (PADs).

The procedure is the following:

- These devices take a BACnet message intended for a device on a remote network
- They look up in a local table for the address of a corresponding PAD on the distant network
- They encapsulate the BACnet message in an IP packet.
- They send the packet to a standard IP router on the local network.
- The process is reversed at the remote PAD and the message is forwarded to its ultimate destination.

11.2.6 Application Layer in BACNet

BACnet operates on two key fundamentals. Firstly the devices' representation of the soft objects, which is a data structure, and the properties of these objects, application layer is a set of objects with its properties, services and prioritization of the services. The internal design and configuration of a BACnet device depends upon the individual vendor and are very proprietary in nature. BACnet system translated the devices into these software objects and provides a way of identifying and accessing information in such a way that it doesn't require any knowledge of the details of the devices.

11.2.6.1 Objects in BACNet

The typical objects are listed in the table below. However one can create a new table. Each object can be understood by its name. However some of them require explanation. Calendar objects represents a list of time and dates which is used for scheduling of any operation of mechanical equipment. Whenever a sequence of starting up of several devices is needed, a command object is used. This is used whenever a multi-action command procedure is used. The general information about vendor name, model name, protocol version, etc., is mentioned in device objects. Upon occurrence of alarm, the concerned objects to be notified are handled through Event Enrollment object. Group objects are used to read several values in one request. Loops can be used to represent any feedback control loop, which is some combination of proportional, integral, or derivative control. Notification Class provides a way to manage the distribution of alarm or event notifications that are to be sent to multiple devices.

11.2.6.2 Application Services in BACNet

While objects are the soft representation of building automation device, the "application services" is the set of commands for accessing and manipulating the information

Table 11.3: Number of Objects in BACNet

Analog Input	Calendar	Loop
Analog Output	Command	Multi-state Input
Analog Value	Device	Multi-state Output
Binary Input	Event Enrollment	Notification Class
Binary Output	File	Program
Binary Value	Group	Schedule

along with some additional functions. There are 35 application services. Some of the services need acknowledgement while others don't require acknowledgement. There is a common theme of the services: either properties are read from or written into the objects. Thus ultimately, services are used by a device to get information to other devices or let other devices know the same.

Table 11.4: BACNet Application Services

Alarm and Event Services	Object Access Services	Remote Device Management Services
AcknowledgeAlarm	AddListElement	DeviceCommunicationControl
ConfirmedCOVNotification	RemoveListElement	ConfirmedPrivateTransfer
ConfirmedEventNotification	CreateObject	UnconfirmedPrivateTransfer
GetAlarmSummary	DeleteObject	ReinitializeDevice
GetEnrollmentSummary	ReadProperty	ConfirmedTextMessage
SubscribeCOV	ReadPropertyConditional	UnconfirmedTextMessage
UnconfirmedCOVNotification	ReadPropertyMultiple	TimeSynchronization
UnconfirmedEventNotification	WriteProperty	Who-Has
	WritePropertyMultiple	I Have
Virtual Terminal Services	**Security Services**	Who Is
VT-Open	Authenticate	I-Am
VT-Close	RequestKey	**File Access Services**
VT-Data		AtomicReadFile
		AtomicWriteFile

11.3 The HART System

HART Communication is a bi-directional industrial field communication protocol used to communicate between intelligent field instruments and host systems. HART is the global standard for smart process instrumentation and the majority of smart field devices installed in plants worldwide are HART-enabled. The global installed base of HART-enabled devices is the largest of all communication protocols at more than 20 million. HART technology is easy to use and very reliable [4] and [5].

The machines on the assembly lines and manufacturing floors talk to each other, using field bus, which is a digital industrial automation protocol. However every industry implements field bus protocol in its own way. The HART protocol provides standards which can be followed and implemented for a unified solution across industries; this avoids the diversities in implementation. HART has been in use for the last 20 years and in its 7th revision the Wireless HART that was proposed, takes care of wireless communication on top of the conventional HART backbone. Like ZigBee, Wireless HART also uses 2.4GHz ISM radio Band and IEEE 802.15.4 compatible DSSS radios with channel hopping on a packet to packet basis. Summarily the proposed chapter will consist of the following: HART System, Wireless HART System, Characteristics of HART specific IEEE 802.15.4 compatible Physical and MAC Layers, Different modes of messaging in Wireless HART, Channel Hopping in Wireless HART, QoS in HART Communication, Security issues in HART Communication, ZigBee versus Wireless HART, is Wireless HART competing with ZigBee or complementing it in industrial automation.

11.3.1 Motivation

When the devices and processes in the industrial automation system follow HART communication protocol, it gets tremendous benefits. The benefits are reflected in all phases of Plant Life Cycle and Maintenance. When properly utilized, the intelligent capabilities of HART-smart devices are a valuable resource for keeping plants operating at maximum efficiency. Real-time HART integration with industrial automation process, safety and asset management systems gives all values of connected devices. This extends the benefits of these devices interacting with each other. The key benefits are:

- 4-20mA compatibility with simultaneous digital information available
- Easy to use and understand
- Low risk - highly accurate and robust
- Cost-effective implementation for both users and suppliers
- Available in a wide variety of device types
- Supported by most industry device and systems suppliers
- Fully interoperable and reliable

Almost all World leading vendors of industrial automation solution support HART protocol in their devices. It is available in HART enabled hardware interfaces and application software for real time monitoring, diagnostics and prognostics. One of the core motivation factor is that the system reduces costs by improving the efficiency and minimizing the disruptions and unplanned shutdown.

11.3.2 Architecture of HART

Frequency Shift Keying is used for modulation of 4-20mA signal with the carrier signal. HART technology is a master/slave protocol which can be used in various modes such as point-to-point or multi-drop for full duplex communicating with smart field instruments and central control or monitoring systems.

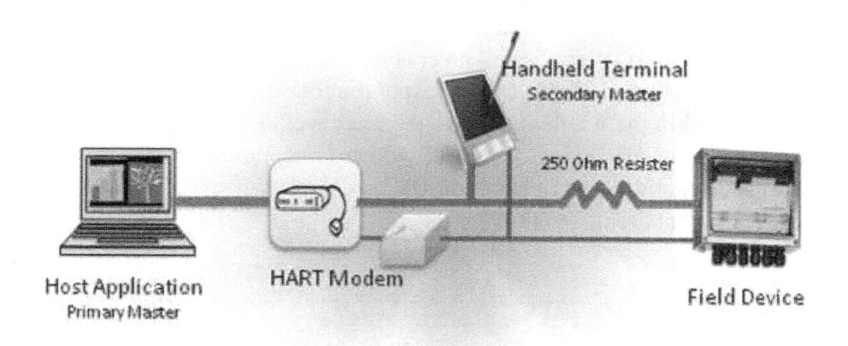

Figure 11.3: The HART system.

HART system can leverage the capabilities of intelligent device data for the improvement of operations. This can be used further to gain early warning to variances in a device, product and process performance speed and the troubleshooting time between the identification and resolution of problems. It can be used to continuously validate the performance of loop and control system strategies. Summarily it can be concluded that using HART system, the devices of the same control and asset management system will be available for more amount of time by proper communication and collaboration.

Further with the HART system in place, device and process connection problems or any other previously undetectable problems are detected in real time. This minimizes the impact of deviation by gaining new and early warning. This can be further used to avoid the high cost of unscheduled shutdown or process disruption which results in reduced maintenance cost, inventory and device management cost.

11.3.3 Wireless HART Protocol

Wireless HART has been designed to connect field devices, PLCs, maintenance tools and asset management systems of automation industry to a centralized control and monitoring unit. It has been ensured that these devices are compatible with conventional HART system and at the same time they get the benefits of the wireless system.

It has been ensured that there is no need to upgrade the software and communication stack of conventional LAN. Thus wireless HART has all other features of HART with the added advantage of the wireless network, such as better communication and collaboration among the devices leading to increased productivity.

Wireless HART protocol is similar to the ZigBee in the sense that modification of OSI layer has been done in the same way of ZigBee. The physical and MAC layers are the same IEEE 802.15.4. Network layer also uses the mesh topology like ZigBee and application layer is according to the requirement of conventional HART system. The transport layer, session layer and presentation layers are absent and their essential functionalities are built up in rest of the four layers. However ZigBee being universally used in the wireless domain there are well defined application layer (see Figure 11.4). However, wireless HART which is supposed to be used in specific domain of automation and control industry have their own specific requirements and they are built up at the different layers.

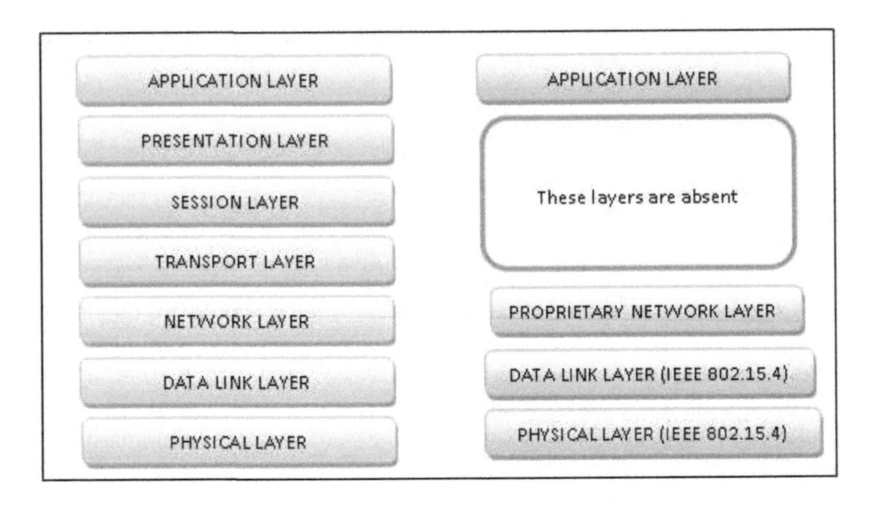

Figure 11.4: Layered architecture of HART.

11.3.4 Physical and Data Link Layer of Wireless HART

There are three main characteristics of Wireless HART. The first one is reliability. The HART system cannot afford to lose any packet. Therefore, the standards have been designed to put maximum reliability in the system. Secondly, the wireless communication is purposefully made secure by putting proper encryption and authentication process. This is a mesh networking technology operating in the global license free 2.4GHz ISM radio band. Wireless HART utilizes IEEE 802.15.4 compatible DSSS radios with channel hopping on a packet by packet basis. It has been ensured that this wireless system can coexist with other wireless and sustain the interferences by other channel. This is resulted in a provision in which field devices of two vendors can communicate seamlessly without compromising the system operation.

In the conventional HART system, token passing technique has been used at the data link layer. In the wireless HART, since it uses the mesh topology for better coordination among the field devices, the physical layer follows the IEEE 802.15.4 standard. TDMA technique is being used in the wireless HART to coordinate between network devices. At the data link layer the two devices establish links specifying the timeslot and frequency to be used for communication between the devices. The link may be dedicated or shared depending upon the requirement.

At the MAC layer, provision is there for long and short addresses. It follows the EUI standards for addressing based on HART unique ID. There is an additional wireless HART data link layer which is used to identify the packet types. The time slots of TDMA are organized into super frames of 100 timeslots per second. The devices of the wireless HART support multiple super frames for mixing the fast and slow network traffic. Super frames can be enabled or disabled based on bandwidth demand. All communication happens in designated timeslots specified by TDMA. The acknowledgement of the receipt of the packet also contains timing information to synchronize the devices with global time.

The links of the wireless HART may be dedicated or shared between multiple sources using contention based access. The decision of frequency hopping is taken on a message to message basis. There are four levels of priority of the packets. Command to the devices has highest priority and packets containing event payloads have minimum priority. Upon occurrence of congestion, the lowest priority messages are discarded and re-routing takes place to minimize further effect of congestion.

11.3.5 Network Layer of Wireless HART

It is specifically designed to suit the requirement of automation and control industry. It has been ensured that this supports a wide range of automation and control units starting from gas pipelines to pharmaceutical companies. It is directly compatible with conventional HART system, applications and tools. Essentially it follows the physical and MAC layer of IEEE 802.15.4. However this is highly secured by using AES 128 block ciphers with individual join and session keys and Data Link Level network Key.

The communication is highly reliable and the network is self healing. This standard supports multiple messaging modes including one way publishing of process and control parameters. It also supports real time and random request and response. If the dataset is large, it has the policy of auto segmentation. All messages have a well defined priority which ensures the QoS message delivery. It uses dedicated and shared bandwidth depending upon the requirement. The dedicated bandwidth is used to ensure the guaranteed delivery of priority message in bounded time and shared bandwidth is also there to provide QoS for bursty message.

Features of network layer are:

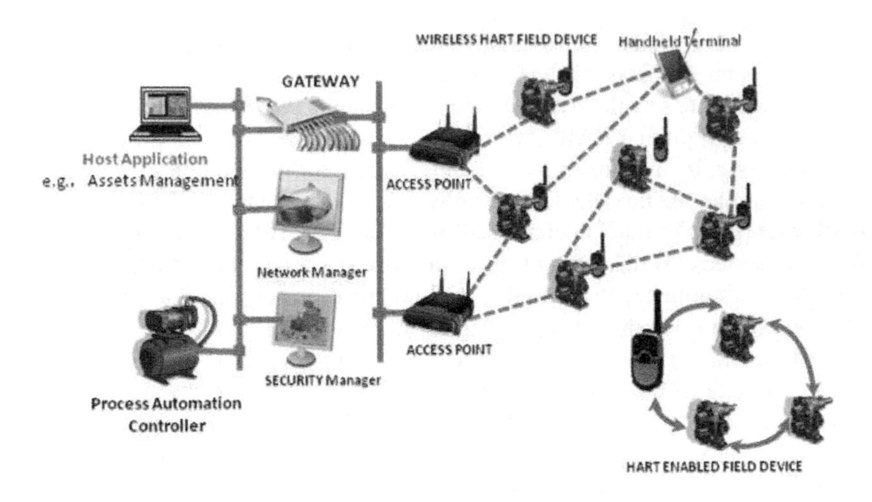

Figure 11.5: Wireless HART.

- It supports full wireless mesh networks. As opposed to the typical ZigBee set up, where there are multiple types of network devices (e.g., full functional, reduced functional, end devices) all network devices of wireless HART are full functional and each device can work as a router for other devices.
- As the network is formed, the multiple redundant communications paths are established and continuously verified. The reliability of network is greater than six sigma (99.9999998).
- There are redundant numbers of routes and upstream and downstream graph routing is used and supports broadcast, multi-cast and unicast transmissions.
- At the time of configuration of field devices, bandwidth is requested or released to satisfy field device communication requirements. However, if the field devices are configured prior to joining the network, communication bandwidth is allocated during the process of joining the network.
- In the shared slots one base bandwidth and one elastic bandwidth are allocated. While base bandwidth remains the same throughout the operation, the elastic part gets changed depending upon the network traffic.
- To transmit high volume of data, high bandwidth is allocated and it is released when a block transfer is complete.
- It also provides un-acknowledged and end to end acknowledgement

Clear Channel Assessment (CCA), channel hopping, blacklisting, and adjustable transmit power are supported to maximize coexistence between Wireless HART networks and other ISM band equipment.

11.3.6 QoS in Wireless HART

The sampling interval of the Wireless HART is such that it doesn't make any compromise with performance of closed loop control. To achieve real time control experience, the feedback control is generally 4-10 times faster than the process response time. Since the sensor and other measurement system cannot be synchronized with the control system, the sensing and measurement frequency is 2-10 times faster than what the process can respond to. The frequent transmission of measured value could result in reduced lifetime of the battery of the field devices; however, the capability of wireless transmitter to schedule the data transmission at a specific moment of time avoids this possibility.

Delays and jitter can affect negatively in timely access to measurement and control information. However wireless HART is a protocol that is time synchronized. Time accuracy is 1 millisecond for all devices. This capability is not available in many other protocols.

The propagation delay and interference also introduce another type of latency and jitter when data is communicated over the wireless channel. However the transmission rate in wireless HART is faster than the traditional wired network. For example in wired network transmission, the rate is 31.25 kbps and delay is 32 microsecond per bit. However in HART the transmission rate wireless is 250 kbps and therefore propagation delay is reduced to 4 microseconds. Transmission and acknowledgement process takes 10 milliseconds and therefore delay and jitter introduced due to wireless channel is compensated.

Wireless channels are prone to failure of message delivery. However in most of the cases, wireless HART networks will try to retransmit the failed message in the next time slot. Failure of delivery increases with a greater number of hops between devices and access points, which in turns increases the propagation delay. In a typical set up in a real plant, 30% of devices are in the range of access points and communicates directly and typical latency is 10 milliseconds. 50% of the devices are one hop away with 20 milliseconds of latency. The rest 20% of devices are 3-4 hops away and average latency is 30 milliseconds.

Thus each communication from the end device to the gateway would take less than 100 milliseconds. On an average the control loop will execute in less than 200 milliseconds thus total operation time will be below the required value of 500 milliseconds.

11.4 RFID

RFID is the technology by which objects can be identified and tracked uniquely [6]. RFID stands for radio frequency identification. A small tag is attached with the object or person or anything which needs to be tracked. This tag can be read with the help of a RFID reader which acts as a sink for all the tag readings. RFID is not a new technology. RFID was first used by the allied forces during World War II to identify friendly planes and enemy planes. This technology has gone through a huge change since then to reach the form it is today. There is a misconception that the barcode and RFID are the same. They are not. The black and white bars which are seen on products are bar codes, whereas a RFID tag is a small device which is attached to the product. RFID tags can be in any form factor. It can be attached to the body of the object or it can even be present inside the object. RFID is a comparatively new technology as compared to a bar code. But there are many advantages that RFID has over the traditional bar codes. Details about this and also the drawbacks will be discussed in the subsequent sections.

Most RFID tags operate in the 13MHz or 900MHz frequency range. 13MHz range is useful when the tag will be surrounded by many objects. It can penetrate through, but the distance is compromised. On the other hand 900 MHz can cover a greater distance, but has low power of penetration. Any kind of data can be written into and stored in the RFID tag keeping in mind the size in which it is available. To write data into RFID tags, which are usually empty when manufactured, we need to use an encoder.

11.4.1 The RFID Architecture, Advantages of Using RFID

There are two types of RFID tags: active tags and passive tags. Tags which can send data without any external intervention are known as active tags whereas tags which need external simulation are known as passive tags. A typical RFID tag has two parts, namely the integrated circuit and the antenna. The integrated circuit is used for processing and storing information as well as for some other specialized functions. Antenna, on the other hand, is responsible for transmitting and receiving data. If the RFID tag is an active tag, it will also contain a battery. The power for the active tags is generated from the battery while the scanning antenna (reader) sends the power to the passive tags so that it has enough power to generate the message and transmit it. When the tag is moved away from the scanning antenna, it again stops sending any data. A scanning antenna can be of any form factor based on the usage. They are usually affixed to walls of the area they are expected to monitor. But all scanning antennas need not be affixed to the wall. The scanning antenna is responsible for waking up passive RFID tags.

The data that is read from the tags by the reader needs to be processed for a sensible representation to the end user. The reader by itself cannot process the data and make it available to the end user, and hence a middleware is needed to do the processing and send the processed data to the end application. So the reader is connected to a processing unit, such as a PC that hosts the middleware needed to process the data and forward it to the end application which in turn can make the data available to the end user in a variety of ways. Middleware does a lot of filtering in order to avoid data repetition.

There are several advantages of using RFID, but interestingly enough, some of these advantages can also become the disadvantages for various reasons. The disadvantages are explained in Section 11.4.2. RFID is easy to install and use. A simple tag has to be attached to the device that needs to be tracked. It can easily be implanted in animals as well as in humans. In the case of animals, it not only helps in tracking the animal, but also helps in managing them as the data about vaccination; age, etc., can be maintained in the tags.

RFID uses the radio frequency. Thus tags do not have to be in the line of sight to be detected by the reader. So the range of reading can be quite large. This was a requirement in the case of bar codes where the object had to be in the line of sight in order to be read. RFID did not adopt that route. Since the tags can be read without being in the line of sight and the exact location of the object can be determined using various standard algorithms.

Further compared to the barcode, RFID can uniquely identify a product. A bar code can identify the category of products, not the unique identification of the product. This is because RFID tags have a much higher storage capacity as compared to bar codes.

RFID at the installation stage might appear to be a costly technology, but there is a sure return on investment. Multiple read/write operations can be performed on a tag.

Even though RFID is prone to some kind of damage, it is better in harsh conditions as compared to bar codes. The data that can be stored and retrieved from a RFID tag is a lot more than what can be stored in a traditional bar code.

11.4.2 Challenges in RFID

As discussed in the previous section, there are lots of advantages to using RFID technology, but like everything else, it also has some drawbacks. For example, the first and foremost challenge in RFID is the cost. While RFID tags are not very expensive, the reader is. So the price of the whole setup goes up drastically when a reader is present. Also RFID installation and software maintenance needs a person to be present which adds to the cost.

The tags are prone to damage in harsh environmental conditions such as high temperature or humidity. Even a high power magnet can damage the tags. Since they are expensive, damaged tags add up to the overall cost. There are also security concerns about RFID even though there are advantages over barcodes. Unlike barcodes which use line-of-sight, RFID works on radio frequency. So it is relatively easy for an intruder to tap the details of the tag without anyone even realizing it and later use the data for personal benefits. An example of this is using RFID in the passports. A mugger can get all the data of the passport holder and use it in wrong ways.

It is possible that there will be dead zones present in a RFID installation. Tags in the dead zones cannot be read. To avoid this problem, multiple readers have to be installed. This increases the cost of the installation. Also if two or more tags are read at exactly the same time, their data can get garbled and the end application receives the data as if the data has come from another tag which is not the same as any of the tags which were read. This is a very rare phenomenon, but it can happen. Another problem in reading multiple tags at the same time is that many of the tags may remain unread. The range of RFID tag falls drastically when it is placed in liquids and metals. Even any object between the reader and the tag reduces the range a lot. This disadvantage is used in the passport problem where the e-passports are lined with a thin strip of metal so that they cannot be read by anyone else other than the officer who is verifying the passport.

11.4.3 RFID Applications

There is a plethora of applications which utilize RFID. Some of them are listed here.

Tracking and monitoring As evident, RFID is used for tracking and monitoring. There can be various other uses such as asset tracking, human tracking, habitat monitoring, etc. A very common example is in retail stores where an RFID tag is attached to each object. There are RFID readers placed on the billing counters and at the exits. So when the tag is brought near the RFID reader, all the details for the billing are filled up automatically. At the billing counter the tag is removed from the object. If the tag is left attached to the object by mistake or a person tries to shoplift, then the readers at the exit send a signal and an alarm is raised. RFID is also used for habitat monitoring. If the behavior of a certain species has to be studied, then RFID tags with a large range are attached to many of the creatures in that species. With the help of a reader, their habitat can be easily monitored.

Some retail stores that have RFID tags attached to their merchandise have implemented a system, where if a shelf displaying a certain type of product is getting exhausted, then the manager of the store is intimated to replenish the objects for the particular shelf. There are many more tracking and monitoring applications for RFID.

Identification and authentication RFID is used in various places for authentication. Many buildings have secure zones which have restricted access. A person who is authorized to enter the restricted access zone is handed over a RFID tag. When the person wants to enter the zone, the RFID reader authenticates the person and if the person is authorized, then the person is allowed to pass. Very small passive RFID tags may be implanted (microchip implant) on pets to make sure that they do not get lost. In case a pet gets lost, the pet can always be traced back. This method also resolves disputes regarding the ownership of pets. Many countries make it compulsory for animals which are imported to be implanted with a microchip so that they can verify whether the animal is vaccinated or not.

Transport payments Many roads have implemented the noncontact payment system in case of toll payment. Previously toll roads had check points where a person would sit and the user of the road had to stop and pay the toll collector to move ahead. Using RFID this whole process can be made easy. RFID tag's can be purchased and installed on the dashboard of the vehicle. When the toll booth detects the car, it deducts the money from the tags present value and lets the user pass through. The same system has also been implemented in many pay-and-parks. Many public transport systems have started using the smart card. The user needs to purchase the smart card and needs to swipe the card at the entrance of the public transport vehicle. The door opens and the amount is again deducted from the card.

Passports Many countries have put RFID in passports. This is also known as e-passport. In the e-passport, along with the various personal details of the passport holder, the passport holder's picture and finger print may be present. There are a lot of controversies about using RFID in passports. Some of the people who oppose this say that details in the passport can be easily stolen by someone in the vicinity.

Location based services Many leading retail giants have implemented the LBS system in their stores. When a shopper is in front of a shelf, the shopper's location is detected and based on the location, relevant advertisements are displayed. The display could be anything right from a screen to a tablet PC which the shopper is carrying. This tactic is usually used to acquaint the shopper with the discounts and the free offers. Such offers are always lucrative for the shopper.

Sports In races such as a marathon, it is a hectic job to have people to monitor the total time of each athlete. Chances of inaccuracy are very high. To tackle this problem, each athlete is given a RFID tag. There is a RFID reader at the start and end point. So when the athlete crosses it, the time is noted down thus giving an accurate result. The same is applied for short distance races where on top of the start and end time, even the lap time has to be monitored.

11.4.4 Standards for RFID Application

Unlike many universal organizations there is no single global authority to govern the regulations of RFID. According to requirements, every country decides its own operational methodology, hence rules and regulations for RFID change from country to country. For example, federal communication commission in the United States regulates the RFID. IEEE 802.15.4f is a standard which provides physical layer and MAC layer for wireless personal area network. RFID is a form of WPAN and intended to follow recommendation of IEEE 802.15.4f. This standard ensures the implementation is cost effective.

IEEE802.15.4 Standards: IEEE 802.15.4, mostly used in RFID, was conceptualized in early 2003. Since then there have been multiple amendments in this standard. The latest amendment has been done in March 2010. Task Group of 802.15.4f (TG4f) recommended the ultra wideband physical layer, ultra high frequency physical layer and 2.4 GHz physical layer. Ultra Wide Band Air Interference recommends three frequency bands for global use, uses 1MHz PRF base and OOK modulation scheme. It has three symbol mapping modes, namely, the base mode with one pulse per symbol, enhanced mode with 3 pulses per symbol and long range mode with 8 to 32 pulses per symbol.

The MAC layer defines the technique to transmit MAC frames through the designated physical layer. The most important part of the MAC layer is the management interface which manages lower physical layer and upper network layer. It also manages frame validation, guarantees time slots and handles node associations. It also provides interfaces for security services. RFID Readers work on listen before talk. This restricts the system and performance is degraded. The North American UHF standard is not accepted in France as it interferes with its military bands. There is no regulation of UHF in China and Japan. In Australia and New Zealand there is no need of license to operate between 918MHz and 926MHz; however, there is upper limit on the transmission power.

Irrespective of the above open environment, the return signal from the RFID tags interfere with other wireless equipment. There has been some standard defined for RFID for specific cases.

ISO 14223/1 : - Radio frequency identification of Animals.

ISO 14443: This standard is used in RFID enabled passport at popular High Frequencies (13.56 MHz) under ICAO 9303.

ISO 15693: This is being used in credit cards at popular HF (13.56 MHz) standard for HighFIDs widely used for non-contact payment.

ISO/IEC 18000: This is being used in item management and mostly used by information technology industries.

ISO 18185: This is being used in e-seal, i.e., electronic seal.

EPCglobal: Under the umbrella of Electronic Product Code EPC global, is the standard completely driven by industry of RFID. It includes the industry leaders of this domain. The goal of the EPC global is to increase efficiency in information flow in supply chain management.

Class 1 Generation 2 UHF Air Interface Protocol Standard: This is known as Gen-2 standard. This defines the physical and logical arrangement of RFID tags and readers operating at frequency 860-960MHz. The interface protocol extends the item level tagging capabilities of UHF Gen-2. There are three additional features: there is an indicator to show the availabilities of formatted data in user memory, there is a block level locking of memory to protect the contents that is already available and whether a RFID tag is decommissioned is indicated by extended protocol control bits.

The readers transmit information to a tag by modulating RF signal. The RFID tag received both information as well as operating energy from RF signal. The RFID tags respond by modulating the reflection coefficient of its antenna, thereby backscattering the information signal to the interrogator. Tags modulate their antenna reflection coefficient with an information signal, only after being directed to do so by a reader. The communication is half duplex and thus a reader and the tag do not talk to each other simultaneously.

11.4.5 Data Management in RFID

The number of RFID tags deployed in supply chain management in almost all products creates huge volume of data. If proper process is not defined, these data are agnostic to losses. Storing data is one aspect of the data management and making them available for IT applications is another non trivial aspect. Data can be compressed to minimize its size. However well defined business processes should be there for proper management of data. One can leverage existing architecture and frameworks for data integration or can define new architecture. There is no standard for this purpose. However while using any process, the consideration of business dynamics related to data ownership and privacy must be accounted for, for the benefit of all stakeholders. It is also very important to ensure data is available in right format. Therefore whether to store data in user's form or in raw form depends on application. Summarily, there are massive data generated by RFID tags and there is no standard system defined for data management, thus it is very challenging for developers to develop applications and architecture of individual applications becomes most important.

11.4.6 Comparison between RFID and ZigBee

RFID is not a competitor for ZigBee. In fact there are cases where RFID does the detecting and identification and the ZigBee protocol is used for RFID to communicate with non-RFID devices. A list of a few major differences in the two technologies follows. The operating frequency for ZigBee is either 900 MHz or 2.4 GHz whereas it's either 13 MHZ or 900 MHz in case of RFID. This is the reason for ZigBee having a much larger range as compared to the RFID technology.

The storage capacity of the ZigBee enabled devices is comparatively much larger as compared to the RFID devices. Security of RFID devices is less stringent compared to ZigBee as there are many security protocols for ZigBee. RFID devices do not have the ability to perform two way communications. Thus it is very difficult to authenticate a device. Since ZigBee devices can do a two way communication, ZigBee is comparatively safer as compared to RFID. Cost of a ZigBee set up is less as compared to the cost of an RFID setup. Also ZigBee is always active, whereas there are passive tags present in RFID.

Table 11.5: Comparison between ZigBee and RFID

	ZigBee	**RFID**
Operating frequency	900MHz/2.4GHz	13MHz/900MHz
Range	Larger than RFID	Less than ZigBee
Storage	More	Less
Security	High	Low
Nature	Always Active	Both Active and Passive possible
Cost	Less	More

11.5 6LoWPAN

With so many real life WSN solutions being deployed, it is necessary to have a standard protocol for sending their data from the source to the destination. IP or Internet protocol is already present to do this function, but the devices in the WSN deployment have many limitations such as low power radio, low memory, low data rate, limited power, etc. So IP cannot be used directly on these systems. IETF (Internet Engineering Task Force) 6LoWPAN was formed with the purpose of successfully using IPv6 over IEEE 802.15.4 devices [7]- [10]. The term 6LoWPAN was coined by combining IPv6 and Low power Wireless Premise Area Network.

By using 6LoWPAN, WSN devices or any other RFD (reduced functionality devices) will have IP communication capability. But in a normal IPv6, the MTU is 1280 bytes which cannot be directly used in the devices with limited capability. To address this problem, 6LoWPAN introduces an adaptation layer above the data link layer. The adaptation layer is responsible for header compression, fragmentation and reassembling packets, routing etc. Addresses in IPv6 use 128 bits as compared to 32 bits used in IPv4. With such a huge address space, it is possible to have 667 x 1021 addresses per square meter on the earth's surface and thus it can be used easily for a dense deployment of the miniature sensors.

11.5.1 Motivation behind the 6LoWPAN Standards

It has always been required to have a standard protocol for data exchange between two nodes in a wireless sensor network. There have been a lot of suggestions about the standard. But IETF 6LoWPAN came up with the argument: why reinvent the wheel when IP is already there? IP has been a standard protocol for so many years. There are lots of people who specialize in programming, administrating and even debugging IP based networks. IP will remove the requirement for anyone working on this system to learn a new protocol, which can be a tedious job at times. There are lots of tools already present for IP. So recreation of these tools is not required. They can be used directly. All these factors will speed up development and deployment process. Using IP will remove the use of a translator at the gateway to connect the WSN nodes to the Internet.

11.5.2 Architecture of 6LoWPAN System

Figure 11.6 depicts the architecture for a 6LoWPAN. There can be three kinds of networks, namely the normal 6LoWPAN network, the extended 6LoWPAN network and the ad-hoc 6LoWPAN network. When a network has no Border Router (6LBR), it becomes an ad-hoc network. When there is one 6LBR for the network, it is a normal network and when there are more than one 6LBRs, then it becomes an extended network.

The Backbone router is a normal IPv6 router which is the intermediate node between the 6LBR and the Internet. It might be required for the backbone router to translate the IPv4 packets into IPv6 and vice versa.

The Border router (6LBR) is used to communicate between the 6LoWPAN network and the backbone router. The 6LBR has an interface for both IPv6 and 6LoWPAN. The 6LBR is responsible for Router Advertisement (RA), Duplicate Address Detection (DAD), Address Resolution, maintaining a whiteboard of the IPv6 addresses, interacting by forwarding packets to other 6LoWPANs, stripping off the adaptation layer in case of interaction with backbone router and adding the adaptation layer in case of sending a packet to the 6LoWPAN. Details of some of these topics are discussed in the section on neighbor discovery. The 6LoWPAN Router

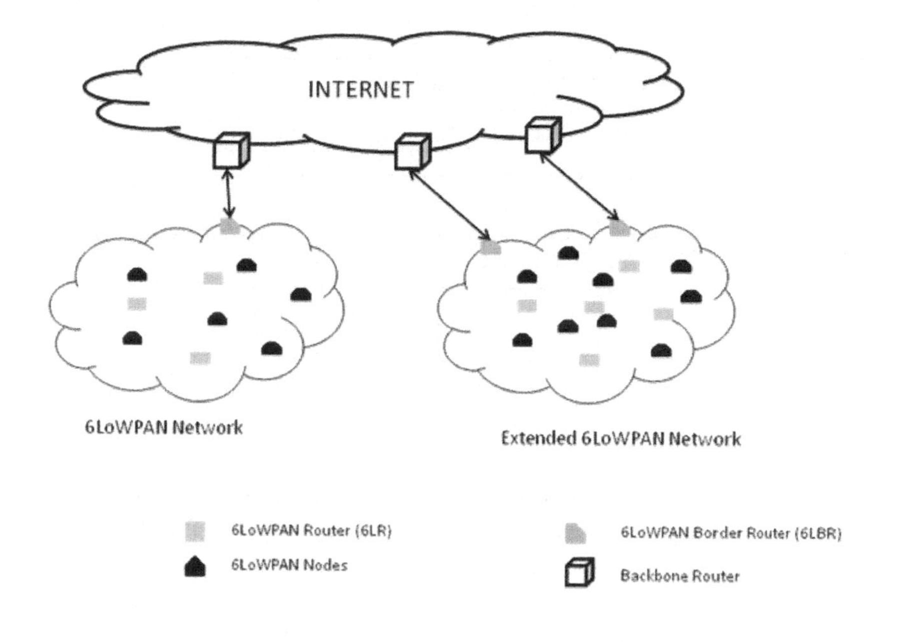

Figure 11.6: 6LoWPAN architecture.

(6LR) is responsible for interaction between the 6LBR and the nodes. They perform the Router Advertisement (RA) and respond to the Router Solicitation (RS). The routing protocol which will be running on the 6LRs will be defined by the IETF ROLL Working Group.

Headers Compression Technique: The number of bytes required in an IPv6 header is 40. This is huge when used in an 802.15.4 RFD. It is not possible to use the header as it is. So the header has been compressed in 6LoWPAN. The compression in 6LoWPAN is to the extent that data can be transmitted by using a header as small as 4 bytes. A typical 6LoWPAN header will have a datagram in containing mesh (L2) addressing, hop-by-hop options, fragmentation and payload.

Table 11.6: A Typical 802.15.4 Frame with the 6LoWPAN Header

Preamble	Start of frame	length	Destination source	Network header	Payload chk

The network header is where the IPv6 datagram goes into. Some possible cases that are possible are explained in the table below.

The HC1 header allows the compression of IPv6 header. There are many fields which are not state dependent and can be reconstructed in the destination or intermediate nodes without them being present in the header. The reconstruction is

Table 11.7: Uncompressed IPV6 Network Header

IPv6 dispatch	IPv6 header	payload

Table 11.8: Dispatch Header Format

00	xxxxxx	Not a 6LoWPAN frame
01	000001	Uncompressed IPv6 address
01	000010	IPv6 HC1 compressed encoding
01	111111	Additional dispatch bytes follow

done by using the link level information in the 802.15.4 frame. A typical IPv6 header is of 40 bytes, but it can be brought down to 2 bytes including the HC1 header. The advantage of using a stacked header is that a new protocol can be introduced any time.

The header has a type identifier which is the first two bits of the dispatch header. The dispatch header is used to define the type of header. IPv6 has kept a provision to handle non 6LoWPAN packets. 00 in the first two bits of the dispatch header means that the frame is not of 6LoWPAN, while 01 means that it is a 6LoWPAN frame. The remaining 6 bits indicate whether it is an uncompressed IPv6 address or HC1 header.

Mesh networking (multi hop) is also supported by 6LoWPAN. In case of mesh networking, the header will be as follows:

Table 11.9: Uncompressed IPV6 Network Header

IEEE 802.15.4 header	Mesh header	Dispatch	Payload

The bits in the mesh header will be: V indicates whether the originator address is a IEEE extended address or a short address. F indicates whether the final address is a IEEE extended address or a short address.

There are cases when the data cannot be transferred in one packet. In such a case the data is sent in multiple packets. This is known as fragmentation and a fragmented header will look like this

Table 11.10: Number of Bits in Mesh Header

Bits	2	1	1	4	..	
	10	V	F	Hops left	Origin address	Destination address

This format will be followed in the first fragmented header. In case of the subsequent headers, the format will be

Table 11.11: Fragmented Header

IEEE 802.15.4 header	Fragmentation Header	Dispatch	Payload

It is not compulsory that all the headers should be present in the same packet. Using 6LoWPAN, the header may or may not contain the mesh header or the fragmented header. This makes the header size variable.

11.5.3 The Physical and MAC Layer in 6LoWPAN

The physical layer as well as the MAC layer is based on IEEE 802.15.4 standards. The physical layer has a data rate of 250Kbps in the 2.4GHz band. The maximum payload is 127 bytes. The function of the MAC layer is to provide a reliable link for communication. The transmission must utilize the channels specified by IEEE 802.15.4 so that a broadcast can be heard by all devices in the network. There is no hard and fast rule that one particular link layer protocol has to be followed in order to be compatible for 6LoWPAN.

11.5.4 Adaptation Layer of the 6LoWPAN

The adaptation layer is the most significant change by IETF 6LoWPAN in the IP. The adaptation layer has many functions. The main purpose of having the adaptation layer is to let IP packets and 6LoWPAN packets understand each other. This layer makes the datagram in such a way so that 802.15.4 networks handle IPv6 packets. There are many other functionalities of the adaptation layer which will be discussed here.

The first major function of the adaptation layer is to compress the TCP/IP header so that it can be used in RFDs. Using a 40 bytes header (which IPv6 header uses) in 802.15.4 will leave hardly any space to transmit the payload. Also other than the IP header there are many other headers that have to be incorporated in the packet so that the final packet is meaningful. So to tackle this problem, the adaptation layer compresses the header as much as possible. The header size varies based on the type of usage the node wants to use it for. So effectively if the node wants to perform some complex functionality, such as mesh networking, then the amount of space left for payload reduces whereas a simple single hop, unregimented packet will have more space for payload.

The next functionality of the adaptation layer is to handle fragmentation. The packets that come in from the IPv6 network have a MTU of 1280, which cannot

Table 11.12: Bits in Fragmented Header

Bits	2	3	11	16
	11	000	Datagram size	Datagram tag

Table 11.13: Bits in Subsequent Fragmented Header

Bits	2	3	11	16	8
	11	000	Datagram size	Datagram tag	Offset

be handled by a normal 6LoWPAN network. So the packets need to be fragmented in such a way so that it is understood by the 6LoWPAN network. So the packets are fragmented into multiple link-level frames so that it can handle IPv6's minimum MTU requirement.

The next functionality of the adaptation layer is routing. It should be possible to interchange data packets between normal Internet, which uses IPv6 and the wireless sensor network, which uses the 6LoWPAN as well as routing within the wireless sensor network group. There are many protocols to help achieve routing. Some of the common ones include LOAD which is based on AODV, DYMO-LOW and Hi-LOW.

The next functionality of the adaptation layer is multicast support. To achieve this, the adaptation layer carries the link-level addresses.

11.5.5 Network Layer of 6LoWPAN

Implementation of network layer has been very challenging due to the characteristics of 6LoWPAN, availability of limited resource in its devices and Internet Protocol. There are different application scenarios and it is very difficult to propose a single architecture that is uniquely optimized for all application scenarios. Thus for different application different routing protocol has been proposed by the working committee.

IETF ROLL Working Group: A working group was formed to address the issues of Routing Over Low power and Lossy (ROLL) networks. Low power and lossy networks require energy efficient systems, their traffic is usually not unicast and they have small frame size. These characteristics force LLNs to have a very specific routing. The ROLL working group evaluated the existing routing protocols and found that the routing protocols do not satisfy the routing requirements. The application area of LLNs is very wide. The working group is focused on finding a routing solution for each of these areas. The routing solution is based only on IPv6. A lot of characteristics such as reliability, computation power, etc., is considered before coming up with the routing protocol. Special attention is given to security and manageability.

6LoWPAN Network Architecture: The nodes in the 6LoWPAN network do not perform any kind of routing. They send the packets to the 6LR which performs the routing. When initialized, the nodes perform the RS and they periodically renew their registration with one or more edge router. A globally unique IPv6 address has to be assigned to the node if it is required to interact with the external world.

6LoWPAN neighbor discovery: When the node is initialized, it sends a Router Solicitation (RS) message. This is a multi cast message. A 6LR responds back to this message with a RA message. The RA message contains the Authoritative Border Routing Option (ABRO) and may or may not contain the 6LoWPAN Context Options (6CO) and Prefix Information Options (PIO). The host then registers its address with one or more routers using the Address Registration Option (ARO). In the ARO, the host chooses its lifetime of the registration. If the router detects that the address is already in use, it sends back the ARO with the status field set to a non-zero value. The host performs the ARO till it receives an acknowledgement for the address registration.

Routing: The nodes used in sensor networks have a lot of limitations. To perform routing even with these limitations, numerous routing protocols have been developed. Based on which layer, the decisions about routing are taken; the protocols can be divided into two categories, namely mesh-under and route-over. In case of mesh-under, the routing decisions are taken care of in the adaptation layer, whereas it's handled in the network layer in case of route-over. Following are the techniques of routing:

(i) Mesh-under Scheme: In this scheme, the adaptation layer performs the routing. Routing and forwarding is done at the link layer based on the 6LoWPAN header. The link layer sends the packet to a neighbor node and the neighbor node passes it on to its neighbor till the packet reaches the destination. A set of multiple hops at the link layer is equivalent to one IP hop. The adaptation layer fragments a large packet into a number of fragments. These fragments may be delivered to the destination by mesh routing. It is not necessary that all packets will follow the same route. The destination gathers all the fragments and the adaptation layer of the destination assembles them back to create the original packet. In case any fragment doesn't reach the destination, all the fragments are retransmitted. The mesh-under scheme doesn't have any 6LRs. Just nodes and 6LBRs exist.

(ii) Route-over: In the route-over scheme, the network layer performs the routing. Each node acts as a router. Each hop to the neighbor is a complete IP hop in itself. For forwarding the packet, the routing table of the node is used. The adaptation layer fragments the IP packet and sends it to the next node based on the routing table. In the next node, the adaptation layer again assembles the fragments to create the IP packet and sends it to the network layer. If all the fragments are not received, then it will request for retransmission. On successful creation of the IP packet, the network layer checks if the packet is meant for this node. If it is, then the packet is sent to the transport layer, else it is again forwarded to the next node based on the routing table.

ARO:

Type	Length	Status	Reserved
Registration Lifetime			
EIU-64			
Registered Address (optional)			

Type	8 bits	To be decided
length	8 bits	Length of the ARO in units of 8 octets
status	8 bits	zero if registration is successful, else non zero
reserved	8 bits	Unused – set to zero
Registration lifetime	32 bits	Amount of time in seconds the router needs to keep the entry in its cache
EIU-64	64 bits	Unique identification the interface of registered address
Registered Address	128 bits	Carries the host address. Shouldn't be filled by the host

6CO:

Type	Length	Context length	res	C	CID
Valid Lifetime					
Context Prefix					

Type	8 bits	To be decided
length	8 bits	Length of the 6CO in units of 8 octets
Context Length	8 bits	Number of valid leading fields in Context prefix which are valid
Res	3 bits	Unused – set to zero
C	1 bit	Indicates if field is valid for compression
CID	4 bits	Context identifier – configured by the 6LBR
Valid Lifetime	32 bits	Amount of time in seconds that the context is valid
Context prefix	Multiple of 8 bytes	IPv6 prefix or address corresponding to the CID

ABRO:

Type	Length=3	Reserved
Version Number		
6LBR Address		

Type	8 bits	To be decided
length	8 bits	Length of the ABRO in units of 8 octets. Always set to 3
Reserved	16 bits	Number of valid leading fields in Context prefix which are valid
Version Number	32 bits	Version number for the specific set of information in RA
6LBR Address	64 bits	Address of the 6LBR which originated this version

Figure 11.7: Various options in 6LoWPAN.

11.5.6 Transport Layer in 6LoWPAN

TCP is a sliding window protocol which provides a mechanism of guaranteed delivery of message. The window size decides how many bytes can be sent before an acknowledgement is received. Window size is configurable and depends on application scenarios. TCP is based on full duplex virtual connection between receivers and transmitter, known as end points. TCP has two interesting properties: firstly it provides end to end acknowledgement and hence reliability in message delivery and secondly messages are ultimately received in sequential order. However, the implementation of this protocol is very complex and final code size is huge.

The devices of 6LoWPAN are small with low computing capabilities, mostly run on limited power sources with a low RAM size. Therefore it is very challenging to implement TCP protocol in this system. UDP on the other hand do not require sequential ordering and there is no end to end acknowledgement mechanism. Thus the complexity of implementation and code size of UDP is really very small. Therefore UDP is potentially a candidate for 6LoWPAN.

11.5.7 Application Layer in 6LoWPAN

The application layer receives the data from the transport layer via protocols such as UDP, TCP, etc. There are various applications that are possible to build. Some common applications which are build include environment monitoring, tracking, home/building automation, etc. This is just a small list. There are a huge number of applications that can be built for wireless sensor networks.

6LoWAPP: IP is being deployed in these devices so that they can interact directly with other IP based devices. Issues related to the integration of these nodes to the present system should be minimum. Today's 6LoWPAN nodes prefer to use UDP over TCP as TCP is quite complex compared to UDP and it is not suitable for lossy networks. Some form of acknowledgement and sequence number should be present which is not supported by UDP. This is done by the application developers at the application layer. Using UDP makes HTTP less desirable, but on the other hand MIME or URI can be used easily. Some of the possible additions which might be desirable include connection splitting at 6LoWPAN Border Routers (6LBRs), full support of Web services, including deployment of small Web accessible resources and enabling SNMP services. 6LoWAPP should be able to interface with various other standards which are mentioned in this chapter and beyond like ZigBee, ISA, etc.

Constrained RESTful environment (CoRE): CoRE working group was formed from the 6LoWAPP activities to target sensor networking or similar applications. CoRE intends to provide a framework for applications running on IP networks in a constrained environment. The system need to operate in small code size, low RAM and limited power environment. The framework created by the CoRE working group is supposed to do sensor data monitoring, actuation and device control.

Devices in CoRE are responsible for all computation and data transfer. Each of these devices are proposed to have one or more resources (e.g., sensors and actuators). The devices are supposed to send instructions for query and for making the change in attached resources on other Devices. They send notifications to all subscribed devices, if the resource values are changed. As a part of the framework, the working group would define a Constrained Application Protocol (CoAP) for resource manipulation on these Devices. CoAP would target the environment defined in ROLL and 6LoWPAN. CoAP would be used for inter device communication within the same network or with devices in other networks. It will also be used for communication between devices and normal nodes in the Internet. The working group will also define a mapping from CoAP to HTTP REST API. CoAP will also support various forms of caching. The initial work of the working group includes:

1. Create, read, update and delete resources on device
2. Publish values to subscriber Devices and subscribe to other devices
3. Support multi-cast
4. Proper working of UDP on the CoAP implemented devices.
5. A definition of how to use CoAP to advertise about or query for a Device's description
6. How to use the HTTP REST API for device communication
7. Finding a way to state if the device is powered on or not

The working group is not supposed to focus on developing a reliable multi-cast solution and developing a service discovery solution.

11.5.8 6LoWPAN Devices

6LoWPAN can be used in both full function device (FFD) as well as reduced function device (RFD). FFDs are those devices which can forward link-level packets to other FFDs or RFDs. RFDs are those devices which cannot forward packets. By using 6LoWPAN, the need to use a translator as a gateway is removed. So a normal router or bridge which is used for IP can be used in 6LoWPAN.

11.5.9 Security in 6LoWPAN

There is no protocol which is fool proof against intrusion. The 6LoWPAN uses the normal IP at higher layers. There are a lot of secure protocols which have been implemented for IP at various levels. All these protocols can be used in 6LoWPAN. The 6LoWPAN uses a normal bridge. So there is only one intrusion point, i.e., at the bridge/router. For other protocols, there can be two intrusion points, namely the router/bridge as well as the gateway that is used to make the protocol available to the internet user.

However using a security protocol means using of bandwidth. 6LoWPAN datagram are small in size and thus adding a security protocol adds to the overhead. The protocols used in the link layer by 802.15.4, which 6LoWPAN uses are not very strong. So intrusion at these levels is possible.

11.5.10 QoS in 6LoWPAN

Quality of service in 6LoWPAN can be varied according to the criticality of the application. There are certain applications in which data has to be delivered in a highly reliable manner and as fast as possible such as hospital storage monitoring (e.g., monitoring of oxygen cylinders in the storage and checking if the number has fallen below a critical threshold) or patient health monitoring system. There are certain systems in which QoS can be compromised such as home automation (e.g., switching on or off lights in the house) or agriculture monitoring.

Most of the protocols used in 6LoWPAN are the same as used in IPv6. There are various protocols in various layers of the IP stack which have some advantage and disadvantage. So a mix and match of the various protocols at various layers can be used based on the purpose of the application.

11.5.11 ZigBee versus 6LoWPAN

There are lots of advantages of using 6LoWPAN instead of ZigBee. The most obvious one is that 6LoWPAN uses the Internet protocol. So it is easy for a person to connect to the device directly using the Internet as compared to using some translator to convert the ZigBee packet into a format that is understood by a device connected to the Internet. A normal bridge or router does the job in case of 6LoWPAN. IP is widely used and so it will be accepted easily.

The size of the header in 6LoWPAN can be as small as 2 bytes whereas at least 8 bytes are needed in case of ZigBee. Even with such a large header size the number of devices that can be present in a ZigBee network is very small compared to the network size of 6LoWPAN. ZigBee does not follow any standard transport layer protocol whereas 6LoWPAN implements the UDP or TCP protocol. This way connecting to the Internet becomes easy in case of 6LoWPAN.

Table 11.14: Comparison between ZigBee and 6LoWPAN

	ZigBee	**6LoWPAN**
Protocol	ZigBee Specific	IP
Header Size	At least 8 bytes	As small as 2 bytes
Number of devices supported	Less than 6LoWPAN	Large
Internet Connectivity	Converter needed	Direct

11.6 Z-Wave

Z-Wave is a technology developed by Danish company Zensys. It has been designed for wireless communication in home automation. It is a low bandwidth, half duplex protocol. Z-Waves operate at a frequency lower than 1 GHZ and so it is impervious to interference from devices such as a Wi-Fi router, cordless phone or any Bluetooth device. The disadvantage of not being in the 2.4 GHz range is that the range might not be free in some countries and so licensing is required. The Z-Wave technology is designed for low data flow. Typically it is used to send command messages such as on-off, high-low, etc. The standard is available under the non disclosure agreement [11] and [12].

11.6.1 Motivation behind the Z-Wave Standards

For home automation applications, the devices used should consume low power, shouldn't break down easily and the communication should be reliable. Keeping all these points in mind, Zensys came up with Z-wave which appealed to many and now there is an alliance which is called the Z-wave alliance which has over a hundred members.

11.6.2 The Architecture of Z-Wave System

The Z-Wave protocol consists of two types of devices, namely the controller device and the slave device which together form a mesh network. Controller devices are the ones that send commands to other nodes. Slave nodes either follow the command or forward it to the other devices which are not in direct range of the controller.

Controller: As stated earlier, controller devices are the ones which send commands in the network. Controller has the complete routing table of the network and hence it can send a command to any device in the network. A Z-wave network can have multiple controllers but the controller which is used to create the network becomes the primary controller. There can be only one primary controller in the network. The difference between the primary controller and any other controller in a network is that only the primary controller has the right to include or exclude devices from the network.

Slave: Slave devices in the Z-Wave network are the ones which receive commands from the controller and take the necessary actions. Slave nodes cannot forward a message unless the controller has asked them to do so.

11.6.3 Physical Layer in Z-Wave

The physical layer in the Z-wave consists of a radio frequency medium. This medium has been designed for low bandwidth data flow, typically around the 900MHz range. Data usually transmitted in a Z-wave network is on-off or some very small amount of data. The 900 MHz range makes it interference free. The end - to - end solution empowered by Z-wave is not expensive because Z-wave devices consume very little power.

11.6.4 MAC Layers Protocol of the Z-Wave

The MAC layer controls the RF media. Data stream is Manchester coded and is sent in the little endian format. MAC layer requires either access to the data frame or to the signal received. The signal can be either a decoded bit stream or Manchester coded. The frame data that is passed on to the transfer layer from the MAC layer looks like this:

Table 11.15: Frame Passed to Transfer Layer

Bits	2	3	11	16	8
Preamble	Start of frame	Data	...	...	End of frame

The MAC layer has a collision avoidance (CA) mechanism. CA is achieved by keeping the nodes in the receive mode when they do not have any data to transmit. If there is data ready to be transmitted, the MAC layer is sensed to check if data transfer is already going on. It waits for a random amount of time before checking again. This is in the order of milliseconds. When the channel is idle, the packet is transmitted.

11.6.5 Transfer Layer of the Z-Wave

Transfer layer is responsible for data transmission between two nodes. It takes care of acknowledgement and does retransmission if ACK is not received. It takes care of the checksum in the packet to check corrupted packets.

There are various types of frame transfer. They are:

Single cast frame type: Frame is sent to one specific node and the node acknowledges the receipt of the frame. If acknowledgement is not received, the frame is transmitted again. By default single cast frames are used with acknowledgement, but it can also be used without acknowledgement.

Multicast frame type: A frame sent to a specific number of nodes is known as the multicast frame type. This type does not support acknowledgement. It is similar to sending the same email to multiple users in one go by filling their names in the TO list, instead of sending it to each user individually.

Broadcast frame type: Broadcast frames can be heard by all the nodes in the network and there is no acknowledgement involved.

In case of broadcast and multicast, reliability is less as compared to single cast. In a network where reliability is of high importance, single cast should be used.

11.6.6 Routing Layer in the Z-Wave

Any node which has a static position and is in the listening mode can participate in the routing. This layer is responsible for sending the frames with the correct forwarder list. This layer is also responsible for maintaining the routing table in the controllers. In the controller, the routing table maintains the network topology. The cells in the tables are filled with 0s and 1s to show a node is not connected to another or a node is connected to another respectively. The primary controller creates the routing table at the time of network establishment.

11.6.7 Application Layer in Z-Wave

The application layer is responsible for decoding and executing the commands sent by the controller. Application layer is implementation specific and can change from one application to another. Application layer is responsible for filling the home id and the node id in the frame that is transmitted. Home id is the unique 32 bit id which each network has and node id is the unique id of a node in the home network.

11.6.8 Z-Wave Devices

Z-wave system is designed and used for home automation system. They can be attached to any device in the home to make them intelligent. Devices can be right from the lights and fans to the thermostats. Z-wave connects all the devices at home without involving any extra wiring. Most of the devices at home can be connected to a Z-wave network by just plugging them to a Z-wave module. With the intelligent Z-wave devices it is also possible to turn on or off one device when something happens in the other device. For example if the door connecting the stairs is opened, then the lights in the stair case turn on.

11.6.9 Security in Z-Wave

Security in Z-wave devices is very low. They should be used primarily for automation of lights, thermostat, etc., and should be avoided for systems which require high security such as home security system.

Z-wave devices are controlled by the primary controller as discussed in section 11.4.1.1. They also have a unique home id. If a malicious controller comes in the range of the existing Z-wave network and it has the home id, it can control all the devices present in the Z-wave network.

Another concern is flooding of network. If a device is introduced in the Z-wave network and it keeps flooding the network with malicious packets, then the response time of the devices will increase drastically. Z-wave technology has been designed in such a way that it can handle only low data rate.

11.6.10 QoS in Z-Wave

Z-waves are used primarily in home automation and controlling. So a high data network is not required. The commands sent by the controller need to reach the end device without loss even though there might be a little time delay.

11.6.11 ZigBee versus Z-Wave

ZigBee is backed by the ZigBee alliance whereas z-wave is backed by the z-wave alliance. Both of these technologies have a lot of big players in the market backing them. ZigBee is backed by major players like Freescale, TI, etc., and is aligned with the IEEE 802.15.4 standards whereas Z-wave, developed by Zensys is backed by companies like Intel, Cisco, etc.

The purpose of Z-wave is home and light industrial monitoring and automation, whereas ZigBee has a huge number of applications in home and light industrial monitoring as well has applications in wireless personal area networks.

Z-wave is proprietary to Zensys whereas ZigBee is based on IEEE 802.15.4 and works both on the 900 MHz free band as well as the 2.4 GHz band. Z-waves on the other hand operate only on the 99 MHz free band. The data rate achieved in ZigBee is 250kbps as compared to 40 kbps achieved in z-waves.

ZigBee devices have a range between 10 to 75 meters without multihop. Z-wave devices on the other hand have a range of about 30 meters without multihop. Both these technologies support multihop and so the final achieved range using multihop is quite large in both the technologies.

Price wise Z-wave devices are more expensive when compared to devices using ZigBee. The ZigBee standard is comparatively older than the Z-wave technology.

Both these technologies have a fair deal of advantages and disadvantages. They should both move forward together instead of trying to cut down each other. The two technologies are not compatible with each other.

Table 11.16: Comparison between ZigBee and Z-Wave

	ZigBee	**Z-Wave**
Applications	Large	Mainly home and small industry automation
Data rate	250kbps	40kbps
Frequency	2.4GHz/900MHz	99MHz
Range	10-75m	30m
Cost	Cheaper	Expensive than zigbee
Nature	Openly Available	Proprietary to zensys

11.7 ISA

The ISA100 is a committee of automation professionals from around the world. This committee was formed in 2005 to develop industrial wireless standards. There are more than 400 automation professionals from nearly 250 companies in this committee. These professionals are here to give their expertise as they all come from various backgrounds [13].

ISA has more than 30000 members worldwide besides the committee members. They have been working on various solutions which are addressed as a different protocol based on its usage. For example Wireless Backhaul Backbone Network is ISA100.15 and People and Asset Tracking and identification is ISA100.21.

The aim of the committee is to provide a single integrated wireless platform in any industry and the new network will communicate between all the existing and yet-to-come standards and protocols wirelessly. The advantage the end user gets is that there is only a single technology for various kinds of applications. The committee members come from various backgrounds, be it security or instrumentation or any other technology. This makes the end user more confident about using the standard specific field. The end user should also be happy about using one single network for a wide range of applications. This reduces the investment, maintenance and operation cost for the end user too.

There are a lot of companies around the world which support ISA100. ABB, Freescale, Honeywell, Siemens are a few of the companies in the huge list of companies. Nivis has come up with the first ISA100.11a standard based system. This is built on the NISA100.11a stack.

11.7.1 Motivation behind the ISA100 Standards

With so many different wireless technologies, it becomes difficult to control each one of them. There needs to be a standard which all these technologies need to comply with in case of any industrial application. In industries there are also requirement for interoperability between two technologies. ISA100 is here to address that as well. ISA will make sure that there won't be any interference between two different technologies and both of them can interoperate.

By using the ISA100 standard, the end user will reduce time and cost in selecting a technology which best suits the industrial requirement. Also the user will be sure that the new technology introduced will be compatible with the existing wireless technologies and if required can communicate between the two technologies. The user will get technical support for installation of this technology from the ISA. In case of a new technology coming in and it is compliant with ISA100, introducing it becomes easier. Also the developer of the new technology can get support from the ISA while developing the technology to meet ISA standards.

11.7.2 The Architecture of ISA100.11a System

The ISA100.11a architecture fits into the 7 layered OSI model. The Physical layer is compliant with the IEEE802.15.4 standard. The Data Link Layer (DLL) consists of an IEEE 802.15.4 MAC sub layer. There is an upper DLL which provides TDMA, channel hopping and mesh routing. The network layer is present to provide inter network routing, i.e., it provides mesh to mesh routing. The network layer frame is in accordance to IETF RFC 4944. The functions of the network layer includes addressing, routing etc. It is also responsible for the security as well as the segmentation of packets so that they can be transferred using the underlying layers. The application layer is where the end application will reside. Along with the end application the application layer also takes care of interoperability between various technologies.

ISA100.11a standard is built keeping in mind the presence of various wireless communication devices being used in the same industry. Many of the devices use the same frequency range. So there will be interference. The committee found various ways of avoiding this interference. The first way was to avoid channels which are already congested. The next way was to do frequency hopping. Data is sent over multiple channels to avoid the load on one channel. The next way was to spread the energy so that it looks like noise to other systems in the same network. The next way was to keep the communication short which in turn helps in reducing the congestion.

11.7.3 ISA100 .11a Scope

ISA100.11a is being created as an open standard. This means that anyone can use the standard without joining a certain group. The standard will be available on the Internet which can be downloaded freely. The standards are being written in as simple manner as possible so that it becomes easy for the end user to understand and implement the standard.

The standard will make sure that the standard can be easily used for industry automation and it will support the same. The network size will be small (LAN or PAN) and the radio that will be used will be in the free frequency range so that it can be deployed world wide. The technology should support class 1 through class 5 of the applications mentioned by ISA100. Further discussion on classes is present in subsequent sections in this chapter.

The standard also keeps in mind the security issue. Security is a high priority for the ISA100.11a working group. The devices complying with the standard will be able to coexist with each other even if they are following a different technology.

11.7.4 ISA100.12: Wireless HART Convergence

The ISA100.12 committee was formed with the goal of achieving coexistence between networks following the ISA100.11a standards and the wireless HART network in the same industry without hindering each other. The committee is supposed to evaluate the benefits of both ISA100.11a and wireless HART and come up with a document stating the advantage of using the technologies in various applications. The future goal of this committee is to converge the two technologies. To achieve this they will keep the network and lower layers the same for both the technologies and the upper layer will be different in case of any of the two technologies. So for the end user, it will look the same. While doing this process, the subcommittee has to make sure that whatever benefit there is each of the two technologies remains intact. They also have to state the changes that have to be made in both the technologies so that they are interoperable at device level. While converging the two technologies, the sub committee also has to make sure that the new converged technology is backward compatible. The subcommittee is supposed to do all this without hindering the schedule of any other ISA100 standard.

11.7.5 ISA100.15 Wireless Backbone Backhaul

The ISA100.15 working group's focus is in the field of wireless backhaul backbone. The wireless backhaul backbone system is used in wireless communication in receiving and sending data to any device in their wireless network.

The purpose of the ISA100.15 group is to use one or more wireless backhaul network to support one or more than one technology to run multiple applications. This will create the interoperability between backhaul wireless network and other networks. ISA100.15 group has defined an interface for the interoperability between the wireless backhaul network and networks such as ISA100.11a, wireless HART, ZigBee, etc. There are many other standards for which the interface is being created. It should also be able to control various control protocols. While doing this entire thing, the group should keep in mind that the quality of service (QoS) is met for various applications. The group must also consider the security issues and ensure that the network layer to the upper layers is interoperable with other similar technologies.

11.7.6 ISA100 Devices

There is no compulsion that ISA100 will perform in certain prescribed devices. It can be used in any kind of device, be it fixed or portable or even moving device. ISA100 standard can be used in low data rate wireless devices as well. ISA100 devices can have high latency (order of 100ms) and thus can be used in less critical applications. The aim of the ISA100 committee itself is to create a hassle free wireless network which will take care of the interoperability and coexistence of the device with any other device present in the network.

11.7.7 ISA100.11a and ISA100.12 Network

ISA100.11a, developed by ISA, has been designed to address proper operation of ISA100 devices wirelessly in harsh environment conditions such as presence of interference in industries. This standard is focused on monitoring and automation performance in industries. Other than addressing the need for proper operation, it also makes sure that the security and management of these communications are proper.

ISA100.11a addresses low powered devices. It will also create a platform for the co existence of ISA100.11a devices with other standard devices as well as make sure that there is interoperability between various ISA100 devices.

ISA100 standard is based on the IEEE 802.15.4 for the MAC and physical layer. This standard is used for secure wireless operation in case of non critical applications. ISA100 has been divided into 6 classes (0 to 5) where class 0 is for safety, class 1 through 3 is for control and class 4 and 5 are for monitoring. ISA100.11a is suitable to address the lesser critical classes (class 1 through 5). It is not suitable to handle critical cases such as safety.

ISA100.12 subcommittee takes care of the wireless HART convergence. This committee has to evaluate both wireless HART specification as well as ISA100.11a specification and has to come up with a way for coexistence of both of these in a single industry. This committee also has to find a vendor who will converge wireless HART and ISA100.11a in the future. The convergence specification needs to be specified as well while keeping in mind not to lose the advantages of any of the two technologies. The committee also has to make sure that this converged technology is compatible with only wireless HART as well as only ISA100.11a networks.

11.7.8 ISA100.21 the Asset Tracking

ISA100.21 standard addresses tracking. Tracking can be tracking of assets or human movement tracking. There are various methods of tracking which already exist. ISA is not defining any new standard for tracking. In fact it is seeing the pros and cons of the various technologies that are already present and then recommending it to the industry. Recommendation is done based on the ease of coexistence of the tracking technology with the other wireless technologies that are already present in the industry or chances of the wireless technology which might be introduced in the industry.

There are various tracking technologies as well as mathematical implementations which are already present. Some of the standard tracking methods include:

GPS - abbreviation for global positioning system. Here the location is based after calculating the signal received from various satellites. In case GPS is used inside a building, AGPS system is required.

Cellular based tracking - tracking of the cellular phone based on the antenna which the phone is near to and to whom it is reporting at the present time. The accuracy as well as the coverage area in this is a lot lower than that of GPS, but tracking can be done easily.

RFID - tracking using the classic RFID. Details of this are covered in the RFID subtopic.

There are other standard tracking methods also and ISA has considered each one of them. Apart from the tracking methods, there are various mathematical ways of determining the location. The most common method is triangulation. Here the RSSI of the device received at three or more antennas is considered and then the location is determined based on calculations. Other methods include angle of arrival, time of arrival and time difference of arrival, etc. Discussing each of these techniques is out of scope for this chapter.

11.7.9 Handling of QoS in ISA100.11a

As discussed earlier, ISA100.11a is not meant for critical applications. In ISA100.11a, applications are usually meant for monitoring, controlling and tracking. These applications fall under class 1 through class 5 of the ISA classes. Class 0 applications are cases where action needs to be taken urgently such as in health care cases. But in case of the other classes, there can be a little delay between the time of actual event and the time it has been reported.

The ISA100.11a devices have to be interoperable with other devices which are following some other technology or standard and are being used in the same industry. This is a high priority thing for ISA100.11a and it cannot be compensated at any cost. As stated earlier, the devices also have to perform even in harsh environment conditions such as interference or any other harsh condition which is common in industries.

11.7.10 ZigBee versus ISA100

ZigBee had come into the picture as a need to address the battery life issue in case of devices. ZigBee has been successful in doing that, but so has ISA100. ISA100 provides features which have been missing in the ZigBee protocol stack.

ISA100 is much more predictable, reliable and secure compared to the ZigBee stack. The use of various standard protocols has helped ISA100 in achieving a better reliability as compared to ZigBee. ISA100 has been defined keeping in mind the industrial standards and the architecture is for wireless sensing in industries. ZigBee on the other hand is not specific for wireless sensing and monitoring in industries. Because of being for industrial applications, ISA100 can handle various industrial protocols such as Wireless HART, field bus, profibus, etc.

ISA100 standard is less affected by interference compared to the ZigBee standard. ISA100 has been built in such a way as to reduce these interferences.

11.8 Comparison of Different Standards

11.8.1 Comparison of ZigBee and Other Standards for Industrial Automation

In Industrial automation system HART, wireless HART, ISA-100 and ZigBee are all being used. However HART has been always the best choice for industries due to its well defined interoperability, and high number devices available from different vendors. Since HART is also very old, and one can easily convert HART into wireless HART therefore, replacement of entire system is not required. Moreover, work is still going on in ISA-100 therefore it less likely that it will compete with wireless HART.

11.8.2 Comparison of ZigBee and Other Building Automation Standards

In building automation space BACNet, LoanTalk, Z-wave and building automation application profile of ZigBee are competing with each other. ZigBee is a wireless protocol which has defined network and application layer. BACNet system is

Table 11.17: ZigBee versus Other Industrial Automation Standards

	ZigBee	HART	ISA-100
Architecture	Open	Open	Open
Medium	Wireless	Primarily wired but wireless also	wireless
Status	Highly active	Highly Active	Less Active
Availability	2004	1986	2005
Wireless Technology	IEEE802.15.4	IEEE802.15.4	not available

originally wired system however wireless has been introduced recently. BACNet and ZigBee both use IEEE 802.15.4 physical and data link layer. ZigBee and BACNet are highly active due to their open architecture. However LonTalk is Proprietary and therefore a lesser number of vendors participates in developing automation solution using LonTalk. Z-wave is used for very short range with relatively low number of devices in the network.

Table 11.18: ZigBee versus Other Building Automation Standards

	ZigBee	BACNet	LonTalk	Z-wave
Architecture	Open	Open	Proprietary	Properietary
Medium	Wireless	Primarily wired but wireless also	Wired	wireless
Status	Highly active	Highly Active	Less Active	Less Active
Availability	2004	1995	1999	2003
Wireless Technology	IEEE802.15.4	IEEE802.15.4	Not available	Subgiga hertz band around 900MHz

There is no winner protocol in the absolute term. It depends upon the requirement that a particular system will be used. ZigBee being one of the widely accepted standard for wireless in automation domain (both industrial and building automation) has a very bright future.

References

[1] Members of ISO 7498, "Information processing systems - Open Systems Interconnection - Basic Reference Model" International Organization for Standardization 1984.

[2] ATA/ANSI 878.1, "ARCNET Local Area Network Standard. American National Standards Institute", New York, NY 1992.

[3] ISO 8802-3, "Carrier sense multiple access with collision detection (CSMA/CD) access method and physical layer specifications", Information processing systems, Local area networks - Part 3, International Organization for Standardization 1993.

[4] Tutorial on HART, "http://www.analogservices.com/about_part0".

[5] HART Communication, "http://www.hartcomm.org/".

[6] "http://www.rfidinc.com/tutorial.html".

[7] Mulligan, G "The 6LoWPAN architecture" In Proceedings of the 4th Workshop on Embedded Networked Sensors (Cork, Ireland, June 25 - 26, 2007). EmNets '07. ACM, New York, NY, p. 78-82. DOI= http://doi.acm.org/10.1145/1278972.1278992.

[8] Kushalnagar, N., Montenegro, G., Schumacher, C. "6LoWPAN: Overview, Assumptions, Problem Statement and Goals" http://www.ietf.org/mail-archive/web/ietf-announce/current/msg03549.html

[9] IETF draft of 6LoWPAN, "http://tools.ietf.org/html/draft-ietf-6lowpan-format-13".

[10] Jonathan W. Hui and David Culler, "Extending IP to Low Power Personal Area Network", IEEE Internet Computing, July-August 2008.

[11] http://www.z-wavealliance.org/.

[12] Abhay Gupta and Michael Tennefoss "Radio Frequency Control Networking", Echelon Corporation white paper, http://www.echelon.com/support/documentation/documents/005-0171a_rf_white_paper.pdf.

[13] http://www.isa.org//source/ISA100_Brochure_2008_Oct.pdf.

Index